Method and Technology of Green Coal Mining

Method and Technology of Green Coal Mining

Editors

Shengrong Xie
Dongdong Chen

MDPI

Basel • Beijing • Wuhan • Barcelona • Belgrade • Novi Sad • Cluj • Manchester

Editors
Shengrong Xie
China University of Mining & Technology (Beijing)
Beijing, China

Dongdong Chen
China University of Mining & Technology (Beijing)
Beijing, China

Editorial Office
MDPI
St. Alban-Anlage 66
4052 Basel, Switzerland

This is a reprint of articles from the Special Issue published online in the open access journal *Energies* (ISSN 1996-1073) (available at: https://www.mdpi.com/journal/energies/special_issues/Green_Coal_Mining).

For citation purposes, cite each article independently as indicated on the article page online and as indicated below:

Lastname, A.A.; Lastname, B.B. Article Title. *Journal Name* **Year**, *Volume Number*, Page Range.

ISBN 978-3-0365-9316-6 (Hbk)
ISBN 978-3-0365-9317-3 (PDF)
doi.org/10.3390/books978-3-0365-9317-3

Contents

About the Editors

Shengrong Xie

Shengrong Xie, male, Han nationality, member of the Communist Party of China, born in Nanjing, Jiangsu Province, in January 1981, is currently a Deputy Secretary, Professor, Doctor and Yueqi young scholar at the Department of Mining Engineering, School of Energy and Mining Engineering, China University of Mining and Technology (Beijing). Xie is mainly engaged in resource mining theory and technology, mine pressure and rock strata control, mine disaster prevention theory and technology research and teaching work. Xie has overseen nearly 30 projects of the National Natural Science Foundation of China, the Youth Foundation of the National Natural Science Foundation of China, and key projects of the National Natural Science Foundation of China, and won two first-class prizes and eight second-class prizes for scientific and technological progress at the provincial and ministerial level, applied for / authorized more than thirty national invention patents, published three books and one textbook. More than 100 of Xie's papers have been published in the *Journal of China Coal Society*, *Chinese Journal of Rock Mechanics and Engineering*, *International Journal of Mining Science and Technology*, *Tunnelling and Underground Space Technology*, *Scientific Reports*, *International Journal of Coal Science and Technology*, among which more than 50 papers are published with Xie as the first author/correspondent author SCI or EI.

Dongdong Chen

Dongdong Chen is male, from Suzhou, Anhui Province, a party member, postdoctoral fellow, postgraduate tutor, associate professor, and an outstanding doctoral graduate in Beijing. Chen is mainly engaged in the research and teaching about roadways surrounding rock control, stope rock slab structure fracture theory, coal mining and mine disaster prevention. As the person in charge, he has overseen more than ten projects of National Natural Science Foundation for Youth Fund, China Postdoctoral Fund Special Project, China Postdoctoral Fund General Project, National Natural Science Foundation Sub-projects and Horizontal Projects. As the person in charge, he has completed dozens of sub-projects of the key projects of the National Natural Science Foundation, and horizontal projects entrusted by enterprises (four of which have reached the international leading level). He has won two first prizes for provincial and ministerial level scientific and technological progress and ten second and third prizes; four monographs; and he has authorized more than 50 national patents (more than 30 first inventors), one patent invention award, and nearly one million yuan for patent transformation (presiding). As the first author/correspondent, he has had 47 SCI/EI papers published in important academic journals such as *Journal of Rock Mechanics and Engineering*, *IJCST*, *TUST* and *RMRE*.

energies MDPI

Article

Control Techniques for Gob-Side Entry Driving in an Extra-Thick Coal Seam with the Influence of Upper Residual Coal Pillar: A Case Study

Shengrong Xie *, Fangfang Guo and Yiyi Wu

School of Energy and Mining Engineering, China University of Mining and Technology-Beijing, Beijing 100083, China; gffever@163.com (F.G.); 15662045768@163.com (Y.W.)
* Correspondence: xsrxcq@163.com

Abstract: In multi-seam mining, the residual coal pillar (RCP) in the upper gob has an important influence on the layout of the roadway in the lower coal seam. At present, few papers have studied the characteristics of the surrounding rock of gob-side entry driving (GED) with different coal pillar widths under the influence of RCP. This research contributes to improving the recovery rate of the extra-thick coal seam under this condition. The main research contents were as follows: (1) The mechanical parameters of the rock and coal mass were obtained using laboratory experiments coupled with Roclab software. These parameters were substituted into the established main roof structure mechanics model to derive the breakage position of the main roof with the influence of RCP, and the rationality of the calculation results was verified by borehole-scoping. (2) Based on numerical simulation, the evolution laws of the lateral abutment stress in the lower working face at different relative distances to the RCP were studied. FLAC3D was used to study the whole space-time evolution law of deviatoric stress and plastic zone of GED during driving and retreating periods with various coal pillar widths under the influence of RCP. (3) The plasticization factor P was introduced to quantify the evolution of the plastic zone in different subdivisions of the roadway surrounding rock, so as to better evaluate the bearing performance of the surrounding rock, which enabled a more effective determination of the reasonable coal pillar width. The field application results showed that it was feasible to set up the gob-side entry with an 8 m coal pillar below the RCP. The targeted support techniques with an 8 m coal pillar could effectively control the surrounding rock deformation.

Keywords: residual coal pillar; gob-side entry driving; extra-thick coal seam; coal pillar size; surrounding rock control

Citation: Xie, S.; Guo, F.; Wu, Y. Control Techniques for Gob-Side Entry Driving in an Extra-Thick Coal Seam with the Influence of Upper Residual Coal Pillar: A Case Study. *Energies* **2022**, *15*, 3620. https://doi.org/10.3390/en15103620

Academic Editor: Sergey Zhironkin

Received: 18 April 2022
Accepted: 12 May 2022
Published: 15 May 2022

1. Introduction

Gob-side entry driving (GED) with a small coal pillar (3~8 m) is widely used in China's mines due to the simple development processes, high resource recovery rate, and the ability of the coal pillar to isolate gangue, harmful gases, water, and fire in the adjacent gob [1,2]. China has many thick and extra-thick coal seams, and the GED has been successfully applied in many mines with extra-thick coal seams [3,4]. For the mining of extra-thick coal seams (>8 m), the characteristics mainly include [5–8]: (1) Strong-dynamic disturbance of abutment stress. Fully mechanized top coal caving mining is generally applied to the mining of extra-thick coal seams, with a large mining space and intense roof activity, resulting in severe abutment stress on the coal mass. (2) A thick top coal over the roadway. The roadway is usually developed along the floor line of the coal seam, the roof of the GED is composed of weak coal masses with a large thickness. The weak properties of the top coal mass seriously increase the difficulty of roadway control. (3) Disturbed by multiple mining-induced stresses. The GED is not only disturbed by the lateral abutment stress of the previous working face, but also advanced abutment stress of the present working face.

Therefore, it is of great significance to study the selection principles of reasonable coal pillar width and surrounding rock control techniques of the GED in extra-thick coal seams.

Presently, many scholars around the world have conducted extensive research on the design of coal pillar width using various methods such as theoretical analysis, numerical simulation, and similar simulation. UDEC can effectively study the macro-mechanical behavior and crack propagation law during the rock failure process, so it is frequently applied to the analysis of the roadway failure mechanism. Bai et al. [9] studied how the failure mechanisms of a 7 m coal pillar width during the formation process caused a large deformation. They investigated the propagation of shear and tensile cracks in the coal pillar of various widths and optimized the coal pillar width and support measures. Shi et al. [10] investigated the crack evolution mechanism of the gob-side entry for different conditions and proposed optimized-support parameters combined with roof-cutting measures. Gao [11–13] carried out a series of numerical simulations using the UDEC Trigon approach to focus on the roadway damage caused by squeezing failure and shear failure, and the effects of rock bolts in the roadway support were also evaluated. FLAC3D can simulate the mechanical behavior of geological materials and geotechnical engineering effectively and is widely used in underground coal mining activities [14–16]. He et al. [17–21] studied the stress distribution characteristics and plastic zone of roadway with different coal pillar widths in the process of coal mining and proposed a support technology for setting a reasonable width of coal pillars. The specific coal pillar width was designed and applied in engineering practices, which realized a good control effect of the roadway surrounding rock. Ma [22] conducted research on the stress distribution characteristics of the narrow coal pillar with different top coal heights of gob-side entry, and they believed that the thick top coal was not conducive to control of the roadway. Jiang [23] presented an approach for evaluating, designing, and optimizing EDG and yield pillar based on the results of numerical simulations and field practice. Han et al. [24–26] considered special geological situations to drive an entry along the gob, such as in an isolated island working face, in a deep soft-broken coal seam, or in a working face adopting roadway side sealing technology. The targeted control techniques of the surrounding rock were proposed and successfully applied in the field practices. These studies had made countless valuable explorations on the width of coal pillars and failure mechanisms and made certain innovations in the research methods. However, there were very few studies on the layout of the GED with the influence of a residual coal pillar (RCP) during the mining of multiple coal seams.

China has many multiple coal seam mining areas, such as Datong, Xinwen, Pingdingshan, etc. [27]. The mining of two adjacent coal seams tends to interact with each other due to the close distance [28,29]. The mining of the upper coal seam generally renders many section pillars, and lower coal seam roadways are usually designed with larger section coal pillars (usually over 20 m) to avoid safety hazards. That results in the waste of massive coal resources, especially for extra-thick coal seams. This paper focused on the layout selection of GED below the RCP. The lateral breakage position of the main roof under the influence of RCP was calculated by theoretical derivation and verified the rationality of the results using borehole-scoping. Under the influence of RCP, the whole space-time evolution laws of deviatoric stress and plastic zone of GED were studied during driving and retreating periods with different coal pillar widths using numerical simulation. The quantified index for the plastic zone of the surrounding rock of GED, plasticization factor P, was proposed. A coal pillar width of 8 m was finally determined. Based on the extent of the plastic zone of GED obtained by borehole-scoping, the targeted roadway support scheme was proposed. That provided a reference for promoting the coal recovery rate in similar geological conditions.

2. Project Overview

2.1. Geological Conditions

Nanyangpo coal mine is located in Shanyin County, Shanxi Province, China. The main mining seams are No. 4 and No. 6 coal seam with a spacing of 32 m. The average thickness of the No. 4 coal seam is about 3.0 m, which has been mined out, but 15 m section coal pillars are left between working faces. The current main mining seam is the No. 6 coal seam with an inclination of 3~5° and an average thickness of 9.6 m. The mining method is fully mechanized top coal caving mining of an extra-thick coal seam, the machine mining height is 3.5 m, and the caving height is 5~10 m. The geological column chart of the No. 6 coal seam is illustrated in Figure 1. The panel 26,102 is near the northern boundary of the mining field, and next to the panel 26,104 and 26,106 that have been mined out with a 20 m coal pillar left. To avoid wasting resources, Nanyangpo coal mine plans to develop the 26,102 tailgate along the gob with a small coal pillar. The layout of the panels is given in Figure 2.

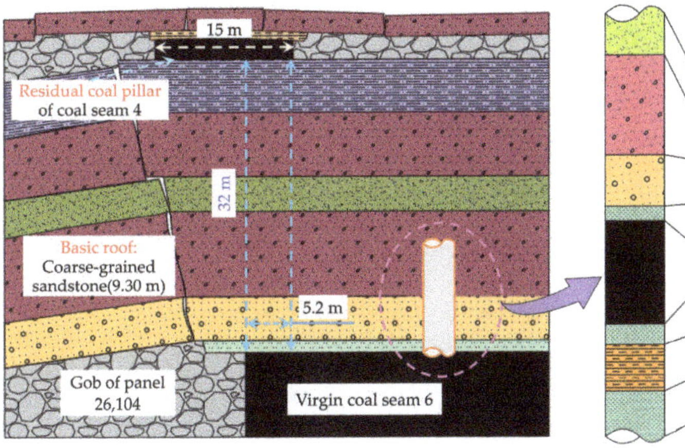

Lithology	Depth (m)	Thickness (m)	RQD (%)
Medium-grain sandstone	233.76	3.82	27
Coarse-grained sandstone	243.06	9.30	30
Gravelly coarse sandstone	247.80	4.74	19
Fine-grained sandstone	249.13	1.33	21
Coal seam 6	258.73	9.60	25
Fine-grained sandstone	260.58	1.85	23
Mudstone	264.90	4.32	46
Fine-grained sandstone	269.21	4.31	30

Figure 1. Generalized stratigraphic column of the test site with the panel layout.

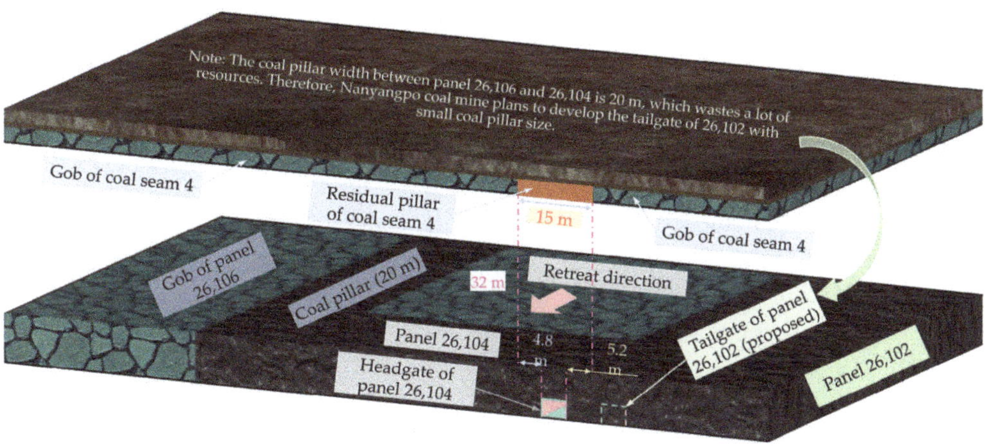

Figure 2. Detailed panel layout of the test site.

2.2. Rock Mass Properties

The physical and mechanical properties of rock mass are an important basis for the design of roadway support. The parameters obtained in the laboratory are also key data for further study on the theoretical model calculation and numerical simulation. The mechanical parameters of coal and rock mass in panel 26,102 are shown in Table 1.

Table 1. Properties of rock mass in panel 26,102.

Lithology	σ_{ci} (MPa)	σ_{ti} (MPa)	c_i (MPa)	φ_i (°)	E_i (GPa)	ν_i
Coal seam 4	26.37	1.02	7.63	26	1.53	0.39
Medium-grained sandstone	89.45	6.86	18.79	29	34.21	0.17
Coarse-grained sandstone	93.56	8.07	24.10	31	38.27	0.23
Gravelly coarse sandstone	47.63	3.34	14.62	25	32.04	0.27
Coal seam 6	34.35	1.27	8.14	24	1.72	0.34
Fine-grained sandstone	98.63	9.34	20.61	31	40.71	0.21
Mudstone	40.37	2.17	11.92	35	11.20	0.31

Note: σ_{ci} is the uniaxial compression strength, σ_{ti} is the tensile strength, c_i is cohesion, E_i is elastic modulus, φ_i is internal friction angle, and ν_i is Poisson's ratio.

It is difficult to get the mechanical parameters of coal and rock samples measured in the laboratory to reflect the actual physical and mechanical properties in engineering sites due to the absence of the original environment and structural characteristics of rock mass [30,31]. E. Hoek and E. T. Brown et al. [32,33] obtained the Hoek–Brown failure criterion by a large number of rock mechanics tests as well as field tests of rock masses after continuous revision and improvement. The parameters of coal and rock samples obtained in the laboratory were imported into Roclab software based on the Hoek–Brown strength criterion for calculation, and the revised parameters were obtained in Table 2, which were more in line with the engineering reality.

Table 2. Revised properties of rock mass by Hoek–Brown strength criterion.

Lithology	*GSI*	m_i	*D*	E_{rm} (GPa)	σ_c (MPa)	σ_t (MPa)	φ (°)	*c* (MPa)	ν
Coal seam 4	56	7	0.7	0.24	1.06	0.08	22.0	0.94	0.39
Medium-grained sandstone	73	18	0.7	13.06	12.59	0.44	38.0	6.26	0.17
Coarse-grained mudstone	77	16	0.7	16.77	17.63	0.74	38.8	7.04	0.23
Gravelly coarse sandstone	71	8	0.7	11.30	5.80	0.44	30.0	5.79	0.27
Coal seam 6	59	6	0.7	0.33	1.73	0.14	22.7	1.26	0.34
Fine-grained mudstone	78	17	0.7	18.39	19.98	0.80	39.8	7.71	0.21
Mudstone	62	7	0.7	2.52	2.54	0.19	25.1	1.68	0.31

Note: m_i is the constant of the intact rock, *GSI* is the constant evaluating the fractured rock mass, and *D* is the disturbance factor.

3. Breakage Position of the Main Roof

3.1. Influence of Upper Residual Coal Pillar

The stress redistribution in the roof strata after mining of the upper coal seam resulted in the RCP bearing a larger load. Thereby, a certain range of stress elevation area was formed at the floor of the RCP. That meant the different positions of the lower working face will make the GED in different stress environments. Based on numerical simulation, the evolution laws of the lateral abutment stress in the lower working face at different relative distances to the RCP were as shown in Figure 3.

Taking the centerline of the upper RCP as the base point, the peak values of lateral abutment stress were 23.2 MPa and 25.5 MPa when the edge of the working face (EWF) was −37.5 m and 37.5 m from the centerline of the RCP separately, with an increase of 0.9% and 10.9%. The peak positions of lateral abutment stress were both about 6 m from the gob. When EWF was −17.5 m and 17.5 m from the centerline of the RCP separately, the peak values of lateral abutment stress were 25.7 MPa and 24.7 MPa with an increase of 11.7% and 7.4%, and the peak positions of lateral abutment stress were both about 6 m from the gob. When EWF was −7.5 m and 7.5 m from the centerline of the RCP separately, the peak values of lateral abutment stress were 29.9 MPa and 29.3 MPa with an increase of 30.0% and 27.4%, and the peak positions of lateral abutment stress were about 7 m and 6 m from the gob. When EWF was −2.5 m and 2.5 m from the centerline of the RCP separately, the peak values of lateral abutment stress were 28.3 MPa and 28.0 MPa with an increase of 23.0% and 21.7%, and the peak positions of lateral abutment stress were about 11 m and 7 m from the gob. The peak values of lateral abutment stress of working face showed an "M-shaped" tendency, and the peak depth first increased and then declined.

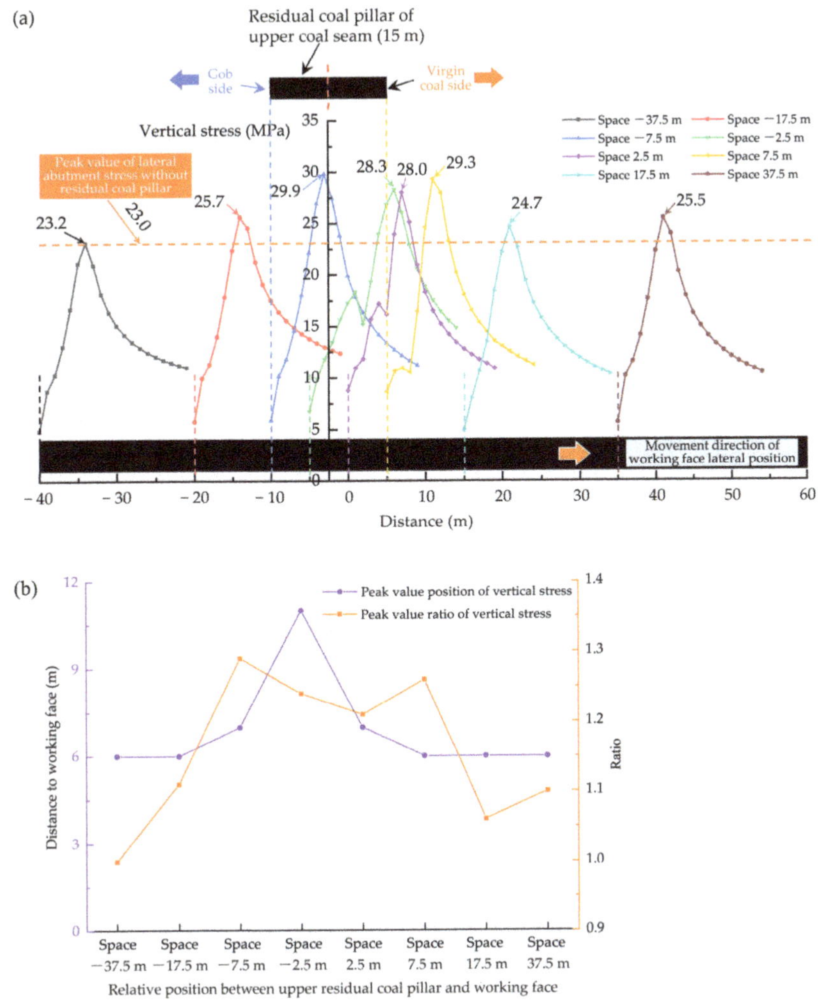

Figure 3. The evolution laws of the lateral abutment stress in lower working face at different relative distances to the RCP. (**a**) Overview under different distances. (**b**) Peak stress ratio and distribution of peak locations. (Note: the gob side and virgin coal side in the figure were both relative to the lower working face).

The above distribution laws indicated that the lateral abutment stress was most significantly affected when the edge of the lower working face was located directly below the residual coal pillar, and the peak growth coefficient was 1.20~1.30. Additionally, when EWF of the lower coal seam was 2.5 m away from the centerline of RCP, the peak growth coefficient of the lateral abutment stress was 1.22. The influence of the stress disturbance of RCP was relatively diminished when a small coal pillar was set up to drive the entry along the gob in such conditions. Consequently, the superimposed disturbance impact of RCP and mining-induced stress on GED should be fully considered to prevent the roadway from instability.

3.2. Mechanical Model of the Main Roof

The main roof controls the upper weak strata of the coal seam. The fracture morphology, the hinge status, and the stability of key blocks after main roof breakage greatly impact the stress distribution of the surrounding rock [34,35]. There is a 15 m coal pillar left in No. 4 coal seam at 32 m above the edge of 26,104 gob in Nanyangpo coal mine. The disturbance of RCP on the stress of floor will inevitably affect the load distribution pattern of the main roof in No. 6 coal seam. Based on the distribution regulation of the lateral abutment stress obtained by numerical simulation in lower working face under the influence of RCP, the mechanical model of the elastic foundation beam is established, as shown in Figure 3. The load curve shows the abutment stress before the lateral breaking of the main roof. The main roof above the virgin coal area is simplified as the elastic foundation beam under the pressure of overburden rock, and the hanging part of the main roof in gob is assumed as the cantilever beam structure.

The rock-beam of the overhanging part in the gob is taken for forces analysis as in Figure 4b. According to $\sum F_y = 0$:

$$\begin{aligned} F &= \int_0^{l_1} \left(\tfrac{(q_1-q_2)x}{l_1} + q_2\right)dx \\ &+ \int_{l_1}^{l_2} \left(q_1 - \tfrac{(q_1-q_c)x}{l_2-l_1}\right)dx + \int_{l_2}^{l_3} q_c dx \end{aligned} \tag{1}$$

Then, the shear force Q_0 and the bending moment M_0 at $x = 0$ are

$$\left.\begin{aligned} Q_0 &= F = \int_0^{l_1} \left(\tfrac{(q_1-q_2)x}{l_1} + q_2\right)dx \\ &+ \int_{l_1}^{l_2} \left(q_1 - \tfrac{(q_1-q_c)x}{l_2-l_1}\right)dx + \int_{l_2}^{l_3} q_c dx \\ M_0 &= \int_0^{l_1} \left(\tfrac{(q_1-q_2)x}{l_1} + q_2\right)xdx \\ &+ \int_{l_1}^{l_2} \left(q_1 - \tfrac{(q_1-q_c)x}{l_2-l_1}\right)xdx + \tfrac{q_c(l_3-l_2)(l_3+l_2)}{2} \end{aligned}\right\} \tag{2}$$

where l_1 is the horizontal distance from the center of RCP to the edge of the lower virgin coal; l_2 is the influence range of RCP on the stress distribution of the main roof; and l_3 is the hanging length of the main roof in the gob, which is approximately equal to the periodic weighting step of the working face.

$q_1 = K_1\gamma H$, $q_2 = K_2\gamma H$, where K_1 and K_2 are the stress increase coefficient, H is the buried depth of the main roof stratum. q_c is the uniform load on the overhanging part of the main roof, the expression is as follows:

$$q_c = \frac{E_1h_1^3(\gamma_1h_1 + \gamma_2h_2 + \cdots + \gamma_nh_n)}{E_1h_1^3 + E_2h_2^3 + \cdots + E_nh_n^3} \tag{3}$$

The beam structure model with immediate floor, virgin coal, and immediate roof as the elastic foundation is shown in Figure 4c. Taking the edge of virgin coal as the origin, the relationship between subsidence y and stress p is as follows:

$$p = -ky \tag{4}$$

$$k = \frac{E_i}{(1-\nu_i{}^2)h_i}, \ \frac{1}{k} = \sum_i^n \frac{1}{k_i} \tag{5}$$

where k is the foundation stiffness coefficient, which is related to Poisson's ratio ν_i and thickness h_i of the weak rock layers below the main roof.

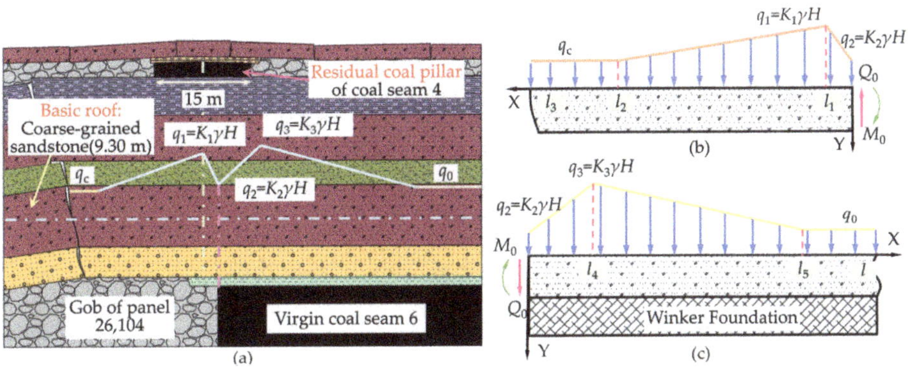

Figure 4. Mechanical model of elastic foundation beam. (**a**) Structural diagram before basic roof breaking. (**b**) Analysis of rock-beam of the overhanging part. (**c**) Analysis of rock-beam with the elastic foundation.

The elastic foundation beam differential equation is

$$\frac{d^4y}{dx^4} + \frac{ky}{EI} = \frac{q(x)}{EI}, \ \beta = \sqrt[4]{\frac{k}{4EI}} = \frac{1}{L} \tag{6}$$

where L is the characteristic length of the beam and EI is the bending rigidity of the main roof rock mass.

Then, the differential equation of elastic foundation beam is

$$\frac{d^4y}{dx^4} + 4\beta^4 y = \frac{q(x)}{EI} \tag{7}$$

According to the calculation of elastic foundation beams [36], the general solution of the deflection equation is

$$\begin{aligned} y &= y_0\phi_1(\beta x) + \theta_0\tfrac{1}{\beta}\phi_2(\beta x) - M_0\tfrac{1}{EI\beta^2}\phi_3(\beta x) \\ &\quad - Q_0\tfrac{1}{EI\beta^3}\phi_4(\beta x) + \left\langle g_{(x)} \right\rangle_t \end{aligned} \tag{8}$$

where the Kralof function, defined to simplify the calculation, is as follows:

$$\left.\begin{aligned} \phi_1 &= ch\beta x \cos\beta x \\ \phi_2 &= \tfrac{1}{2}(ch\beta x \sin\beta x + \sin\beta x \cos\beta x) \\ \phi_3 &= \tfrac{1}{2} sh\beta x \sin\beta x \\ \phi_4 &= \tfrac{1}{4}(ch\beta x \sin\beta x - \sin\beta x \cos\beta x) \\ ch\beta x &= \tfrac{e^{\beta x} + e^{-\beta x}}{2} \\ sh\beta x &= \tfrac{e^{\beta x} - e^{-\beta x}}{2} \end{aligned}\right\} \tag{9}$$

where y_0, θ_0, M_0, Q_0 are the deflection, angle of rotation, bending moment, and shear force at $x = 0$; $\left\langle g_{(x)} \right\rangle_t$ denotes the correction term that should be added when $x > t$.

For the solution of the elastic foundation beam under the loading condition as shown in Figure 4c, the deflection equation of the elastic foundation beam with a correction term section can be given according to the superposition principle:

$$\left.\begin{array}{rl} g & = \frac{q_2}{k} + \left(\frac{q_3-q_2}{kl_1} - \frac{q_2}{k}\right)\frac{1}{\beta}\phi_2(\beta x) - \frac{q_3-q_2}{kl_1}x \\ & + \left\langle\left[\frac{q_2-q_3}{kl_4} + \frac{(q_3-q_0)}{(l_5-l_4)k}\right]x + \frac{2q_2-q_3}{k}\phi_1[\beta(x-l_4)]\right. \\ & \left. + \left[\frac{(q_0-q_3)l_4}{(l_5-l_4)k} - \frac{q_2}{k}\right] - \left[\frac{q_2-q_3}{kl_4} + \frac{(q_3-q_0)}{(l_5-l_4)k}\right]\frac{1}{\beta}\phi_2[\beta(x-l_4)]\right\rangle_{l_4} \\ & \left\langle\frac{(q_0-q_3)l_5}{(l_5-l_4)k} - \frac{(q_0-q_3)}{(l_5-l_4)k}x + \frac{(q_0-q_3)}{(l_5-l_4)k}\frac{1}{\beta}\phi_2[\beta(x-l_5)]\right\rangle_{l_5} \end{array}\right. \quad (10)$$

where the load on the main roof at l_5 length in virgin coal approximately equals the original rock stress, that is $q_0 = \gamma H$, $q_3 = K_3\gamma H$, which is approximately considered that the peak value of the abutment stress of the main roof above the virgin coal with a length of l_4 from the edge of gob. K_3 is the stress disturbance coefficient factor.

The equation for the deflection, angle of rotation, bending moment, and shear force of the main roof before breakage are given:

$$\left.\begin{array}{rl} y & = y_0\phi_1(\beta x) + \theta_0\frac{1}{\beta}\phi_2(\beta x) - M_0\frac{1}{EI\beta^2}\phi_3(\beta x) \\ & - Q_0\frac{1}{EI\beta^3}\phi_4(\beta x) + g \\ \theta & = -y_0\beta\phi_4 + \theta_0\phi_1 - M_0\frac{1}{EI\beta}\phi_2 - Q_0\frac{1}{EI\beta^2}\phi_3 + g' \\ M & = y_0 \cdot 4EI\beta^2\phi_3 + \theta_0 \cdot 4EI\beta\phi_4 + M_0 \cdot \phi_1 + Q_0\frac{1}{\beta}\phi_2 + g'' \\ Q & = y_0 \cdot 4EI\beta^3\phi_2 + \theta_0 \cdot 4EI\beta^2\phi_3 - M_0 \cdot 4\beta\phi_4 + Q_0\phi_1 + g''' \end{array}\right\} \quad (11)$$

The boundary conditions are:

$$\left.\begin{array}{rl} M_0 & = [q_1(4l_1^2 + 5l_2^2 + 5l_1l_2) + q_2l_1^2 \\ & + q_c(l_1^2 - 5l_2^2 + 3l_3^2 + l_1l_2)]/6 \\ Q_0 & = [q_1(l_2 - l_1) + 2q_2l_1 + q_c(2l_3 - l_1 - l_2)]/2 \\ & y|_{x=l} = 0, \theta|_{x=l} = 0 \end{array}\right\} \quad (12)$$

According to the numerical simulation results and the geological conditions of the mine, the engineering parameters are given as follows: the depth of the 26,104 working face is about 250 m; the dip length of the working face is 240 m; and the periodic weighting step is 20~25 m.

The immediate roof, coal seam, and immediate floor are considered as the elastic foundation of the main roof beam. The foundation coefficient k is 0.0380 GPa; the elastic modulus, bending modulus, and bending stiffness of main roof are 21.78 GPa, 67.03 m^4, and 1460 GN·m^2, respectively; l_1 = 2.3 m, $l_2 \approx l_3 \approx$ 20~25 m; $l_4 \approx$ 6 m, l_5 approximately equals to 22 m, $l_5 \geq 3\,L$, taken as 3 L = 60 m; K_1, K_2, K_3 were approximated to 1.2, 0.7, 1.5; q_0 = 5.85 MPa, q_1 = 7.02 MPa, q_2 = 4.10 MPa, q_3 = 8.78 MPa.

Combining (9)–(12) and substituting the data, the breaking position of main roof is 4.4~5.8 m from the edge of the gob.

3.3. Borehole-Scoping in the Main Roof

Borehole-scoping can accurately visualize the lithology, thickness, delamination, cracks, and fractures of the overlying rock strata. To verify the theoretical derivation result of the lateral breakage position of the main roof after panel 26,104 retreated, the borehole-peeping stations were arranged in the test section of 26,102 tailgate to observe the cracks propagation in the roof strata. A total of 23 holes were drilled in 4 groups with a total depth of over 640 m. The site construction is shown in Figure 5.

The obvious crack was defined by a width over 5 mm and a length over 10 mm, and the remarkable crack was defined by a width over 10 mm and a length over 100 mm. The distribution of cracks propagation in the roof strata of 26,102 tailgate is shown in Figure 6, indicated that: (1) most of the cracks were distributed between 10.55 m and 21.43 m above the coal pillar, that is, the range of main roof thickness. (2) Remarkable longitudinal cracks with intersecting circumferential cracks occurred at depths of 19.49 m in BII and 13.93 m in BII, indicating that the main floor had been fractured. The borehole wall within 1 m of the two cracks sites was relatively intact again, so it could be deduced that the fracture line of the main roof was located above the coal pillar. (3) Based on the angles and depths of the boreholes where the two remarkable cracks were observed, the position of fracture line in the main roof was 5~6 m away from the edge of the gob, which was consistent with the theoretical calculation result.

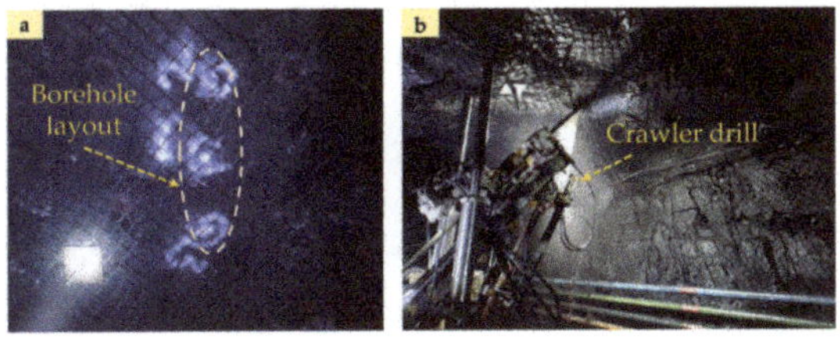

Figure 5. Site construction. (**a**) Layout of part boreholes in roof. (**b**) Crawler drill.

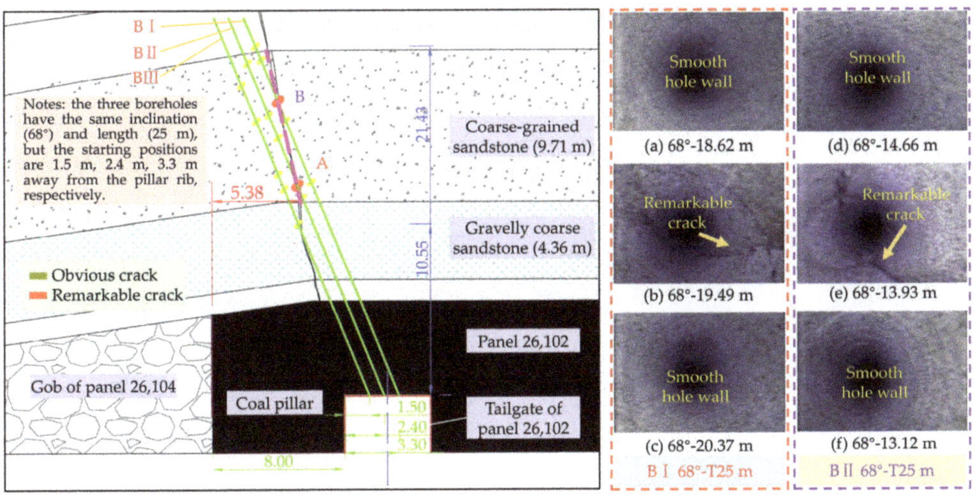

Figure 6. The distribution of roof strata cracks in 26,102 tailgate. Note: taking the representation of 45°-T35 m as an example, 45° denotes the borehole angle, and T35 m represents the full length of the borehole; taking the description of (**a**) 68°-19.62 m as an example, (**a**) denotes the serial number of the camera image of borehole, 68° denotes the borehole angle, and 19.62 m represents the detection depth.

4. Discussion and Analysis of the Simulation Results

4.1. Global Model for the Pillar Width

The model selects the 26,102 tailgate as the test roadway, which is adjacent to the gob of panel 26,104. The *X*-axis is 180 m in the length direction of working face, the *Y*-axis is 120 m in the advancing direction and the *Z*-axis is 110 m in the vertical direction. The roadway section is rectangular, and the dimension is 5.2 m × 3.5 m (width × height). The boundary displacement of the model is constrained in horizontal and bottom. The upper of the model is subjected to a stress of 4.05 MPa equivalent to the self-weight of the overburdened rock, and the lateral pressure coefficient is 1.2. The model is calculated using the Mohr-Coulomb model, while the double-yield model is used for the gob, and the mechanical parameters of each rock formation are taken from Table 2.

4.2. Double-Yield Model of the Gob

With the retreating of the working face, the broken blocks are backfilled in the gob after the main roof periodic fracture. The modulus coefficient of the gangue in the gob will increase significantly after compacted. The densely compacted gob can bear part of the abutment pressure, which effectively weakens the stress concentration in the coal pillar. Therefore, the double-yield model in FLAC3D can be used to simulate the compaction and hardening process of waste in the gob [37]. The overburden pressure parameters and mechanical parameters of gangue in the gob can be obtained by the Salamon formula, expressed as follows [38]:

$$\sigma = E_0\varepsilon / \left(1 - \frac{\varepsilon}{\varepsilon_{\max}}\right) \tag{13}$$

where E_0 is the initial tangential modulus, GPa; σ is the stress of gangue in the gob, MPa; ε is the bulk strain of the compressed gangue in the gob; $\varepsilon_{\max}$ is the maximum bulk strain. The values are given:

$$\left.\begin{array}{l} \varepsilon_{\max} = (b-1)/b \\ E_0 = \frac{1.039\sigma_c^{1.042}}{b^{7.7}} \\ b = (h+h_c)/h_c \end{array}\right\} \tag{14}$$

where b is the crushing expansion coefficient of gangue; σ_c is the gangue compressive strength; h is the mining height of the coal seam; h_c is the height of the roof caving zone in the gob.

According to the engineering situations of Nanyangpo coal mine, the average mining height of coal seam during retreating period is 9.6 m, and the height of the roof caving zone in the gob is about 34.6 m. The values can be substituted into Equations (13) and (14) to derive the double-yield model parameters of the gangue in the gob as shown in Table 3.

Table 3. Double-yield model parameters of gangue in the gob.

Strain	Stress (MPa)	Strain	Stress (MPa)	Strain	Stress (MPa)
0.01	0.10	0.08	1.23	0.15	4.67
0.02	0.22	0.09	1.50	0.16	5.83
0.03	0.34	0.10	1.80	0.17	7.46
0.04	0.48	0.13	2.17	0.18	9.93
0.05	0.63	0.14	2.60	0.19	14.12
0.06	0.81	0.13	3.14	0.20	22.77
0.07	1.01	0.14	3.81	0.21	60.00

The specification of the established unit sub-model of the gob is 1 m × 1 m × 1 m, and a constant velocity of 10^{-5} m/s is applied to the upper surface of the model to determine the mechanical properties of the gob by trial-and-error method. when the parameters are set to the gangue with a density of 1000 kg/m^3, bulk modulus of 11.12 GPa, shear modulus of 5.20 GPa, internal friction angle of 5°, the stress–strain curve of the numerical simulation matches well with that of the theoretical calculation, as shown in Figure 7.

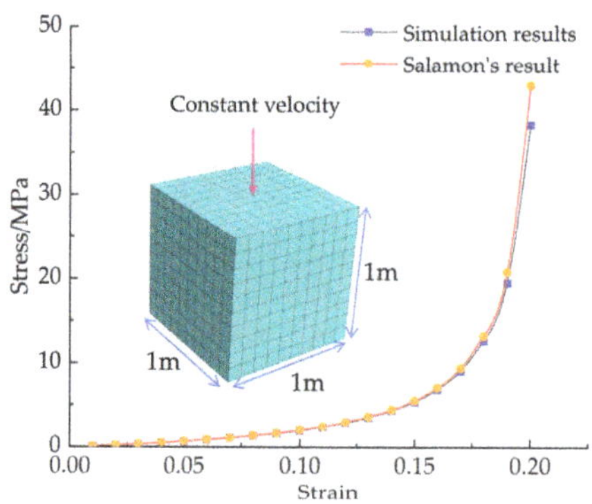

Figure 7. Numerical simulation inversion of goaf parameters.

4.3. Discussion and Analysis of the Simulation Results

4.3.1. Results with Various Coal Pillar Widths

Deviatoric stress is the synthesis of horizontal, vertical, and tangential stresses, and represents the distribution of shear stress in the material subjected to loads, revealing that the essential force of rock failure is mainly caused by shear stress. This stress index is gradually adopted by an increasing number of papers [39,40].

The size of the coal pillar affects the state of deviatoric stress distribution, the failure range of the plastic zone and the deformation extent of the roadway surrounding rock in GED. In this paper, we define the plasticization factor P as a parameter to characterize damage of the surrounding rock in GED, which is given as

$$P_1 = \frac{S_1}{S_p}, P_2 = \frac{S_2}{S_r}, P_3 = \frac{S_3}{S_e}, P_4 = \frac{S_4}{S_e} \tag{15}$$

where P_1, P_2, P_3, and P_4 are the plasticization factors of coal pillar subdivision (I), top coal subdivision (II), virgin coal upper corner subdivision (III), and virgin coal subdivision (IV) in GED, respectively. S_1, S_2, S_3, and S_4 are the areas of plastic zone in I, II, III, and IV. S_p, S_r, and S_e are the cross-section area of I, II, and gob-side entry, respectively.

As shown in Figures 8 and 9, with the increase of coal pillar width, the peak zone of deviatoric stress in the virgin coal area of GED tended to be gradually reduced, which contrasted with the coal pillar area. With the width of coal pillar being 4~6 m, the surrounding rock of gob-side entry was in a lower stress environment. While the plastic zone in II was coalesced with that of the overlying rock and the high stress of the surrounding rock was mostly concentrated in the virgin coal area, indicating that the bearing capacity of coal pillar was poor in such conditions. The range of plastic zone of GED decreased rapidly with a coal pillar of 8 m. The peak zone of deviatoric stress in coal pillar progressively enlarged, while virgin coal side continued to diminish up to similar peak values on both sides. It indicates that the bearing capacity of the coal pillar was enhanced and initiated to sustain the overburden stress in concert with the virgin coal. The coal pillar was in a high stress state with a width of 10~15 m. Additionally, the main bearing body of the overlying load gradually transformed from the virgin coal side to the coal pillar. At this time, the plastic zone of GED was small with a high stability of the surrounding rock, while a significant amount of the coal resources were wasted.

With the increase of coal pillar width, the plasticization factor in four subdivisions of the entry gradually declined. Correspondingly, the peak value of deviatoric stress in the coal pillar area grew swiftly to a stable state, while the value in virgin coal area continued to get smaller. When the width of coal pillar is 8 m, the P were not more than 80% in I and II, and 60% in III and IV. The peak deviatoric stress in coal pillar area was roughly coincident with that of the virgin coal area.

4.3.2. Results with Disturbance of Panel Retreating

Mining-induced stress exerted an essential influence on the stability of the surrounding rock in GED. Exploring the distribution characteristics of the deviatoric stress and plastic zone of the surrounding rock advanced of the working face could provide the necessary basis for the surrounding rock control.

As shown in Figures 10 and 11:

(1) In the vicinity of the working face, the peak and range of deviatoric stress in coal pillar were much larger than that in virgin coal area, and the peak ratio was 1.41, indicating that the stress in the surrounding rock was mainly sustained by the coal pillar. P were more than 90% in I and II, and 100% in III and IV, representing that extensive damage occurred in the surrounding rock of GED, that meant reinforced-supported measures should be implemented to avoid destabilization of the coal pillar.

(2) At 7 m ahead of the working face, the peak deviatoric stress in the coal pillar reached the maximum value and was much smaller than that of virgin coal area with a peak ratio of 0.68, and the mining-induced stress of the working face was mainly undertaken by virgin coal area. *P* were still more than 90% in I and II, and 100% in III and IV. *P* dropped rapidly by 15% in II, and the value in I did not vary significantly compared to that of the vicinity of the working face.

(3) At 30 m ahead of the working face, the discrepancies of peak deviatoric stress on both sides of entry were further reduced with a peak ratio of 0.81. It stated that the mining-induced stress was progressively performed by virgin coal area in cooperation with coal pillar. *P* in III and IV diminished promptly, and *P* were virtually invariable in I and II, indicating that the stress disturbance of the working face was fading quickly.

(4) At 30 m ahead of the working face, the peak ratio of deviatoric stress between coal pillar and virgin coal area was 0.84. *P* in III and IV were further depressed by 9% and 7%, and *P* in I and II kept constant, indicating that the extent of damage to the roadway had steadily stabilized.

The above analysis suggested that the plasticization factor *P* in III and IV were more significantly troubled by mining-induced stress of the working face. *P* firstly dropped from both much more than 100% to 40% and 60% in III and IV, respectively. The deviatoric stress in coal pillar and virgin coal area first increased rapidly to the peak value, and then gradually tended to be stable. That indicated that the main bearing body of overburden stress was "the coal pillar → the virgin coal area → the collaboration of coal pillar and virgin coal area".

4.3.3. Coal Pillar Width Determination

Based on the theoretical calculation and field measurement results, the fracture position of the main roof is located in the range of 5~6 m above the virgin coal from edge of the gob. That means the fracture line of the main roof is above the coal pillar when the width is 8 m, which contributes to maintaining the stability of the surrounding rock of the gob-side entry. The overlying strata load of GED is performed by virgin coal area in cooperation with coal pillar. The plasticization factors of coal pillar subdivision and top coal subdivision are both less than 80%, which are within the control of support. The surrounding rock in GED, advanced 60 m of the working face, is obviously affected after the panel retreated, and the influence on the stress and plastic zone of the surrounding rock tends to be stable. Therefore, it is feasible to set up an 8 m coal pillar with targeted support techniques to maintain the stability of the surrounding rock of GED.

5. Surrounding Rock Control Techniques

5.1. Cracks Distribution of the Coal Body

The results of borehole-scoping in 26,102 tailgate are shown in Figure 12:

(1) On the coal pillar side of GED. The borehole wall within 2.08 m was relatively broken from the horizontal distance of the coal pillar rib of the roadway, indicating that the integrity of coal body was poor. The integrity of the coal body was enhanced within 2.08~6.07 m from the coal pillar rib with certain cracks existing. When the horizontal distance from the gob edge of coal pillar was less than 1.93 m, the borehole wall was seriously damaged or even collapsed, making it impracticable for further observation.

(2) On the virgin coal side of GED. The borehole wall within 1.84 m from the virgin coal rib of the roadway was relatively broken, and the borehole wall integrity was improved with increasing borehole depth. Slightly developed cracks existed in the borehole within 1.84~3.76 m from virgin coal rib. Borehole wall was gradually smooth at a horizontal distance of over 3.79 m from virgin coal rib, that indicated good coal body integrity.

(3) In the roof of GED. The coal body was severely damaged at a depth of 1.00 m from upper corner of coal pillar rib, 1.95 m from the roof of the roadway, and 1.76 m from the upper corner of virgin pillar rib. The coal body gradually turned intact with the distances exceeding 1.76 m, 4.72 m, and 3.97 m, respectively.

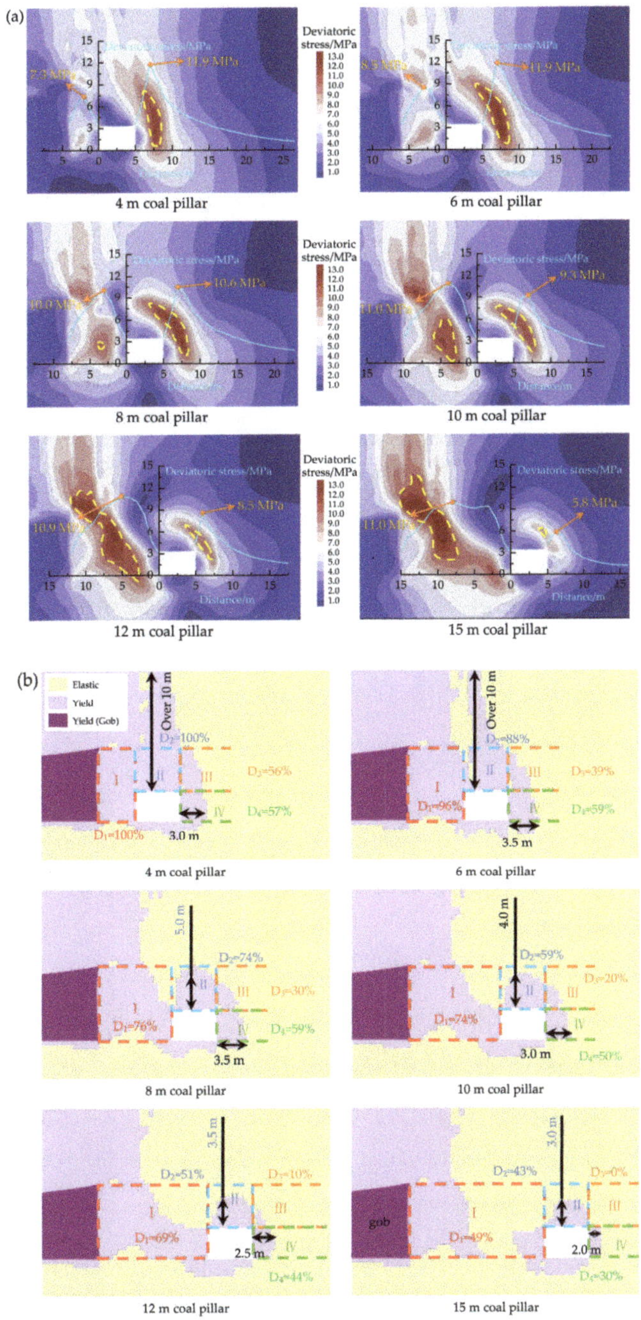

Figure 8. Characteristics of the surrounding rock with various coal pillar widths. (**a**) Deviatoric stress. (**b**) The plastic zone.

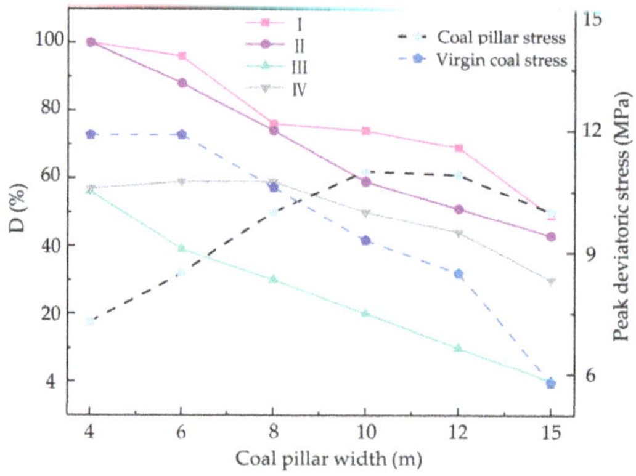

Figure 9. Distribution of the plasticization factor in subdivisions.

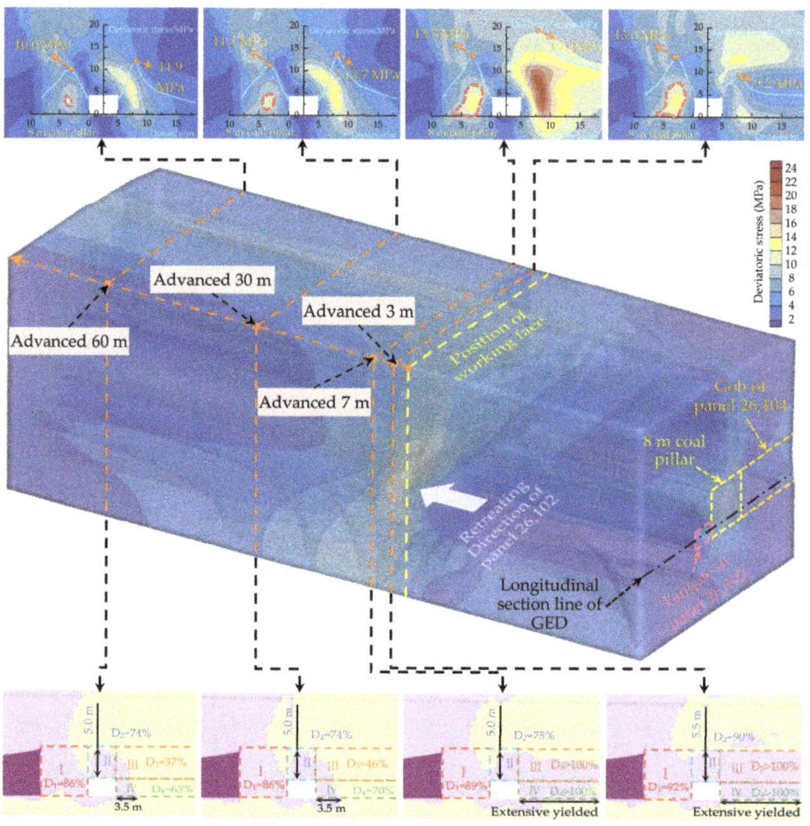

Figure 10. The distribution characteristics of surrounding rock advanced of the working face.

5.2. Support Principles

Based on the crack distribution of the surrounding rock, the compressive stress zone of bolts should cover the fracture zone, and the length of anchor cables should be greater than the depth of plastic zone, which played an effective role in improving the stress situations of the surrounding rock. The control techniques of the surrounding rock with the cooperative bearing of bolts and anchor cables were proposed, and the support principles are shown in Figure 13. The main contents were as follows: (1) Shallow bearing area of bolts. The 2.4 m prestressed bolts were used in the shallow part of the surrounding rock, supplemented by high strength plates and W-shaped steel straps, to provide combined compressive stress for the fractured zone, which contributed to preventing support components from failure caused by the surrounding rock falling of the fractured zone. (2) Deep bearing zone of anchor cables. The 8.0 m high prestressed anchor cables were used in the roof and were embedded in the stable strata by passing through the top coal of about 6 m thick in the roof. The 5.0 m high prestressed anchor cables were used in both ribs to pass through the fracture zone and anchored into the relatively intact coal body. The high pre-stressed anchor cable could increase the shear resistance of the coal body, making the surrounding rock of the roadway to reinstate the state of three-dimensional stress in a certain extent, and exert the bearing capacity of the deep surrounding rock of the roadway.

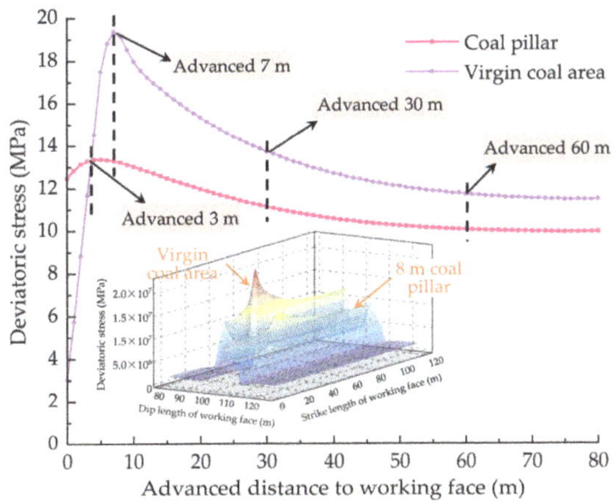

Figure 11. The stress distribution advanced of the working face in GED.

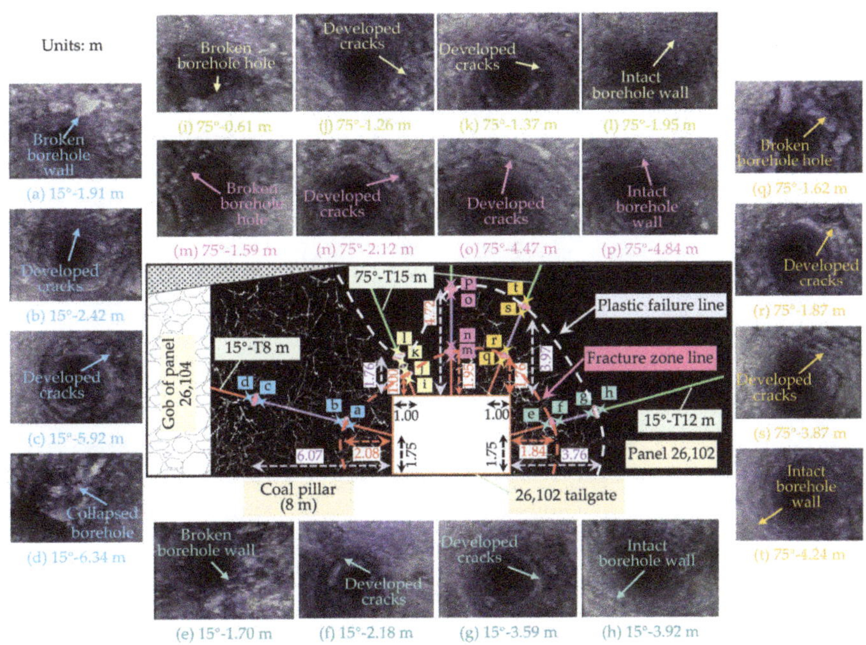

Figure 12. Borehole-scoping for the coal body in 26,102 tailgate.

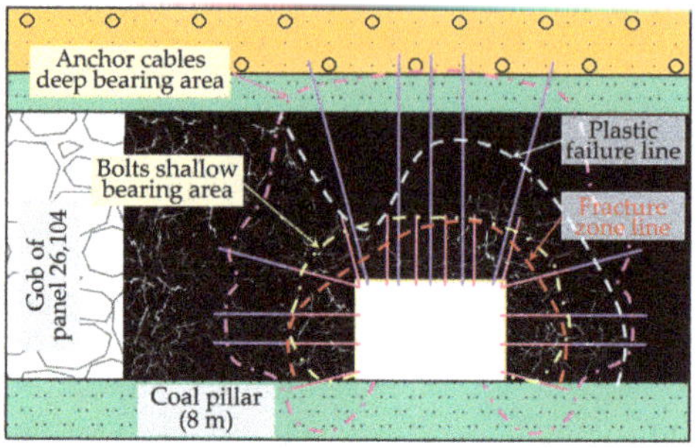

Figure 13. Cooperative bearing structure of bolts and anchor cables.

5.3. Technical Measures

Based on the field geological conditions, the support scheme of 26,102 tailgate is shown in Figure 14. The specific support parameters are as follows:

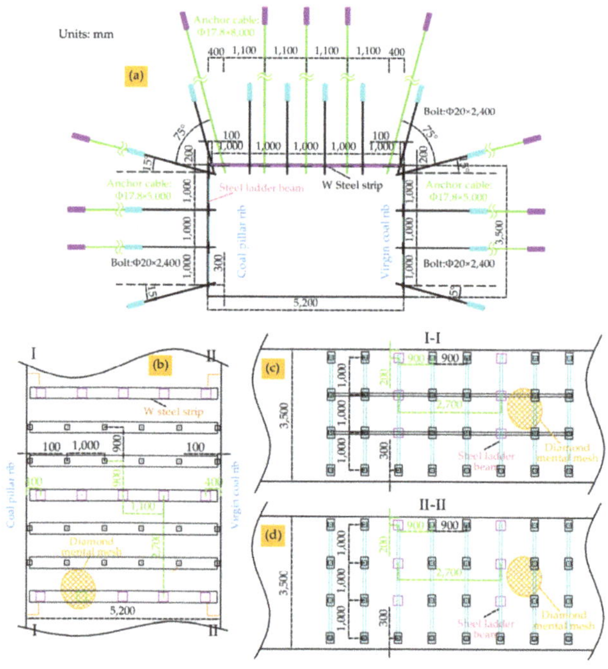

Figure 14. Detailed support parameters. (**a**) Entry section. (**b**) Support pattern in roof. (**c**) Support pattern in coal pillar rib. (**d**) Support pattern in virgin coal rib.

The roof and two ribs were supported by high-strength steel bolts 20 mm in diameter and 2400 mm in length. The inter-row spacing of the bolts was 1000 mm × 900 mm. A row of anchor cables is arranged for every two rows of bolts. The anchor cables 17.8 mm in diameter and were used for roadway support with a length of 8000 mm in roof and 5000 mm in two ribs. The inter-row spacing for the cables was 1100 mm × 2700 mm. The bottom one of each row of anchor cables was replaced by a high-strength steel bolt 20 mm in diameter and 2400 mm in length for both ribs. The bolts and anchor cables were connected with steel ladder beams made of 16 mm round steel in both ribs and with W-shaped steel straps with length × width × thickness of 5000 mm × 300 mm × 3 mm in the roof. Bolts in the roof and rib corners were installed at a 15° incline. Bolt plates of W-shaped steel straps were selected for the bolts in both ribs. A metal mesh 4 mm in diameter was used in the roadway section to prevent broken coal body from falling.

6. Application and Analysis

Field measurements of ground response can effectively and comprehensively reflect the working status of the support system and verify the effect of the roadway support scheme, which contributes to the stability of the roadway support. Three stations were set up in the 200 m test section of 26,102 tailgate, and each station included one group deformation observation of GED and one group of bolts and anchor cables forces monitoring. The station layout is shown in Figure 15.

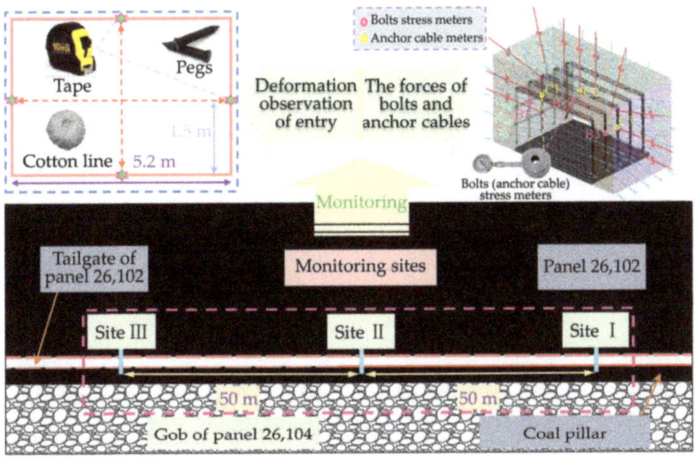

Figure 15. Layout of the field measurements.

Taking the monitoring results of the typical station II as an example, the surface deformation of the roadway surrounding rock is shown in Figure 16. With the combined support of high-strength prestressed bolts and anchor cables, there was no considerable deformation and damage in GED during roadway driving and retreating period. The surrounding rock deformation in the roof, virgin coal rib and coal pillar rib finally stabilized by 140 mm, 96 mm, and 105 mm separately after 25 days of roadway development. During the driving period, the maximum deformation of roadway in roof, virgin coal rib, and coal pillar rib advanced of the working face was 296 mm, 272 mm, and 251 mm, respectively.

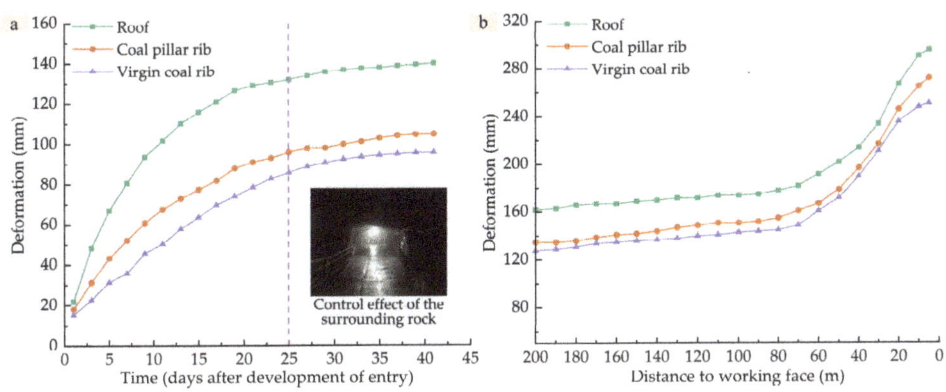

Figure 16. Deformation observation of the surrounding rock. (**a**) During roadway development. (**b**) During working face retreating.

At each monitoring section, the forces of bolts and anchor cables were monitored in the roof and two ribs, respectively. The measured bolts and anchor cables of the coal pillar rib, roof and virgin coal rib were numbered B1, B2, B3 and C1, C2, C3, respectively. The initial prestressing forces of bolts and anchor cables were 72~80 kN and 175~190 kN, which were 40~45% and 34~37% of the breaking load separately. The monitoring results are shown in Figure 17.

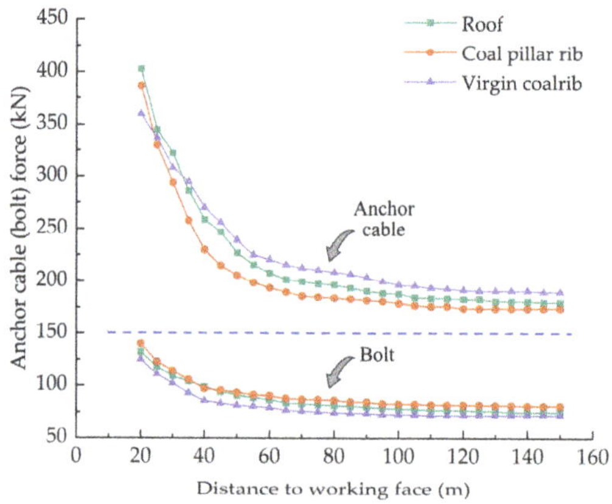

Figure 17. Anchor cable (bolt) force at different distances from the working face.

When the monitoring section was 60 m away from the working face, the forces of B1, B2, and B3 bolts were about 51%, 49%, and 45% of their breaking load (179 kN), and the forces of C1, C2, and C3 anchor cables were about 37%, 40%, and 43% of their breaking load (520 kN). The growth of bolts and anchor cables forces was less than 10%, indicating that the roadway section was weakly affected by the mining-induced stress at a distance of over 60 m from the working face. The disturbance impact of abutment stress was rapidly intensified once the monitoring section was less than 60 m from the working face. The forces of B1, B2, and B3 bolts were about 78%, 74%, and 70% of their breaking load, and the forces of C1, C2, and C3 anchor cables were about 75%, 78%, and 69% of their breaking load with 20 m from the working face. The forces of bolts and anchor cables were far less than its upper breaking limit, indicating that they were in good working condition. The 20 m area of GED advanced of the working face was supported by single hydraulic props, which contributed to avoiding the instability of the surrounding rock caused by intense disturbance of mining-induced stress. Therefore, the monitoring of mining pressure in this area could not be highlighted. Field measurements of ground response showed that the combined control techniques of bolts and anchor cables with an 8 m coal pillar achieved effective control of the roadway surrounding rock under the influence of upper residual coal pillar.

7. Conclusions

Based on theoretical calculation, numerical simulation, and field measurements, the evolution laws of the lateral abutment stress in lower working face at different relative distances to the RCP were studied, as well as the whole space-time evolution law of deviatoric stress and plastic zone of GED during driving and retreating periods with various coal pillar widths under the influence of RCP. The targeted support techniques with an 8 m coal pillar were proposed. The conclusions were as follows:

(1) The above distribution laws indicated that the lateral abutment stress was most significantly affected when the edge of the lower working face was located directly below the RCP, and the peak stress growth coefficient was 1.20~1.30. When the EWF of the lower coal seam was 2.5 m away from the centerline of RCP, the peak growth coefficient of the lateral abutment stress was 1.22. The influence of stress disturbance of RCP was relatively diminished when a small coal pillar was set up to drive the entry along the gob in such conditions. Consequently, the superimposed disturbance impact of RCP and mining-induced stress on GED should be fully considered to prevent the roadway from instability.
(2) The mechanical parameters of the rock and coal mass were obtained using laboratory experiments coupled with Roclab software. These parameters were substituted into the established main roof structure mechanics model to derive the breakage position of the main roof with the influence of RCP, which was verified the rationality of the calculation results using borehole-scoping.
(3) FLAC3D was used to study the whole space-time evolution law of deviatoric stress and plastic zone of GED during driving and retreating periods with different coal pillar widths under the influence of RCP, and the plasticization factor P was introduced to quantify the evolution of the plastic zone in four subdivisions of the roadway surrounding rock. The results showed that when the width of the coal pillar was 8 m, P were not more than 80% in I and II, and 60% in III and IV. The peak ratio of deviatoric stress between coal pillar and virgin coal area was 0.84. Additionally, the plasticization factor P in the virgin coal upper corner subdivision and virgin coal subdivision were more significantly troubled by mining-induced stress of the working face. P firstly dropped from both being much more than 100% to 40% and 60% in III and IV.
(4) Field measurements of ground response showed that the combined control techniques of bolts and anchor cables with an 8 m coal pillar achieved effective control of the roadway surrounding rock under the influence of the residual coal pillar. The maximum deformation of roadway in the roof, virgin coal rib and coal pillar rib advanced of the working face was 296 mm, 272 mm, and 251 mm separately during the retreating period. The forces of bolts and anchor cables were 70%~78% and 69%~75% of their breaking load during the retreating period, and the supporting components were all in good working conditions, which realized a valid effect of surrounding rock control. This provided a reference for promoting the coal recovery rate in similar geological conditions.

Author Contributions: Conceptualization, S.X. and F.G.; methodology, F.G. and Y.W.; software, F.G. and Y.W.; data curation, F.G. and Y.W.; writing—original draft preparation, F.G.; writing—review and editing, S.X. and F.G.; supervision, S.X.; project administration, S.X., F.G. and Y.W. All authors have read and agreed to the published version of the manuscript.

Funding: This work is financially supported by the National Natural Science Foundation of China (Grant No. 52074296), the Fundamental Research Funds for the Central Universities (Grant No. 2022YJSNY18), the National Natural Science Foundation of China (Grant No. 52004286), the Fundamental Research Funds for the Central Universities (Grant No. 2022XJNY02), and the China Postdoctoral Science Foundation (Grant No. 2020T130701, 2019M650895), all of which are gratefully acknowledged.

Conflicts of Interest: The authors declare no conflict of interest.

References

1. Zhang, Z.Z.; Bai, J.B.; Chen, Y.; Yan, S. An innovative approach for gob-side entry retaining in highly gassy full-mechanized longwall top-coal caving. *Int. J. Rock Mech. Min. Sci.* **2015**, *80*, 1–11. [CrossRef]
2. Wu, R.; He, Q.; Oh, J.; Li, Z.; Zhang, C. A new gob-side entry layout method for two-entry longwall systems. *Energies* **2018**, *11*, 2084. [CrossRef]
3. An, Y.P.; Zhang, N.; Zhao, Y.M.; Xie, Z.Z. Field and numerical investigation on roof failure and fracture control of thick coal seam roadway. *Eng. Fail. Anal.* **2021**, *128*, 105594. [CrossRef]

4. Wang, J.C.; Yang, S.L.; Wei, W.J.; Zhang, J.W.; Song, Z.Y. Drawing mechanisms for top coal in longwall top coal caving (LTCC): A review of two decades of literature. *Int. J. Coal Sci. Technol.* **2021**, *8*, 1171–1196. [CrossRef]
5. Zhang, G.C.; Wen, Z.J.; Liang, S.J.; Tan, Y.L.; Tian, L.; Zhao, Y.Q.; Zhao, D.S. Ground Response of a Gob-side Entry in a Longwall Panel Extracting 17 m-Thick Coal Seam: A Case Study. *Rock Mech. Rock Eng.* **2020**, *53*, 497–516. [CrossRef]
6. Yan, H.; Zhang, J.X.; Feng, R.; Wang, W.; Lan, Y.W.; Xu, Z.J. Surrounding rock failure analysis of retreating roadways and the control technique for extra-thick coal seams under fully-mechanized top caving and intensive mining conditions: A case study. *Tunn. Undergr. Space Technol.* **2020**, *97*, 103241. [CrossRef]
7. Yan, H.; Zhang, J.X.; Ding, Z.W.; Huang, Y.L. Surrounding rock deformation mechanism of roadways with extra-thick coal seam. *Disaster Adv.* **2013**, *6*, 226–233.
8. Jiang, B.Y.; Gu, S.T.; Wang, L.G.; Zhang, G.C.; Li, W.S. Strain burst process of marble in tunnel-excavation-induced stress path considering intermediate principal stress. *J. Cent. South Univ.* **2019**, *26*, 984–999. [CrossRef]
9. Wu, W.D.; Bai, J.B.; Wang, X.Y.; Shuai, S.; Wu, S.X. Numerical study of failure mechanisms and control techniques for a gob-side yield pillar in the Sijiazhuang coal mine, China. *Rock Mech. Rock Eng.* **2019**, *52*, 1231–1241. [CrossRef]
10. Shi, X.S.; Jing, H.W.; Zhao, Z.L.; Gao, Y.; Zhang, Y.C.; Bu, R.D. Physical experiment and numerical modeling on the failure mechanism of gob-side entry driven in thick coal seam. *Energies* **2020**, *13*, 5425. [CrossRef]
11. Gao, F.Q.; Stead, D.; Kang, H.P.; Wu, Y.Z. Discrete element modelling of deformation and damage of a roadway driven along an unstable goaf—A case study. *Int. J. Coal Geol.* **2014**, *127*, 100–110. [CrossRef]
12. Gao, F.Q.; Stead, D.; Kang, H.P. Simulation of roof shear failure in coal mine roadways using an innovative UDEC Trigon approach. *Comput. Geotech.* **2014**, *61*, 33–41. [CrossRef]
13. Gao, F.Q.; Stead, D.; Kang, H.P. Numerical simulation of squeezing failure in a coal mine roadway due to mining-induced stresses. *Rock Mech. Rock Eng.* **2015**, *48*, 1635–1645. [CrossRef]
14. Liu, X.; He, M.; Wang, J.; Ma, Z. Research on non-pillar coal mining for thick and hard conglomerate roof. *Energies* **2021**, *14*, 299. [CrossRef]
15. Wang, K.; Huang, Y.; Gao, H.; Zhai, W.; Qiao, Y.; Li, J.; Ouyang, S.; Li, W. Recovery technology of bottom coal in the gob-side entry of thick coal seam based on floor heave induced by narrow coal pillar. *Energies* **2020**, *13*, 3368. [CrossRef]
16. Batugin, A.; Wang, Z.Q.; Su, Z.H.; Sidikovna, S.S. Combined support mechanism of rock bolts and anchor cables for adjacent roadways in the external staggered split-level panel layout. *Int. J. Coal Sci. Technol.* **2021**, *8*, 659–673. [CrossRef]
17. He, W.R.; He, F.L.; Zhao, Y.Q. Field and simulation study of the rational coal pillar width in extra-thick coal seams. *Energy Sci. Eng.* **2020**, *8*, 627–646. [CrossRef]
18. Li, X.B.; Zhao, Y.Q.; He, W.R.; Li, L.; He, F.L. Study on coal pillar width and surrounding rock control of gob-side entry in extra-thick coal seam. *Geotech. Geol. Eng.* **2020**, *38*, 1–14. [CrossRef]
19. Yang, R.S.; Zhe, Y.; Li, Y.L.; Li, W.Y.; Lin, H. Coal pillar size design and surrounding rock control techniques in deep longwall entry. *Arab. J. Geosci.* **2020**, *13*, 1–14. [CrossRef]
20. Qi, W.Y.; Zhang, Q.; Zhang, J.X.; Zhang, J.G.; Zhu, C.L. Design of coal pillars of gob-side entry between mining faces with large differences in mining height in deep mine. *Energy Sources Part A Recovery Util. Environ. Eff.* **2020**, *42*, 2648–2663. [CrossRef]
21. Wang, Q.; Gao, H.K.; Jiang, B.; Li, S.; He, M.C.; Wang, D.C.; Li, W.; Gao, S.; Yang, H.C. Research on reasonable coal pillar width of roadway driven along goaf in deep mine. *Arab. J. Geosci.* **2017**, *10*, 466. [CrossRef]
22. Ma, Z.; Chen, C.; Liang, X.; Chen, A.; Song, W. Field and numerical investigation on the stability of coal pillars of gob-side entry driving with top coal. *Arab. J. Geosci.* **2020**, *13*, 1–11. [CrossRef]
23. Jiang, L.S.; Zhang, P.P.; Chen, L.J.; Hao, Z.; Sainoki, A.; Wang, Q.B. Numerical Approach for Goaf-Side Entry Layout and Yield Pillar Design in Fractured Ground Conditions. *Rock Mech. Rock Eng.* **2017**, *50*, 3049–3071. [CrossRef]
24. Han, C.L.; Yang, H.Q.; Zhang, N.; Deng, R.J.; Guo, Y.X. Zoning control technology of gob-side roadway driving with small coal pillar facing mining in a special isolated island working face: A case study. *Appl. Sci.* **2021**, *11*, 10744. [CrossRef]
25. Wang, E.; Xie, S.R. Determination of coal pillar width for gob-side entry driving in isolated coal face and its control in deep soft-broken coal seam: A case study. *Energy Sci. Eng.* **2022**, 1–12. [CrossRef]
26. Zha, W.H.; Shi, H.; Liu, S.; Kang, C.H. Surrounding rock control of gob-side entry driving with narrow coal pillar and roadway side sealing technology in Yangliu Coal Mine. *Int. J. Min. Sci. Technol.* **2017**, *27*, 819–823. [CrossRef]
27. Xie, S.R.; Pan, H.; Chen, D.D.; Gao, M.M.; He, S.S.; Song, B.H. The Structure and activity laws of interlayer equivalent basic roof in mult-seam mining. *J. China Univ. Min. Technol.* **2016**, *46*, 1218–1225.
28. Zhang, C.L.; Zhang, Y.; Zuo, J.P.; Gao, S.Y. Coal seam mining in a physical model vis-à-vis MATLAB analysis and geological radar. *Min. Metall. Explor.* **2021**, *38*, 897–911.
29. Xie, S.R.; Wu, X.Y.; Chen, D.D.; Sun, Y.H.; Wang, E.; Wu, X.W.; Duan, X.B.; Jiang, Z.S.; Qi, P.Y.; Shi, S.H. Failure mechanism and control technology of thick and soft coal fully mechanized caving roadway under double gobs in close coal seams. *Shock Vib.* **2020**, *2020*, 8846014. [CrossRef]
30. Wang, M.; Bai, J.B.; Li, W.F.; Wang, X.Y.; Cao, S.G. Failure mechanism and control of deep gob-side entry. *Arab. J. Geosci.* **2015**, *8*, 9117–9131. [CrossRef]
31. Yan, S.; Bai, J.B.; Wang, X.Y.; Huo, L.J. An innovative approach for gateroad layout in highly gassy longwall top coal caving. *Int. J. Rock Mech. Min. Sci.* **2013**, *59*, 33–41. [CrossRef]
32. Hoek, E.; Brown, E.T. Practical estimates of rock mass strength. *Int. J. Rock Mech. Min. Sci.* **1997**, *34*, 1165–1186. [CrossRef]

33. Hoek, E.; Carranza-Torres, C.T.; Corkum, B. Hoek-Brown failure criterion–2002 edition. In Proceedings of the Fifth North American Rock Mechanics Symposium, Toronto, ON, Canada, 7–10 July 2002; pp. 18–22.
34. Zhang, G.C.; He, F.L.; Jiang, L.S. Analytical analysis and field observation of break line in the main roof over the goaf edge of longwall coal mines. *Math. Prob. Eng.* **2016**, *2016*, 4720867. [CrossRef]
35. Zhu, C.L.; Zhang, J.X.; Li, M.; He, Z.W.; Wang, Y.Y.; Lan, Y.W. Effect mechanism of strata breakage evolution on stope deformation in extra-thick coal seams. *Alex. Eng. J.* **2022**, *61*, 5003–5020. [CrossRef]
36. Long, Y.Q. *Calculation of Elastic Foundation Beam*; People's Education Presss: Beijing, China, 1981.
37. Wang, M.; Li, W.F. Analysis on the loading mechanism of caving gangue in goaf based on Salamon model. *Coal Sci. Tech.* **2015**, *41*, 50–54. [CrossRef]
38. Salamon, M. Mechanism of caving in longwall coal mining. Rock mechanics contributions and challenges. In Proceedings of the 31st US Symposium. on Rock Mechanics, Golden, CO, USA, 18–20 June 1990; pp. 161–168.
39. Liu, R.; Zhu, Q.J. An analysis of the impact of deviatoric stress and spherical stress on the stability of surrounding rocks in roadway. *Geofluids* **2021**, *2021*, 3095690. [CrossRef]
40. Chen, D.D.; Wang, E.; Xie, S.R.; He, F.L.; Wang, L.; Zhang, Q.; Wu, X.Y.; Jiang, Z.S.; Li, Y.B.; Shi, S.H. Roadway surrounding rock under multi-coal-seam mining: Deviatoric stress evolution and control technology. *Adv. Civ. Eng.* **2020**, *2020*, 9891825. [CrossRef]

energies MDPI

Article

Study on the Long-Distance Gas Pre-Drainage Technology in the Heading Face by Directional Long Borehole

Yunbing Hou, Junqi Cui * and Ruipeng Liu

School of Energy and Mining Engineering, China University of Mining and Technology-Beijing, Beijing 100083, China
* Correspondence: 18339796517@163.com

Abstract: Gas control in the heading face of a coal roadway is an important and difficult point in coal mining in China. On the basis of analyzing the disadvantages of high gas control cost and long drainage period in the existing mine heading face, a long-distance pre-drainage method of long-distance drilling is proposed to control the gas in the heading face so as to improve the tunneling speed. Applied to the engineering geological conditions of Changcun coal mine, the technology is studied in detail. First, a gas migration model considering permeability changing with time is established, and the model is put into the numerical simulation software to study the variation law of permeability and gas pressure under the conditions of single borehole and multi-borehole drainage. The results show that with the increase of drainage time, the permeability around the borehole increases gradually, the gas pressure decreases gradually, and the permeability at the borehole boundary increases the most, reaching 1.2 times the initial permeability. In the process of multi-borehole drainage, there will be mutual influence between boreholes, but with the increase of borehole spacing, the degree of this influence gradually decreases. Second, according to the results of numerical simulation, a reasonable gas drainage scheme is designed and applied in the field. The field application shows that the technology has a good gas drainage effect, the gas drainage concentration and flow are at a high level for a long time, the drilling cuttings quantity is always lower than the critical value, and the excavation length of roadway increases by more than 50 m per month. These results indicate that this technology is a promising method to realize the safe and rapid excavation of a mine coal roadway.

Keywords: directional long borehole; long-distance; heading face; permeability evolution; gas pressure evolution; gas drainage

Citation: Hou, Y.; Cui, J.; Liu, R. Study on the Long-Distance Gas Pre-Drainage Technology in the Heading Face by Directional Long Borehole. *Energies* **2022**, *15*, 6304. https://doi.org/10.3390/en15176304

Academic Editor: Sergey Zhironkin

Received: 16 August 2022
Accepted: 25 August 2022
Published: 29 August 2022

1. Introduction

Gas disaster is one of the main disasters in coal mines in China. The occurrence of gas accidents not only affects the safety production of the mine, but also seriously affects the life safety of workers. Therefore, it is necessary to control the coal seam gas [1–3]. Coal seam gas drainage is an effective method to prevent the occurrence of gas accidents. At the same time, the use of the extracted gas not only increases the total amount of energy, but also is conducive to environmental protection [4,5].

According to the different location of drainage, coal seam gas drainage can be divided into working face drainage and heading face drainage. Among them, the heading face is difficulty of gas control, which often leads to abnormal gas emission accidents due to a poor drainage effect [6,7]. At present, gas drainage in the heading face mainly includes cross-measure borehole gas drainage technology in the floor rock roadway and in-seam borehole gas drainage technology in the heading face [8–10]. Cross-measure borehole gas drainage technology is to extract coal seam gas by drilling in the rock roadway under the coal seam (Figure 1a). This method has the advantages of high safety, good borehole structure stability and long drainage time [11–13]. However, due to the need to dig out a rock roadway, there are some shortcomings, such as high cost, large engineering quantity, long construction period and so on [14]. In-seam borehole gas drainage technology is to extract the coal seam gas by drilling along the coal seam in the heading face (Figure 1b). Compared with the cross-measure borehole technology, this method does not need to excavate a rock roadway, which reduces the construction cost. However, the length of one-time pre-drainage is short, and can only reach 80–100 m. After passing the inspection, the excavation can be carried out, and then the next drainage cycle is carried out, that is "Drilling-Drainage-Inspection-Excavation-Drilling" [15,16], which leads to the slow excavation speed of roadway.

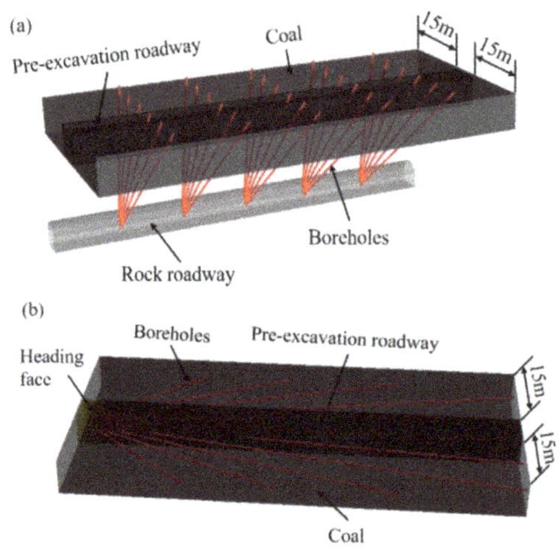

Figure 1. Schematic diagram of gas drainage technology in the heading face. (**a**) Cross-measure borehole gas drainage technology. (**b**) In-seam borehole gas drainage technology.

With the development of technology, directional long borehole gas drainage technology has been gradually applied in coal mines. Wang et al. [17] summarized the application of directional long boreholes in China's coal mines, including pre-mining and post-mining drainage, and found that it can effectively improve the gas drainage efficiency. Lu et al. [18] conducted the gas drainage testing of directional long boreholes in Daning Coal Mine, China, which effectively controlled the gas outburst accident risk of the mine. Compared with the ordinary borehole drainage method, the proportion of coal seam drilling is higher, and the drainage effect is better. Wang et al. [19], Li et al. [20] and Hao et al. [21] studied the gas drainage in goaf by directional long boreholes instead of a separate drainage roadway, analyzed and determined the reasonable horizon of directional long boreholes in the roof, and found that this method can effectively solve the problem of gas overrun in the upper corner of the working face, and can replace the roadway-based gas drainage, thereby saving much work.

Based on the advantages of the directional long borehole and the disadvantages of existing ordinary borehole gas drainage in the heading face, a new method of long-distance gas drainage in the heading face by using the directional long borehole is proposed. Through the construction of boreholds by drilling in the completed roadway (not floor rock roadway), the coal seam gas near the pre-excavation roadway is extracted. The construction diagram is shown in Figure 2. Compared with the cross-measure borehole gas drainage technology, this method does not need to excavate a rock roadway, and has the advantages of low cost. Compared with the in-seam borehole gas drainage technology in the heading face, the drainage distance is longer, which increases the length of what was once the excavation roadway, and can extract the coal seam gas for a long time. Taking the return air roadway of working face 2302 in Changcun coal mine as an example, this paper establishes a gas migration model considering the change of permeability with drainage time, studies the gas migration law and borehole layout of directional long borehole drainage, carries out industrial testing on site, and analyzes the drainage effect.

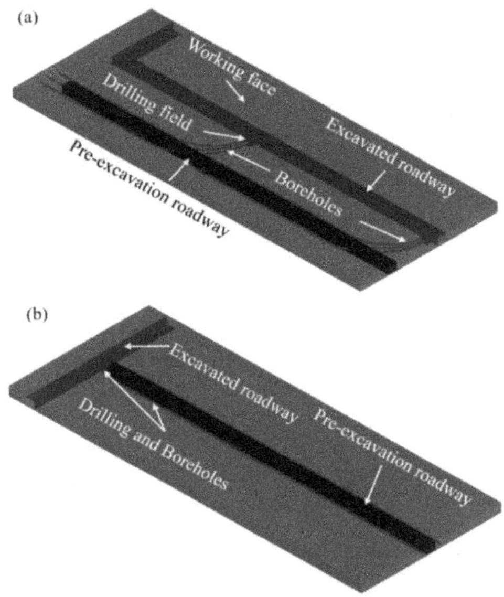

Figure 2. Schematic diagram of gas pre-drainage in the heading face with directional long borehole. (**a**) Construction borehole at the gateway. (**b**) Construction borehole at preparation roadway.

2. Project Summary and Basic Parameter Test

The Changcun coal mine is located in the Changzhi City, Shanxi Province, China. It is a large modern mine of Shanxi Lu'an environmental protection energy development Co., Ltd. (Changzhi, China). The main mining area is 3# coal seam. The geological structure is mainly fold. The strata strike nearly north-south and incline westward with an inclination of 3–6°. In the east, it is mainly a tilt structure with near east-west undulation. In the west, there are nearly north-south folds. The geological structure provides a good environment for coal seam gas storage. The gas content of the coal seam is generally 8–10 m^3/t, and the gas content is large.

2.1. Project Summary

2302 working face is located in No. 23 mining area of the mine, mining 3# coal seam; the thickness of the coal seam is generally 4.84–7.32 m, and the average thickness is 6.09 m. The average depth of the coal seam is about 500 m. The roadway layout adopts the "two air intakes and one return" mode; the 2302 belt transportation roadway and auxiliary transportation roadway provide for entry of the air, and the 2302 return air roadway returns the air. A 2302 belt transportation roadway and auxiliary transportation roadway have been excavated, and 2302 return air roadway is the pre-excavation roadway. The mine location and roadway layout of 2302 working face are shown in Figure 3. According to the field measurement data, the gas content of 2302 working face is 8.5 m^3/t, the gas pressure is 0.35 MPa, and the permeability is 8.09×10^{-8} m^2. The attenuation coefficient of borehole gas flow is 0.1726–0.3025 d^{-1}. According to the difficulty degree of coal seam drainage table (Table 1), the gas drainage in 2302 working face is classified as difficult.

The gas pressure is produced by the thermal movement of free gas. Changcun coal mine has high gas content and low gas pressure, which indicates that the coal seam has strong adsorption capacity, and the adsorption content of coal seam gas accounts for more.

Gas desorption is a slow process. Research shows that under natural conditions, when the coal particle size is 1 cm, the time required for desorption of 90% of the gas is 15 years [22]. Therefore, for the coal seam with strong adsorption capacity, gas drainage should be carried out for a long time even under negative pressure.

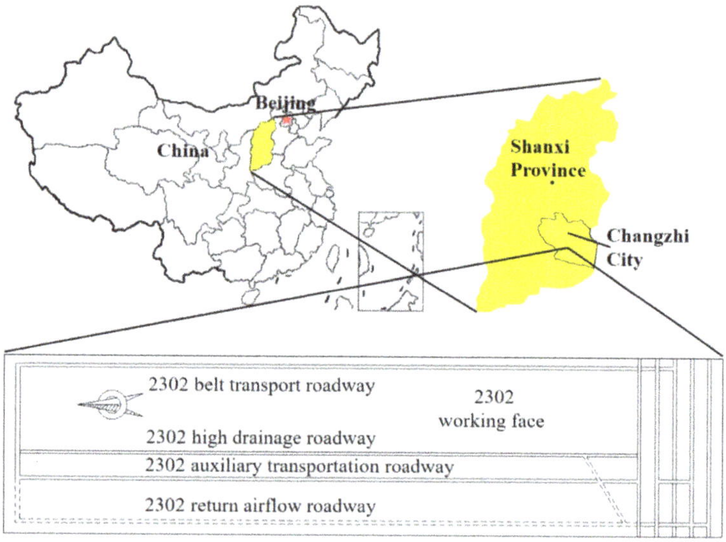

Figure 3. Location of Changcun Coal Mine and roadway layout of 2302 working face.

Table 1. Difficulty degree of coal seam gas drainage.

Classification	Attenuation Coefficient of Borehole Gas Flow/d^{-1}	Coal Seam Permeabilit/m^2
Easy to be extracted	<0.003	>2.5 × 10^{-4}
Can be extracted	0.003–0.05	2.5 × 10^{-6}–2.5 × 10^{-4}
Difficult to be extracted	>0.05	<2.5 × 10^{-6}

2.2. Basic Parameter Test

The coal samples were taken on site, and the coal samples were processed into standard samples and pulverized with different particle sizes, respectively. The coal samples are used to test the mechanical parameters, industrial analysis and gas basic parameters of coal. The test results are shown in Tables 2 and 3.

Table 2. Mechanical parameters of coal samples.

Elastic Modulus (E/GPa)	Poisson's Ratio (v)	Cohesion (c/MPa)	Internal Friction Angle (φ/°)
1.1	0.3	0.8	28

Table 3. Industrial analysis and gas basic parameters.

Water Content (W/%)	Ash Content (A/%)	Porosity (η/%)	Maximum Gas Adsorption Capacity (a/m^3/kg)	Adsorption Constant (b/MPa^{-1})
1.14	7.76	4.29	37.08	0.82

3. Gas Migration Equation in Coal Seam

3.1. Control Equation of Coal Seam Permeability Considering Gas Pressure Variation and Gas Adsorption and Desorption

The research shows that coal seam permeability k can be expressed as an exponential function related to effective stress [23]:

$$k = k_0 \exp\left(-3C_f \Delta\sigma_e\right) \quad (1)$$

where k_0 is the initial permeability, m^2; C_f represents cleat volume compressibility, MPa^{-1}; $\Delta\sigma_e$ is effective stress variation, MPa.

The cleat volume compressibility C_f can be expressed by the following formula [24]:

$$\begin{cases} C_f = \frac{1}{K_p} \\ K_p = \eta \cdot K = \eta \cdot \frac{E}{3(1-2v)} \end{cases} \quad (2)$$

where K_p is pore bulk modulus, GPa; η is porosity of coal seam, %; K is elastic modulus of coal matrix, GPa; E is elastic modulus of coal, GPa; v is Poisson's ratio of coal.

There are three changes in the process of drilling and gas drainage. Firstly, the coal seam is disturbed in the process of drilling, and the stress around the borehole redistributes, resulting in the change of effective stress. Secondly, the gas in a free state is extracted under negative pressure, and the decrease of gas pressure leads to the increase of effective stress, resulting in the compression of coal pores and the decrease of gas flow channels. Thirdly, the decrease of gas pressure promotes gas desorption and coal matrix shrinkage, resulting in the increase of coal fracture channels. Therefore, the variation of effective stress can be expressed as three parts:

$$\Delta\sigma_e = \Delta\sigma_s + \Delta\sigma_p + \Delta\sigma_x \quad (3)$$

where $\Delta\sigma_s$ is effective stress variation caused by stress redistribution, MPa; $\Delta\sigma_p$ is effective stress variation caused by gas pressure drop, MPa; $\Delta\sigma_x$ is effective stress variation caused by gas desorption, MPa.

Coal is a kind of soft elastic-plastic body. After stress redistribution, the tangential stress around the borehole will be higher than the strength of coal, so that the coal will be destroyed and a plastic zone and elastic zone will appear. Assuming that the coal is in the limit equilibrium state in the plastic zone, the tangential stress around the borehole can be described by the following formula [25,26]:

$$\Delta\sigma_t = \begin{cases} c \cdot cot\varphi \cdot \left[\left(\frac{1+sin\varphi}{1-sin\varphi}\right) \cdot \left(\frac{2x}{R_0}+1\right)^{\frac{2sin\varphi}{1-sin\varphi}} - 1 \right], x \leq H \\ \sigma_0 \cdot \left[1 + \frac{4H^2}{(2x+R_0)^2}\right] - \frac{4H^2}{(2x+R_0)^2} \cdot c \cdot cot\varphi \left[\left(\frac{2H}{R_0}\right)^{\frac{2sin\varphi}{1-sin\varphi}} - 1 \right], x > H \end{cases} \quad (4)$$

where σ_0 is initial stress, MPa; c is cohesion of coal, MPa; φ is internal friction angle, °; x is distance from coal body to borehole wall, m; R_0 is borehole diameter, m; H is distance from plastic zone boundary to borehole wall, m. H can be expressed as follows:

$$H = \frac{R_0}{2} \left\{ \left[\frac{\sigma_0 \cdot (1 - sin\varphi)}{c \cdot cot\varphi} + 1 \right]^{\frac{1-sin\varphi}{2 \cdot sin\varphi}} - 1 \right\} \quad (5)$$

According to the coal seam buried depth of 500 m, the initial stress σ_0 is 12.5 MPa, the diameter of long hole is 113 mm. The tangential stress as shown in Figure 4 can be calculated by substituting the cohesion c and internal friction angle φ measured in the laboratory and original stress and borehole diameter into Equations (3) and (4). If the position where the stress changes by 5% is taken as the influence boundary of the borehole, it can be seen from Figure 4 that the radius of the influence range of the directional long borehole is 0.313 m, which is smaller than the thickness of the coal seam. Therefore, the influence of the stress change around the borehole on the permeability of the coal seam can be ignored, and the change of the effective stress can be expressed as follows:

$$\Delta\sigma_e = \Delta\sigma_p + \Delta\sigma_x \tag{6}$$

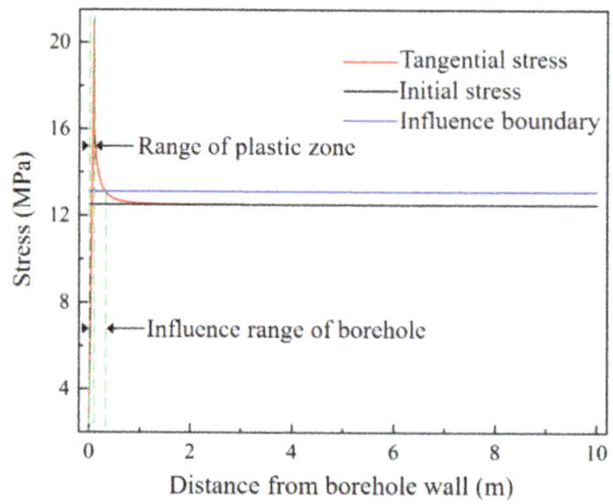

Figure 4. Distribution of tangential stress around borehole.

The effective stress variation caused by gas pressure drop can be expressed by the following formula:

$$\Delta\sigma_p = P_0 - P \tag{7}$$

where P_0 is initial gas pressure of coal seam, MPa; P is gas pressure of coal seam, MPa.

The effective stress variation caused by gas adsorption and desorption can be expressed by the following formula:

$$\Delta\sigma_x = K\cdot\Delta\varepsilon_x \tag{8}$$

where ε_x is the coal bulk strain caused by gas adsorption and desorption, which satisfies the Langmuir equation. It can be expressed by the following formula [24]:

$$\varepsilon_x = \frac{\varepsilon_l\cdot P}{P + P_l} \tag{9}$$

where ε_l is Langmuir volume strain constant; P_l is Langmuir pressure constant.

The control equation of coal seam permeability considering gas pressure variation and gas adsorption and desorption can be obtained by simultaneous Formulas (1), (6)–(9):

$$k = k_0 exp\left\{-3C_f(P_0 - P)\cdot\left[1 - \frac{E}{3\cdot(1-2\nu)}\cdot\frac{\varepsilon_l P_l}{(P+P_l)(P_0+P_l)}\right]\right\} \tag{10}$$

3.2. Control Equation of Gas Flow

Coal seams are porous media, in which gas seepage conforms to the mass conservation equation [27]:

$$\frac{\partial X}{\partial t} + \nabla(\rho V) = 0 \tag{11}$$

where X is the gas content in unit volume coal, kg/m^3; t is the time variable, s; ρ is the gas density in the coal seam, kg/m^3; V is the gas seepage velocity, m/s.

There are two forms of gas in a coal seam, namely free state and adsorption state. Therefore, the gas content in a coal seam includes free gas content and adsorption gas content, which can be expressed by a gas state equation and Langmuir equation, respectively:

$$\begin{cases} X_1 = \eta\rho \\ X_2 = \frac{abp}{1+bp} \cdot \frac{100-A-W}{100} \cdot \frac{1}{1+0.31W} \\ X_3 = X_1 + X_2 \end{cases} \tag{12}$$

where X_1 is the free gas content, kg/m^3; X_2 is the content of adsorbed gas, kg/m^3; a is the maximum gas adsorption constant per unit mass of coal, m^3/kg; b is the adsorption constant of coal, MPa^{-1}; A is ash content of coal, %; W is moisture content of coal, %.

Assuming that the gas is an ideal gas, the gas density in the coal seam is as follows:

$$\rho = \frac{M_g P}{RT} \tag{13}$$

where M_g is molecular weight of gas, 16 g/mol; R is the ideal gas constant, 8.314 J/(mol·K); T is the absolute temperature, K.

It is assumed that the flow of gas in the coal seam conforms to Darcy's law [8,27]:

$$V = -\frac{k}{\mu}\nabla P \tag{14}$$

where k is the permeability of coal seam, m^2; μ is the dynamic viscosity of gas, Pa·s.

The control equation of gas flow can be obtained by simultaneous Formulas (11)–(14):

$$\left[\frac{M_g\eta}{RT} + \frac{ab}{(1+bp)^2} \cdot \frac{100-A-W}{100} \cdot \frac{1}{1+0.31W}\right]\frac{\partial P}{\partial t} - k \cdot \frac{M_g}{\mu RT}\nabla(P\nabla P) = 0 \tag{15}$$

It can be seen from Formulas (10) and (15) that gas drainage and permeability are influenced by each other. With gas drainage, gas pressure decreases and gas desorption occurs, which affects the permeability. Accordingly, the change of permeability will affect the results of gas extraction. The relationship between gas drainage and permeability is shown in Figure 5.

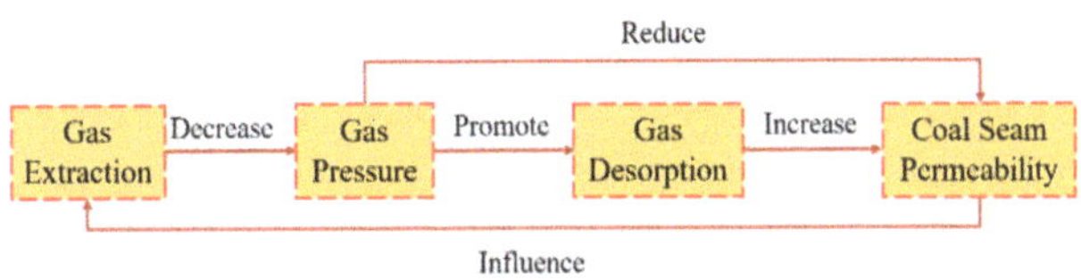

Figure 5. Relationship between gas drainage and permeability.

4. Variation of Permeability and Gas Pressure around Borehole

4.1. Single Borehole Drainage

4.1.1. Geometric Model and Parameter Setting

Since the directional borehole is parallel to the pre-excavation roadway after construction in the design area, the geological conditions of the area through which the borehole passes have little change, so the numerical calculation model can be simplified into a local three-dimensional model. It is assumed that gas is the only flowing gas in the coal seam and the coal seam is isotropic. According to the coal seam thickness and gas parameters of Changcun coal mine, the COMSOL numerical simulation software is used to establish the coal seam gas migration model. In order to reduce the influence of boundary effect, the length of the model is 70 m, the height is 6 m, and the diameter of the borehole is 113 mm. The model diagram is shown in Figure 6. At the same time, a 15 m long monitoring line is arranged from the borehole center to the depth of coal seam, and four monitoring points are arranged at 1 m, 3 m, 5 m and 7 m away from the borehole center to monitor the change of gas parameters. There are no flow boundary conditions around the model, and constant pressure boundary conditions around the borehole simulate the drainage pressure, and the drainage pressure is 20 kPa. The PDE module of numerical simulation software is used to substitute Formulas (10) and (15) into the calculation, and the calculation parameters are listed in Table 4.

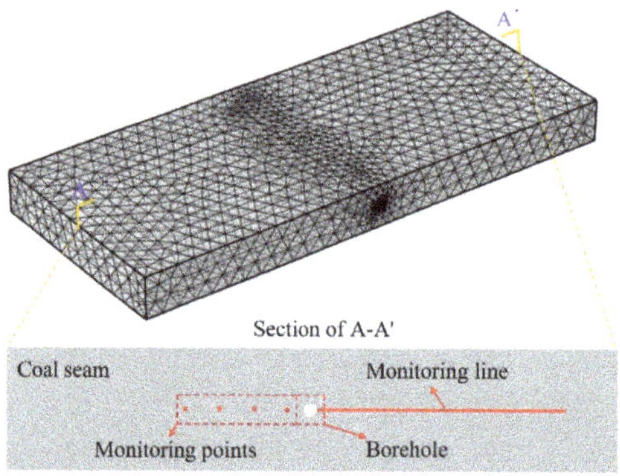

Figure 6. Numerical model.

Table 4. Numerical calculation parameters.

Parameter	Numerical Value
Elastic modulus of coal (E/GPa)	1.1
Poisson's ratio of coal (υ)	0.3
Internal friction angle of coal (φ/°)	28
Porosity of coal (η/%)	1.21
Original gas pressure (P_0/MPa)	0.35
Maximum gas adsorption capacity of coal (a/m^3·kg^{-1})	37.08
Adsorption constant of coal (b/MPa^{-1})	0.82
Ash content of coal (W/%)	1.14
Moisture content of coal (A/%)	7.76
Langmuir volume strain constant (ε_l)	0.01266
Langmuir pressure constant (P_l/MPa)	4.31
Molecular weight of gas (M_g/g/mol)	16
Ideal gas constant (R/J·mol^{-1}·K^{-1})	8.314
Dynamic viscosity of gas (μ/Pa·s)	1.84×10^{-5}
Absolute temperature (T/K)	273
Initial permeability of coal seam (k_0/m^2)	8.09×10^{-8}

4.1.2. Mesh Independence Test of Model

In order to study the influence of model mesh on simulation results, it is necessary to test the mesh independence. The mesh is divided into normal, fine and finer, as shown in Figure 7. According to the theoretical formula, the permeability and gas pressure distribution law of drainage time is 10 d are simulated, as shown in Figure 8.

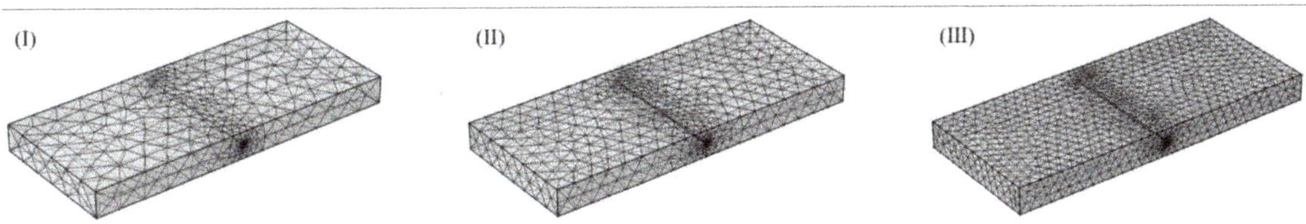

Figure 7. Mesh division types: (**I**) normal; (**II**) fine; (**III**) finer.

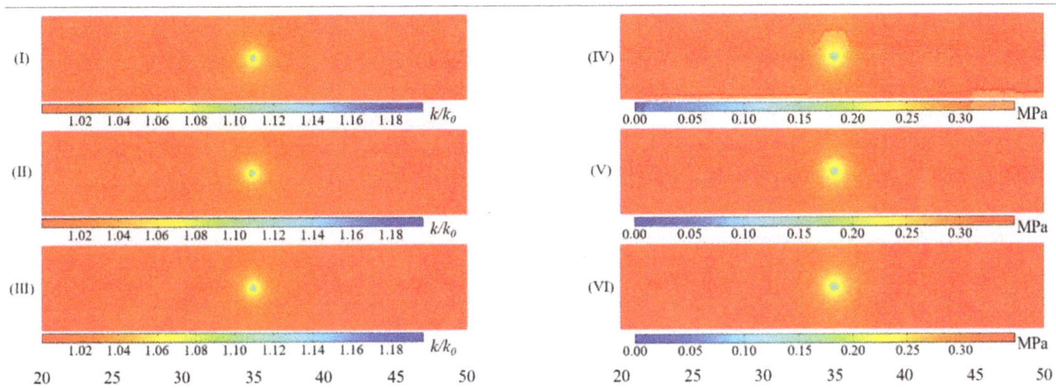

Figure 8. Cloud chart of permeability and gas pressure distribution of different mesh types at 10 d of drainage: (**I–III**) are the permeability distribution of mesh type from normal to finer; (**IV–VI**) are the gas pressure distribution of mesh type from normal to finer.

It can be seen from the figures that the mesh division type is from normal to finer, the permeability and gas pressure distribution around the borehole are almost unchanged. In order to more clearly analyze the influence of mesh division type on simulation results, the data measured by the monitoring line are plotted into a curve, as shown in Figure 9.

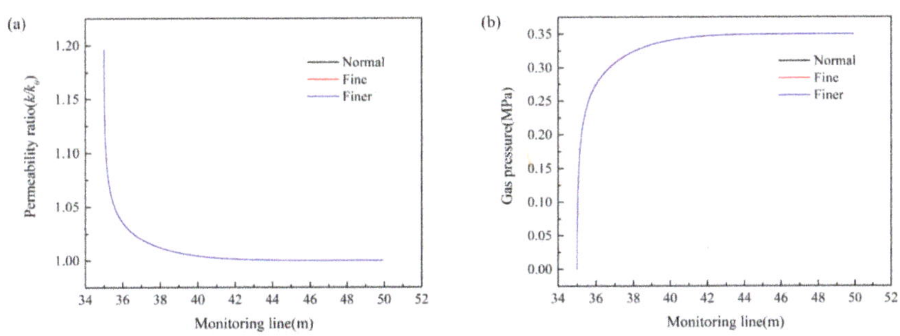

Figure 9. Permeability and gas pressure distribution curves of different mesh types at 10 d of drainage: (**a**) Permeability distribution curve. (**b**) Gas pressure distribution curve.

It can be seen from the figures that the three different mesh types have obtained the same observation curve, so it can be determined that the mesh type has no effect on the simulation results of single borehole drainage. In order to save the time of numerical simulation, the mesh type is selected as normal mode in the simulation of single borehole drainage.

4.1.3. Permeability Variation Law

According to the theoretical formula, the numerical simulation software is used to calculate the distribution characteristics of permeability around the borehole when the drainage time is 50 d, 100 d, 150 d, 200 d, 250 d and 300 d respectively, so as to observe the variation law of permeability around the boreholes. The curves from the monitoring data of monitoring line and monitoring points are shown in the figures below.

Figures 10 and 11 show the change of coal permeability around the borehole at different drainage times. It can be seen from the figure that the permeability around the borehole is greater than the initial permeability. The further away from the borehole boundary, the smaller the permeability, and the closer to the initial permeability. At the same time, it can also be seen that with the increase of drainage time, the range of permeability around the borehole gradually increases. The monitoring data of monitoring line and monitoring points form curves, as shown in Figure 12. It can be seen from Figure 12a that the permeability of the borehole boundary increases by 1.2 times. With the increase of drainage time, the coal permeability also increases, but the permeability at the borehole boundary remains unchanged. It can be seen from Figure 12b that no matter how far the distance from the borehole is, with the increase of drainage time, the permeability gradually increases from the initial permeability, but the increase in the amplitude of permeability decreases with the increase of time. The closer to the borehole, the faster the permeability increases with time in the early stages.

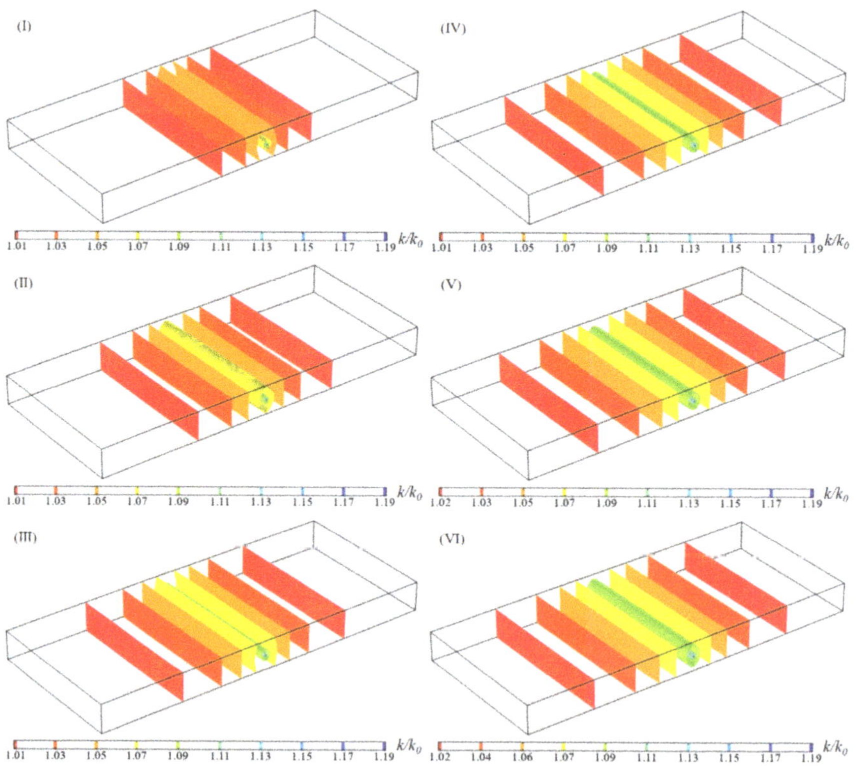

Figure 10. Three-dimensional cloud chart of coal permeability distribution around boreholes at different drainage times: **(I)** 50 d; **(II)** 100 d; **(III)** 150 d; **(IV)** 200 d; **(V)** 250 d; **(VI)** 300 d.

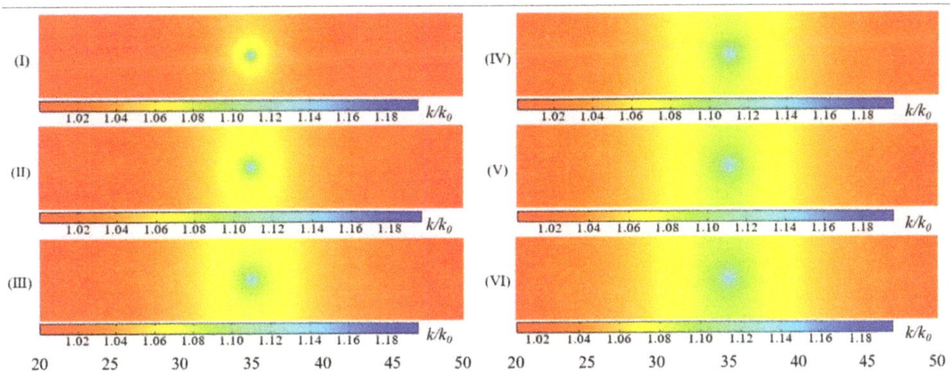

Figure 11. Cloud chart of coal permeability distribution around borehole at different drainage times: (**I**) 50 d; (**II**) 100 d; (**III**) 150 d; (**IV**) 200 d; (**V**) 250 d; (**VI**) 300 d.

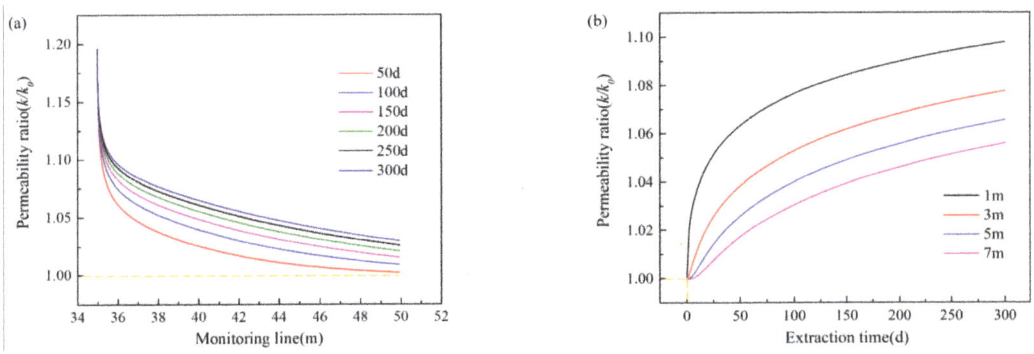

Figure 12. Permeability data monitored of monitoring line and monitoring points. (**a**) Permeability variation with monitoring line at different drainage times. (**b**) Permeability variation with time at different distances from the borehole.

4.1.4. Variation Law of Gas Pressure

Based on the theoretical formula, the numerical simulation software is used to calculate the distribution characteristics of gas pressure around the borehole when the drainage time is 50 d, 100 d, 150 d, 200 d, 250 d and 300 d, respectively, so as to observe the variation law of gas pressure around the borehole and to plot the curves from the monitoring data of monitoring line and monitoring points, as shown in the figure below.

Figures 13 and 14 show the cloud diagram of gas pressure distribution around the borehole at different drainage times. It can be seen from the figure that the gas pressure increases gradually from the borehole boundary to the depth of the coal seam. With the increase of drainage time, the range of gas pressure decrease gradually increases. The monitoring data of monitoring line and monitoring points are plotted into curves, as shown in Figure 15. It can be seen from the figure that with the increase of drainage time, the gas pressure decreases more. The closer to the borehole, the greater the gas pressure drop rate in the early stage of drainage.

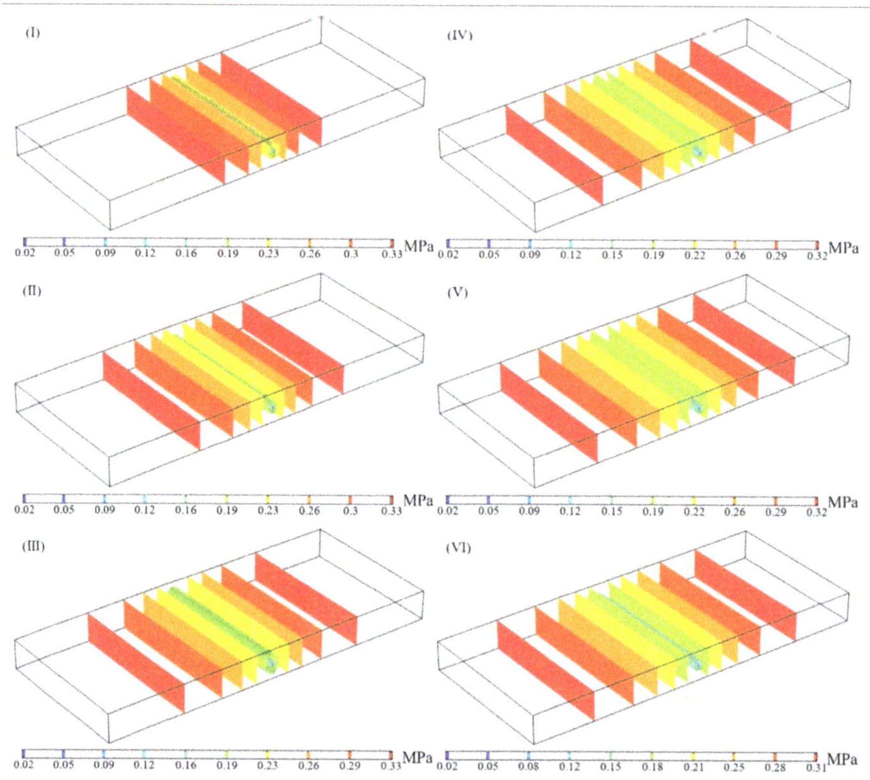

Figure 13. Three-dimensional cloud chart of gas pressure distribution around boreholes at different drainage times: (**I**) 50 d; (**II**) 100 d; (**III**) 150 d; (**IV**) 200 d; (**V**) 250 d; (**VI**) 300 d.

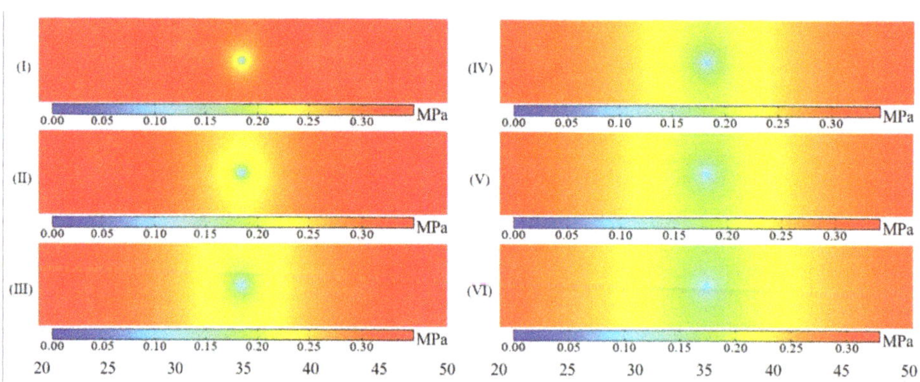

Figure 14. Cloud chart of gas pressure distribution around borehole at different drainage times: (**I**) 10 d; (**II**) 50 d; (**III**) 100 d; (**IV**) 150 d; (**V**) 200 d; (**VI**) 300 d.

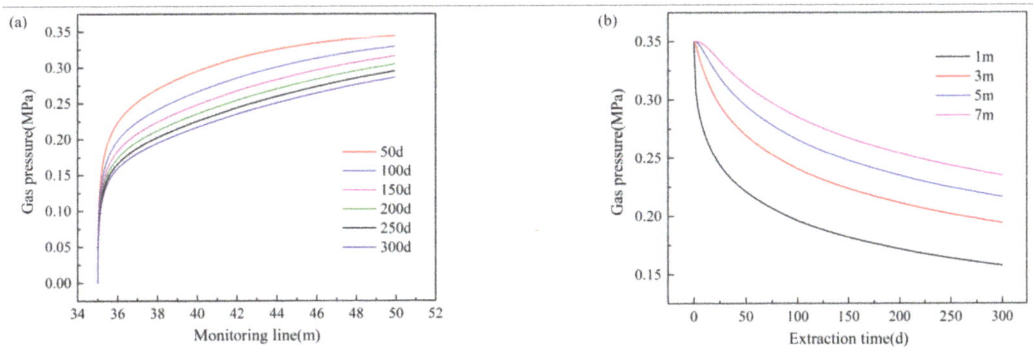

Figure 15. Monitoring data of gas pressure of monitoring line and monitoring points. (**a**) Variation law of gas pressure with monitoring line at different drainage times. (**b**) Variation of gas pressure with time at different distances from borehole.

4.2. Multi-Borehole Drainage

4.2.1. Geometric Model Setting

Based on the single borehole drainage model, a multi-borehole drainage model has been established, as shown in Figure 16. Except for increasing the number of boreholes, other calculation parameters of the model have not changed. Because it is estimated that the drainage time of boreholes is more than 200 d, the gas drainage effect of boreholes with different spacing is studied based on the drainage time of 200 d. The spacing of boreholes is 4 m, 6 m, 8 m, 10 m, 12 m and 14 m, respectively. In the process of borehole drainage, the whole coal seam in the area covered by the borehole should be drained to meet the standard. Therefore, a 60 m long monitoring line is arranged at the top of the coal seam, that is, in the middle of the upper boundary of the model, to monitor the simulation results. At the same time, a monitoring point is arranged at the midpoint of the two boreholes, to monitor the simulation results in the middle of the boreholes.

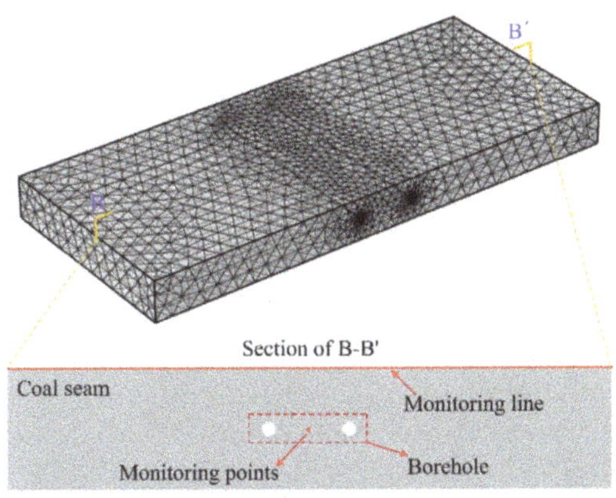

Figure 16. Numerical model.

4.2.2. Mesh Independence Test of Model

In order to study the influence of model mesh on simulation results, it is necessary to test the mesh independence. The mesh is divided into normal, fine and finer, as shown in Figure 17. According to the theoretical formula, the permeability and gas pressure distribution law of the simulation borehole spacing of 6 m and the drainage of 20 d are shown in Figure 18.

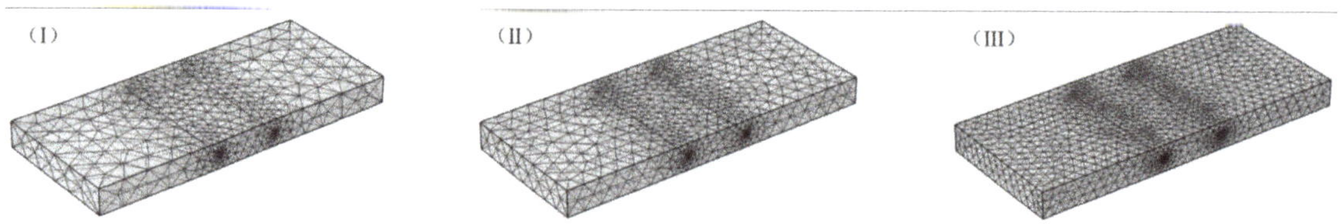

Figure 17. Mesh division types: (**I**) normal; (**II**) fine; (**III**) finer.

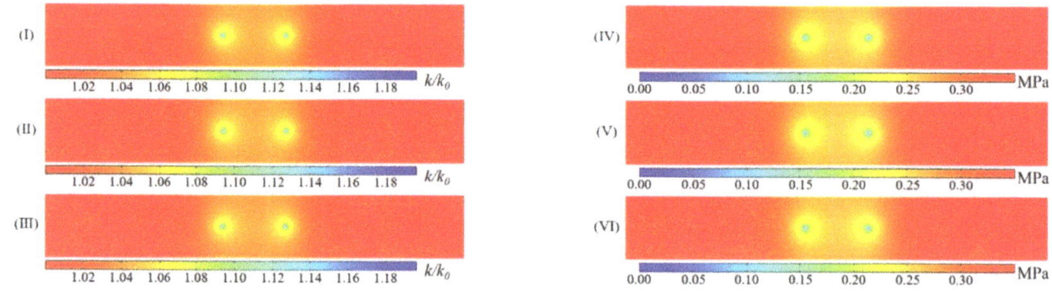

Figure 18. Cloud chart of permeability and gas pressure distribution of different mesh types at 10 d of drainage: (**I–III**) are the permeability distribution of mesh type from normal to finer; (**IV–VI**) are the gas pressure distribution of mesh type from normal to finer.

It can be seen from the figures that the mesh type varies from normal to finer, and there is no obvious change in the permeability and gas pressure distribution around the borehole. In order to more clearly analyze the influence of mesh type on simulation results, the data measured by the monitoring line are plotted into a curve, as shown in Figure 19.

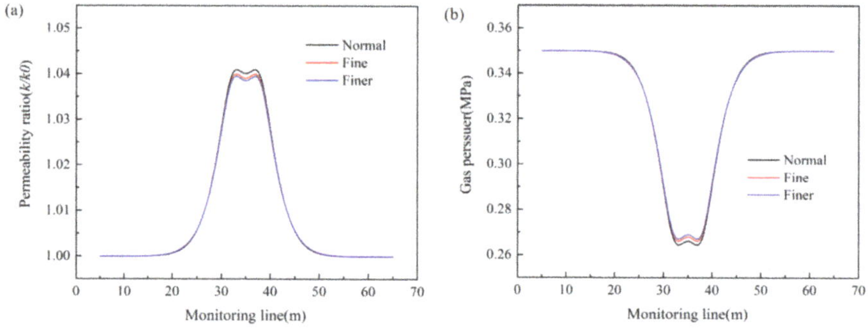

Figure 19. Permeability and gas pressure distribution curves of different mesh types at 10 d of drainage: (**a**) Permeability distribution curve. (**b**) Gas pressure distribution curve.

It can be seen from the figures that the simulation results of three different mesh types are different between boreholes under the multi-borehole drainage. The mesh type is from normal to finer, the permeability between boreholes decreases, and the gas pressure increases. The maximum error is 0.093% from normal to fine mesh types, and 0.049% from fine to finer mesh types. The error of simulation results between different mesh types is small. Since it is necessary to determine the borehole spacing when simulating the multi-borehole extraction, in order to ensure the extraction effect and consider the error, the finer mesh type is selected for simulation.

4.2.3. Permeability Variation Law

According to the theoretical formula, the numerical simulation software is used to calculate the distribution characteristics of permeability around the borehole under different borehole spacing when the drainage time is 200 d, so as to observe the variation law of permeability around the borehole., and to draw the curves from the monitoring data of monitoring line and monitoring point, as shown in the figures below.

Figures 20 and 21 shows the distribution of permeability around differently spaced boreholes. It can be seen from the figure that with the increase of borehole spacing, the range of permeability gradually increases, but the permeability between boreholes decreases with the increase of borehole spacing. The data monitored by the monitoring line and the monitoring point are drawn into curves, as shown in Figure 22. It can be seen from Figure 22a that with the increase of borehole spacing, the permeability between boreholes gradually changes from greater than at the top of the boreholes to less than at the top of the boreholes, indicating that the drainage boreholes will interact with each other, resulting in the increase of permeability between boreholes; however, with the increase of borehole spacing, the degree of interaction between boreholes gradually decreases. It can be seen from Figure 22b that when the drainage time is 200 d, and the spacing between boreholes is 6 m, 10 m and 14 m, that is, the distance between measuring point and boreholes is 3 m, 5 m and 7 m, the permeability ratios are 1.112, 1.100 and 1.088, respectively, and the permeability gradually decreases. Compared with 1.067, 1.056 and 1.046 at the same time and location of single borehole drainage, the increase is 4.22%, 4.17% and 4.02%, respectively, which can be concluded as above.

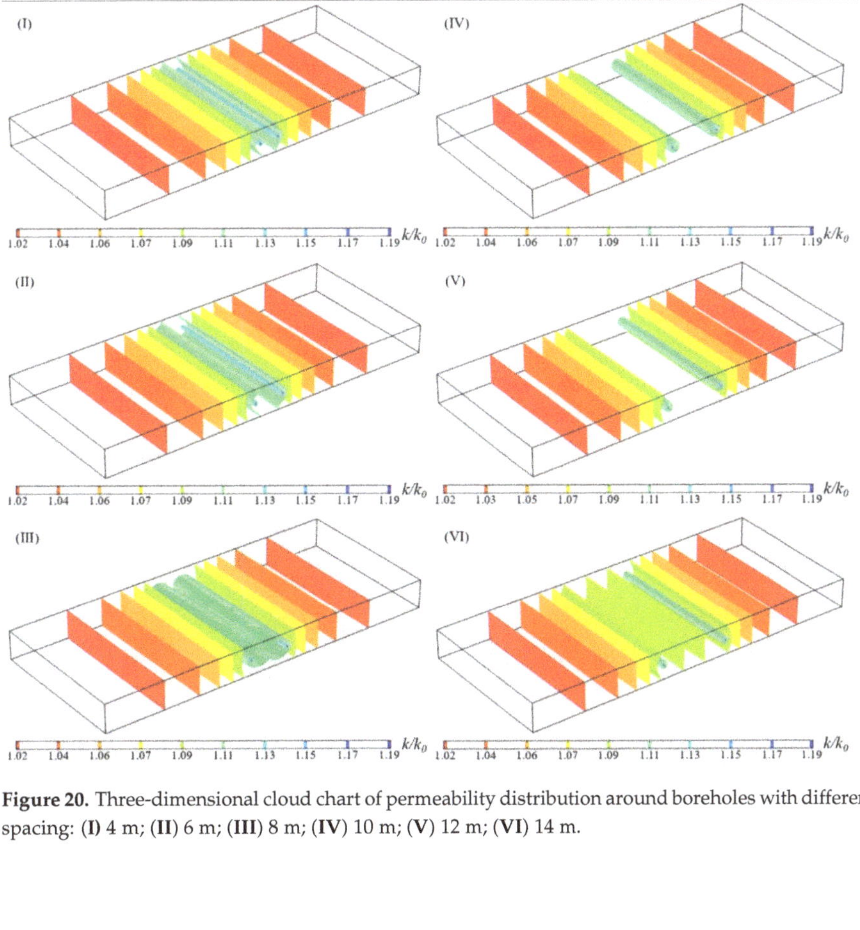

Figure 20. Three-dimensional cloud chart of permeability distribution around boreholes with different spacing: **(I)** 4 m; **(II)** 6 m; **(III)** 8 m; **(IV)** 10 m; **(V)** 12 m; **(VI)** 14 m.

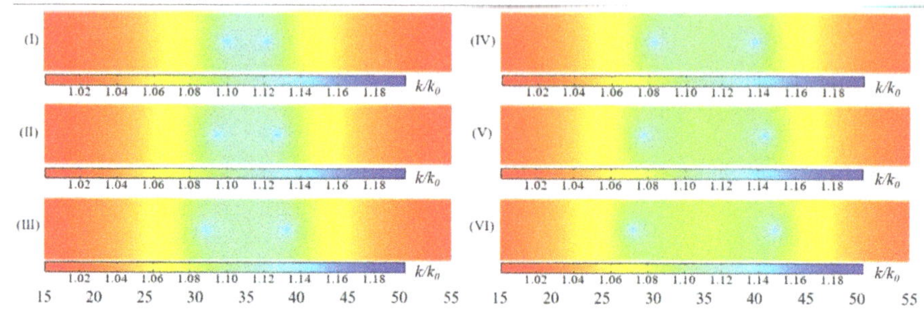

Figure 21. Cloud chart of permeability distribution around boreholes with different spacing: (**I**) 4 m; (**II**) 6 m; (**III**) 8 m; (**IV**) 10 m; (**V**) 12 m; (**VI**) 14 m.

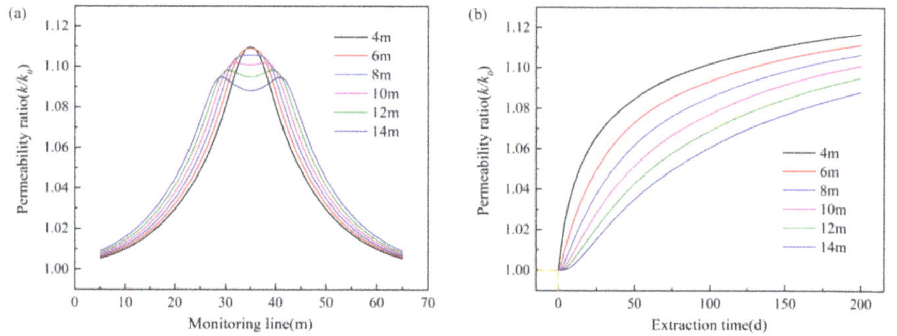

Figure 22. Monitoring data of permeability of monitoring line and monitoring point. (**a**) Variation law of permeability with monitoring line at different borehole spacing. (**b**) The law of permeability variation with time with two boreholes.

4.2.4. Gas Pressure Variation Law

According to the theoretical formula, the distribution characteristics of gas pressure around the boreholes are calculated respectively by using numerical simulation software under different borehole spacing when drainage time is 200 d. The gas pressure variation law around the boreholes is observed, and the data monitored by the monitoring line and the monitoring point are plotted into curves, as shown in the following figure.

Figures 23 and 24 show the distribution of gas pressure around different boreholes according to their spacing. It can be seen from the figure that with the increase of borehole spacing, the gas pressure reduction range gradually increases, but the gas pressure between boreholes increases with the increase of borehole spacing. The data monitored by the monitoring line and the monitoring point are drawn into curves, as shown in Figure 25. It can be seen from Figure 25a that with the increase of borehole spacing, the gas pressure between boreholes gradually changes from less than the top of boreholes to more than at the top of boreholes, indicating that the drainage boreholes will interact with each other, resulting in the decrease of gas pressure between boreholes; with the increase of borehole spacing, the degree of interaction between boreholes gradually decreases. It can be seen from Figure 25b that when the drainage time is 200 d, and the distance between boreholes is 6 m, 10 m and 14 m, that is, the distance between measuring point and boreholes is 3 m, 5 m and 7 m, the gas pressure is 0.133 MPa, 0.154 MPa and 0.175 MPa respectively, and the gas pressure increases gradually. Compared with 0.213 MPa, 0.233 MPa and 0.253 MPa at the same time and position of single borehole drainage, the values decreased by 36.49%, 35.32% and 31.10%, respectively, similar to the findings above.

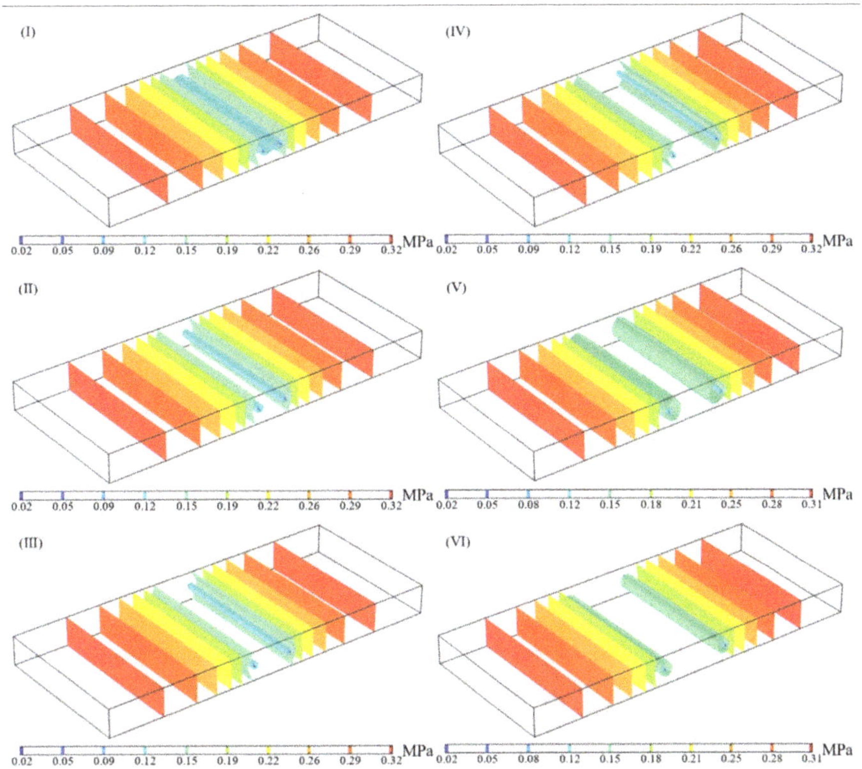

Figure 23. Three-dimensional cloud chart of gas pressure distribution around boreholes with different spacing: (**I**) 4 m; (**II**) 6 m; (**III**) 8 m; (**IV**) 10 m; (**V**) 12 m; (**VI**) 14 m.

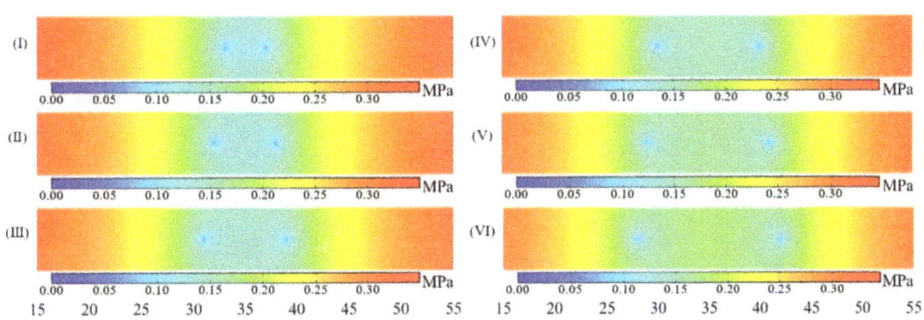

Figure 24. Cloud chart of gas pressure distribution around boreholes with different borehole spacing: (**I**) 4 m; (**II**) 6 m; (**III**) 8 m; (**IV**) 10 m; (**V**) 12 m; (**VI**) 14 m.

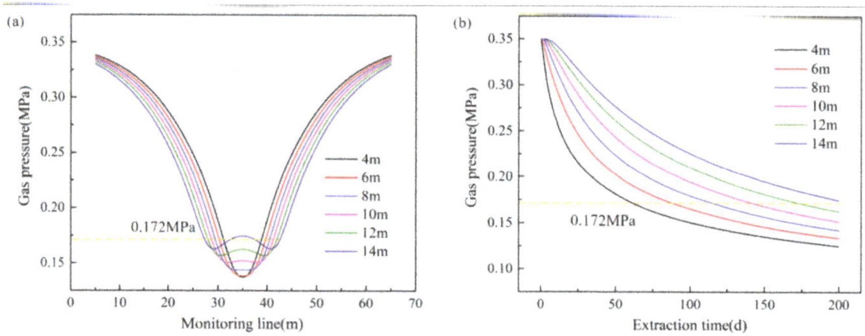

Figure 25. Monitoring data of gas pressure of monitoring line and monitoring point. (**a**) Variation law of gas pressure with monitoring line at different borehole spacing. (**b**) The law of gas pressure variation with time with two boreholes.

For a high gas coal seam with gas pressure lower than 0.74 MPa, the effective range of borehole drainage is often defined by relative pressure, that is, the boundary of the effective drainage range is 51% decrease in gas pressure [28]. Therefore, the effective drainage boundary of gas pressure is 0.172 MPa. It can be seen from Figure 25, when the spacing between boreholes is 14 m, the gas pressure between boreholes will be greater than 0.172 MPa, that is, the drainage between boreholes will not meet the standard. In order to more clearly analyze the effective drainage range between boreholes, the range of gas pressure lower than 0.172 MPa at different borehole spacing is drawn, as shown in Figure 26. It can be seen from the figure that with the increase of borehole spacing, the range of gas pressure lower than 0.172 MPa also gradually increases (the red part in the figure). However, when the borehole spacing is 14 m, there is a non-standard drainage zone between boreholes.

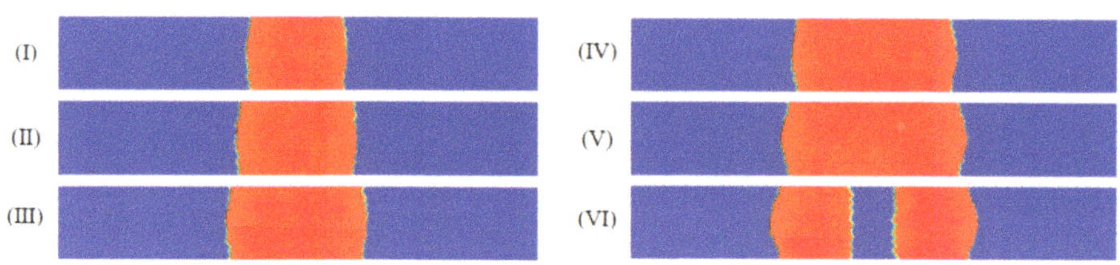

Figure 26. Range of gas pressure lower than 0.172 MPa at different borehole spacing: (**I**) 4 m; (**II**) 6 m; (**III**) 8 m; (**IV**) 10 m; (**V**) 12 m; (**VI**) 14 m.

5. Field Application

5.1. Method Statement

According to the above numerical simulation results, the final borehole spacing is determined to be 12 m. In order to control the coal seam within 15 m on both sides of the pre-excavation roadway, four boreholes need to be arranged. According to the length of roadway, four drilling fields with the size of 8 m × 5 m × 5 m are designed and constructed in the 2302 auxiliary transportation roadway. The distance between drilling fields is 400 m. There are eight boreholes designed and constructed in 1# drilling field and four boreholes in 2–4# drilling fields. The borehole layout plan is shown in Figure 27. The sealing method of "two plugging, one injection and one row" is adopted, and the sealing depth of the borehole is 20 m to ensure tight sealing without air leakage. And each borehole in the drilling fields of 1–4# is equipped with a concentration measuring port and orifice flowmeter, which is convenient for real-time monitoring of gas concentration and gas drainage flow.

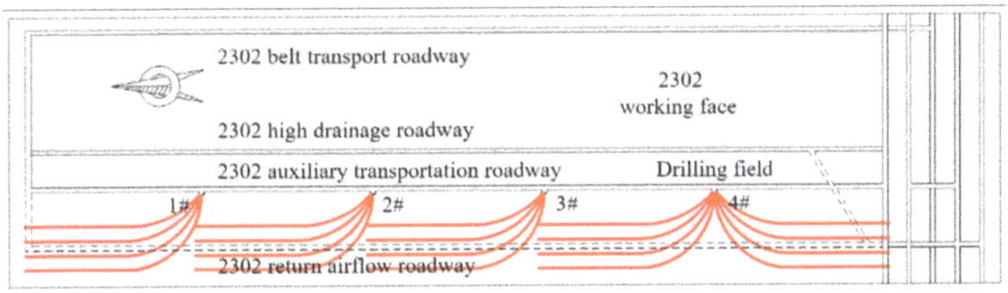

Figure 27. Borehole layout plan.

The drilling equipment used in the construction is shown in Figure 28, mainly including operation console, guidance system console, water pump, directional drill, rotary unit, front gripper, crawler, etc. The operation console controls the hydraulic system and water system of the drilling rig. The guidance system console displays the collected data of the unit in the borehole on the screen. The water pump is used to provide water for the downhole motor. A directional drill is used to control drilling direction. The rotary unit is used to rotate the drill pipe. The front gripper is used to push and dismantle the drill pipe, and also to guide the drill pipe. The crawler can enable the drilling rig to walk freely in the underground roadway.

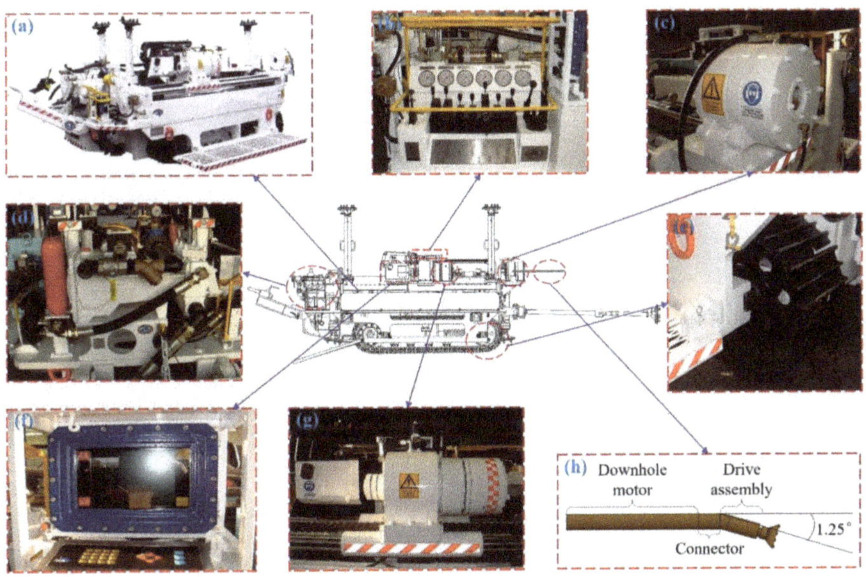

Figure 28. Construction equipment. (**a**) Overall drawing of drilling rig. (**b**) Operation console. (**c**) Front gripper. (**d**) Water pump. (**e**) Crawler. (**f**) Guidance system console. (**g**) Rotating unit. (**h**) Directional drill.

5.2. Effect Analysis

After the completion of drilling construction, record the gas drainage flow and gas concentration after different drainage times. After the completion of drainage, test the effects on the coal seam within the scope of the pre-excavation roadway. At the same time, record the excavation speed of the heading face during the excavation, and analyze the drainage effect.

Figure 29 shows the data record during borehole drainage and 2302 return air roadway excavation. Figure 29a,b show the gas concentration and gas drainage flow of a single borehole in four drilling fields in the first 50 d. It can be seen from the figure that the gas concentration and gas drainage flow change little in the first 35 d, with gas concentration higher than 70% and gas drainage flow higher than 0.38 m^3/min. The gas concentration is still higher than 65% and the gas drainage flow is still higher than 0.33 m^3/min after 50 d of drainage, with little decrease and good drainage effect. Figure 29c shows the measurement of the drilling cutting quantity in the coal seam before the roadway excavation (after each excavation distance of the roadway, it is necessary to predict the outburst in front of the heading face, and the drilling cutting quantity is one of the important indicators). It can be seen from the figure that the drilling cutting quantity is 3.1–4.7 kg/m, which is lower than the critical value of outburst of 6.0 kg/m, and there is no danger of outburst, i.e., there is no gas dynamic phenomenon in the process of roadway excavation. Figure 29d shows the excavation length per month. It can be seen from the figure that the excavation length of the long-distance gas pre-drainage method is more than 80 m per month; compared with the in-seam ordinary borehole gas drainage technology (Figure 1b), the excavation length can be more than 50 m per month, and the excavation speed is greatly increased.

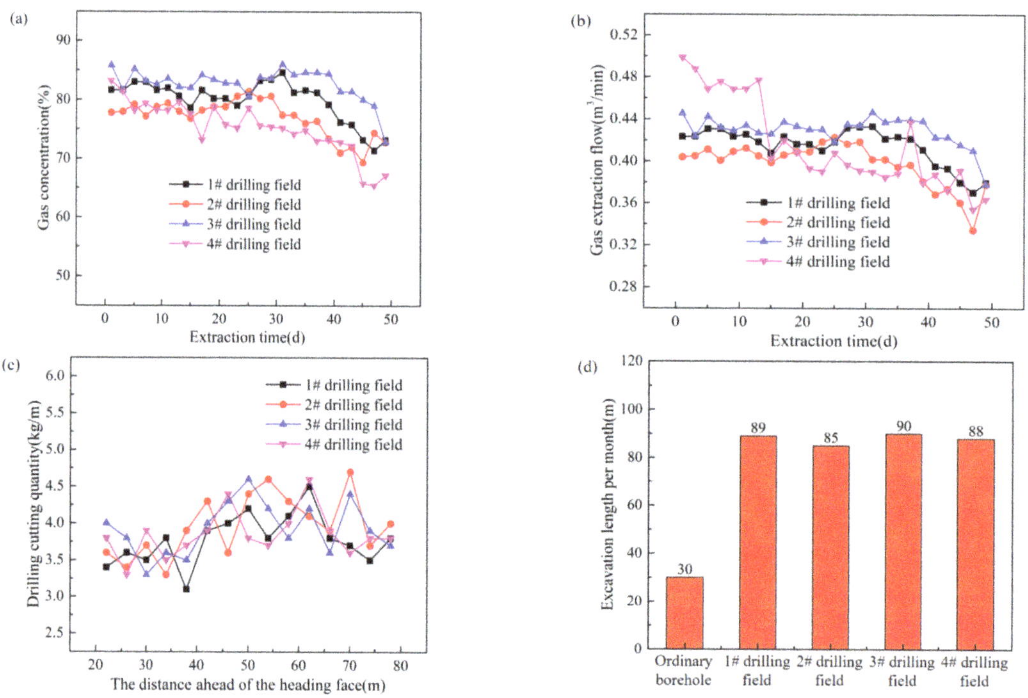

Figure 29. Analysis of gas drainage effect. (**a**) Gas drainage concentration. (**b**) Gas drainage flow. (**c**) Drilling cutting quantity. (**d**) Coal roadway excavation speed.

6. Discussion

This paper introduces and studies the technology of long-distance advance pre-drainage of gas in the heading face, and has achieved good application results in the field. This technology can enable drilling of boreholes to control the gas in the area of the roadway to be excavated by use of the roadway that has been excavated. This technology is applicable to the following situations: (1) The mine is a mine with high gas or coal and gas outburst. (2) The mine needs long-term gas drainage to solve the gas problem. (3) There is a shortage of mining and replacement, which affects the efficient production of mine. This technology can save a large amount of cost compared with the technology of driving bottom drainage, and can save a large amount of time compared with the technology of digging while performing drainage, which has great technical advantages.

7. Conclusions

In view of the shortcomings of the existing commonly used gas drainage methods in the heading face, this paper proposes to adopt the long-distance gas pre-drainage technology in the heading face by directional long borehole. The main conclusions are as follows:

1. Taking the return air roadway of 2302 working face in Changcun coal mine as the research object, a gas migration model considering the change of permeability with drainage time was established. The results show that the permeability decreases gradually from the borehole boundary to the coal depth in the same drainage time. Whether with single borehole or multi-borehole drainage, the range of permeability increases gradually with the increase of drainage time. In the process of multi-borehole drainage, the permeability between boreholes will increase due to the interaction between boreholes. The maximum permeability does not change with drainage time and is 1.2 times the initial permeability.
2. With the increase of drainage time, the gas pressure around the boreholes gradually decreases. When multi-boreholes are used for drainage, due to the mutual influence between boreholes, the decrease in the range of gas pressure between boreholes increases. Compared with single borehole drainage, the effective drainage radius of boreholes increases, but with the increase of borehole spacing, the degree of mutual influence between boreholes decreases. In the field construction, a reasonable borehole spacing should be selected.
3. According to the field conditions and research results, a reasonable gas drainage scheme was designed and implemented. The field application shows that the concentration and purity of gas drainage can maintain a high level for a long time. The drilling cutting quantity index is always lower than the critical value of outburst, and there is no gas dynamic phenomenon in the excavation process. Compared with the in-seam ordinary borehole gas drainage technology, the excavation length of the roadway is more than 50m per month, which greatly increases the excavation speed of the roadway. This technology provides a new way of thinking for solving the problem of mining replacement shortage in high gas or coal and gas outburst mines that need long-time extraction.

Author Contributions: Y.H.: Supervision, funding acquisition, writing the review, and editing. J.C.: Writing the original draft, methodology, and software. R.L.: Writing—review and editing. All authors have read and agreed to the published version of the manuscript.

Funding: This work was financially supported by the National Natural Science Foundation of China (Grant No. 52074296, 52004286), the China Postdoctoral Science Foundation (Grant No. 2020T130701, 2019M650895), the State Key R&D Plan (2017YFC0804303) of the Ministry of Science and Technology of China.

Institutional Review Board Statement: Not applicable.

Informed Consent Statement: Not applicable.

Data Availability Statement: Available from corresponding author.

Conflicts of Interest: The authors declare no conflict of interest.

References

1. Xie, S.; Li, E.; Li, S.; Wang, J.; He, C.; Yang, Y. Surrounding rock control mechanism of deep coal roadways and its application. *Int. J. Min. Sci. Technol.* **2015**, *25*, 429–434. [CrossRef]
2. Hu, S.; Wang, E.; Li, Z.; Shen, R.; Liu, J. Time-Varying multifractal characteristics and formation mechanism of loaded coal electromagnetic radiation. *Rock Mech. Rock Eng.* **2014**, *47*, 1821–1838. [CrossRef]
3. Wang, L.; Cheng, L.; Cheng, Y.; Yin, G.; Xu, C.; Jin, K.; Yang, Q. Characteristics and evolutions of gas dynamic disaster under igneous intrusions and its control technologies. *J. Nat. Gas Sci. Eng.* **2014**, *18*, 164–174. [CrossRef]
4. Hook, M.; Tang, X. Depletion of fossil fuels and anthropogenic climate changed: A review. *Energy Policy* **2013**, *52*, 797–809. [CrossRef]

5. Yuan, L. Strategic thinking of simultaneous exploitation of coal and gas in deep mining. *J. China Coal Soc.* **2016**, *41*, 1–6.
6. Lu, L.; Wang, G. Mine gas control method for mine high gassy roadway heading face. *Coal Sci. Technol.* **2007**, *35*, 30–32.
7. Shen, G. Analysis of the gas treatment technology state and existent problems at the roadway heading face. *Saf. Coal Mines* **2012**, *43*, 127–129.
8. Kong, X.; Wang, E.; Liu, X.; Li, N.; Chen, L.; Feng, J.; Kong, B.; Li, D.; Liu, Q. Coupled analysis about multi-factors to the effective influence radius of hydraulic flushing: Application of response surface methodology. *J. Nat. Gas Sci. Eng.* **2016**, *32*, 538–548. [CrossRef]
9. Lin, B.; Yan, F.; Zhu, C.; Zhou, Y.; Zou, Q.; Guo, C.; Liu, T. Cross-borehole hydraulic slotting technique for preventing and controlling coal and gas outbursts during coal roadway excavation. *J. Nat. Gas Sci. Eng.* **2015**, *26*, 518–525. [CrossRef]
10. Zhang, H.; Cheng, Y.; Liu, Q.; Yuan, L.; Dong, J.; Wang, L.; Qi, Y.; Wang, W. A novel in-seam borehole hydraulic flushing gas extraction technology in the heading face: Enhanced permeability mechanism, gas flow characteristics, and application. *J. Nat. Gas Sci. Eng.* **2017**, *46*, 498–514. [CrossRef]
11. Zhou, H.; Cheng, Y.; Liu, H.; Guo, P.; Wang, L. Permeability improvement technology of array crossing boreholes and its application in outburst coal seam. *J. China Coal Soc.* **2011**, *36*, 1515–1518.
12. Liu, Q.; Guo, Y.; An, F.; Lin, L.; Lai, Y. Water blocking effect caused by the use of hydraulic methods for permeability enhancement in coal seams and methods for its removal. *Int. J. Min. Sci. Technol.* **2016**, *26*, 615–621. [CrossRef]
13. Ma, Y.; Nie, B.; He, X.; Li, X.; Meng, J.; Song, D. Mechanism investigation on coal and gas outburst: An overview. *Int. J. Miner. Met. Mater.* **2020**, *27*, 872–887. [CrossRef]
14. Li, H.; Liu, M.; Hao, G.; Deng, Q.; Dong, Y. Technology feasibility of gas drainage with comb–shaped long borehole in floor instead of trans layer borehole. *Coal Geol. Explor.* **2019**, *47*, 32–38.
15. Cheng, B.; Fan, Z.; He, X. Technology and practice of enhanced seepage enhancement in driving face of high gas coal seam based on hydraulic cutting. *China Min. Mag.* **2020**, *29*, 121–126.
16. Xiang, Z. Coal roadway driving technology in low permeability coal seam based on short hole rapid drainage. *Saf. Coal Mines* **2020**, *51*, 97–100.
17. Wang, F.; Ren, T.; Tu, S.; Frank, H.; Naj, A. Implementation of underground longhole directional drilling technology for greenhouse gas mitigation in Chinese coal mines. *Int. J. Greenh. Gas Control* **2012**, *11*, 290–303. [CrossRef]
18. Lu, S.; Cheng, Y.; Ma, J.; Zhang, Y. Application of in-seam directional drilling technology for gas drainage with benefits to gas outburst control and greenhouse gas reductions in Daning coal mine, China. *Nat. Hazards* **2014**, *73*, 1419–1437. [CrossRef]
19. Wang, G.; Fan, C.; Xu, H.; Liu, X.; Wang, R. Determination of long horizontal borehole height in roofs and its application to gas drainage. *Energies* **2018**, *11*, 2647. [CrossRef]
20. Li, H.; Ma, J. Analysis on gas drainage effect of large diameter roof directional long borehole instead of high drainage rock roadway. *Coal Sci. Technol.* **2020**, *48*, 304–310.
21. Hao, S.; Duan, H.; Mo, H.; Peng, X.; Peng, D. Gas drainage technology and practice analysis of large diameter high position directional long borehole. *Coal Geol. Explor.* **2020**, *48*, 243–248.
22. Guo, J. *Research on Characteristics and Its Mechanism of Intensifying Gas Desorption and Seepage of Anthracite by Electrochemical Method*; Taiyuan University of Technology: Taiyuan, China, 2015.
23. Shi, J.; Durucan, S. Drawdown induced changes in permeability of coalbeds: A new interpretation of the reservoir response to primary recovery. *Transp. Porous Media* **2004**, *56*, 1–16. [CrossRef]
24. Zhang, L.; Zhou, H.; Wang, X.; Rong, T.; Chen, C. Characteristics of deep coalbed gas migration based on the time-dependent effect. *J. China Coal Soc.* **2019**, *44*, 1771–1779.
25. Brady, B.H.G.; Brown, E.T. *Rock Mechanics for Underground Mining*; Springer: Dordrecht, The Netherlands, 2006.
26. Jiang, C.; Yu, Q. *Spherical Shell Instability Mechanism of Coal and Gas Outburst and the Technology of Prevention and Control*; China University of Mining and Technology Press: Xuzhou, China, 1998.
27. Zhang, H.; Liu, J.; Elsworth, D. How sorption-induced matrix deformation affects gas flow in coal seams: A new FE model. *Int. J. Rock Mech. Min. Sci.* **2008**, *45*, 1226–1236. [CrossRef]
28. Liu, S.; Ma, G.; Lu, J.; Lin, B. Relative pressure determination technology for effective radius found on gas content. *J. China Coal Soc.* **2011**, *36*, 1715–1719.

energies

Review

Application of Pre-Splitting and Roof-Cutting Control Technology in Coal Mining: A Review of Technology

Shengrong Xie, Yiyi Wu, Fangfang Guo, Hang Zou, Dongdong Chen *, Xiao Zhang, Xiang Ma, Ruipeng Liu and Chaowen Wu

School of Energy and Mining Engineering, China University of Mining and Technology, Beijing 100083, China
* Correspondence: chendongbcg@163.com

Abstract: According to the development requirements of green mining of coal resources, it is imperative to improve the extraction rate of coal and the application of safe and efficient mining technology. Pre-splitting and roof cutting technology is widely used in reducing residual coal pillars and safe pressure relief mining, which has become the crucial technology for pillar-free mining methods. Therefore, it is essential to review and discuss the research hotspots, cutting-edge methods, principles of action, and application areas of the development of this technology. Above all, the research data on pre-splitting and roof-cutting development in the past ten years are summarized and outlined. The research's hot spots are pressure relief technology and gob-side entry retaining technology. Then, the functional forms of pre-splitting and roof cutting technology are discussed and compared, including explosive blasting (directional energy gathering blasting, liquid explosive blasting, and composite blasting), hydraulic fracturing, liquid CO_2 gas fracturing, and mechanized roof cutting (chain arm saw machine and directional cutting roof rig). Through the analysis of field application cases, the application field is divided into three major areas: non-coal pillar mining (gob-side entry driving with narrow coal pillar, gob-side entry retaining with the filling body, completely gob-side entry retaining, and "N00" construction method), pressure relief at working face (thick and hard main roof cracking and end area hard roof cracking), and pressure relief at roadway (gob-side roadway pressure relief and blasting pressure relief technology for roadways). By detailing the process of each application technology one by one, the principle and mode of pre-splitting in each technology are expounded. Finally, the development prospects of pre-splitting and roof cutting in new technical methods, deep pressure relief mining, intelligent unmanned mining, and green and efficient mining are prospected, providing references for similar projects.

Keywords: pre-splitting blasting; roof cutting technology; hydraulic fracturing; no-coal pillar mining; thick hard roof; pressure relief technology

Citation: Xie, S.; Wu, Y.; Guo, F.; Zou, H.; Chen, D.; Zhang, X.; Ma, X.; Liu, R.; Wu, C. Application of Pre-Splitting and Roof-Cutting Control Technology in Coal Mining: A Review of Technology. *Energies* **2022**, *15*, 6489. https://doi.org/10.3390/en15176489

Academic Editor: Sheng-Qi Yang

Received: 1 August 2022
Accepted: 31 August 2022
Published: 5 September 2022

1. Introduction

In Chinese coal mining, high-intensity mine pressure in the roadway and working face has been the critical factor that endangers the average coal production [1–4]. In the past 30 years, with the rapid development of computer science and sensor technology, the study of mine pressure in the coal mining process has become more profound and transparent [5–8]. Theoretical models of mine pressure in mining fields and roadways have been proposed and gradually recognized by field engineers, such as the structural model of "masonry beam" [9,10], the mechanical model of "plate structure" of the stope [11–15], the theory of slip line of roadway floor [16], and the theory of butterfly plastic zone [17,18]. Many scholars have profoundly and extensively studied the location and causes of stress concentration areas in the mining field and the roadway through pressure monitoring in the field and numerical calculations by computer [19–22]. Among them, the stability and movement of the roof after coal mining are closely related to the regional stress concentration condition of the mining field and the roadway. Especially for the coal mining face with a hard roof, composite roof, and thick roof, when it adopts usual mining methods, many problems such as support crushing, large deformation, and strong disturbance will occur [23–27]. Simultaneously, the increase of the buried depth makes the surrounding rock conditions of the roadway increasingly bad. Under the complex surrounding rock environment, it is difficult to maintain the stability of the roadway only by support, and the application of pressure relief technology is urgently needed [28–30]. Based on this, pre-split or cutting off the roof before working face mining is gradually being implemented to study pressure relief in mining fields and roadways.

Pre-splitting and roof-cutting control technology refers to the directional pre-fracturing or cutting off of the roof of the mining field and roadway by explosive blasting or high-strength hydraulic fracturing [31–33]. As shown in Figure 1, according to its application method, it can be classified into directional blasting roof cutting technology, directional hydraulic fracturing technology, and liquid CO_2 gas fracturing technology [34–36]. According to its application object, it can be classified into pre-splitting control technology of mining field's roof and roadway's roof. According to its application scenario, it can be classified into the technology applied to roof cutting and pressure releasing of gob-side entry driving, roof cutting and pressure releasing of gob-side entry retaining, roof cutting for self-forming roadway technology, hard main roof cracking, and pressure relief on the surrounding rock of the roadway.

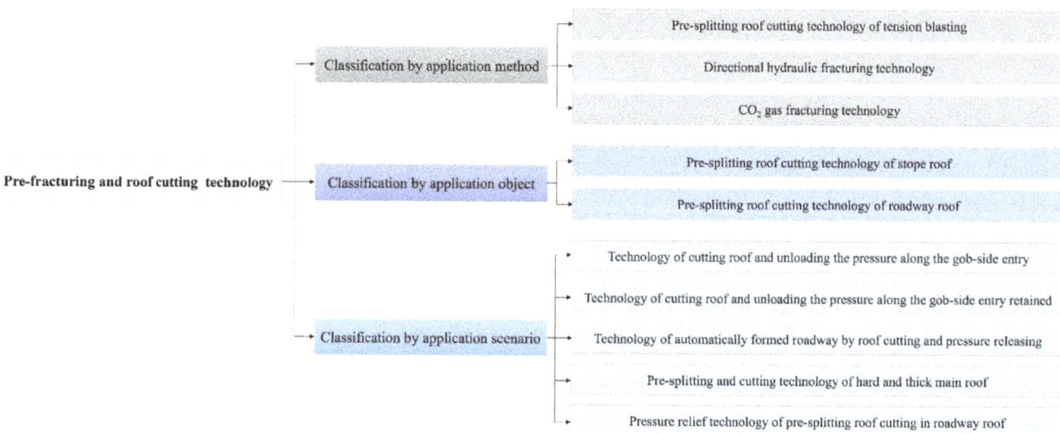

Figure 1. Category of pre-splitting and roof-cutting technology.

With the development of roof cutting and pressure relief theory, pre-split means, and supporting equipment, pre-splitting and roof-cutting technology has become essential to realize pressure relief and no pillar mining. Based on this technology, the recovery rate of coal mining increases and the roof disaster decreases, which is conducive to producing high-quality and efficient green coal mining [37–40]. The research application of pre-splitting and roof-cutting technology in the coal mining field in the past five years is analyzed by VOSviewer, as shown in Figure 2.

By counting more than 300 papers on pre-splitting and roof-cutting in the global mining industry in the past five years, we can get that the research hotspots are: ① Pre-splitting roof applied to the stability study of working face and roadway roof; ② Analysis of the mechanical mechanism of pre-cracked roof effect, including mechanical model and pressure release principle; ③ Process design and optimization of support means of pre-fractured roof technology; ④ Numerical simulation analysis study of pre-fractured roof technology, including the evolution of its plastic zone, stress, and deformation; ⑤ Application of pre-fractured roof technology, including hard-roof cracking and pressure relief, no pillar mining (gob-side entry driving and gob-side entry retaining), and self-forming roadway technology; ⑥ Research on gas and rock explosion in the process of the pre-fractured roof.

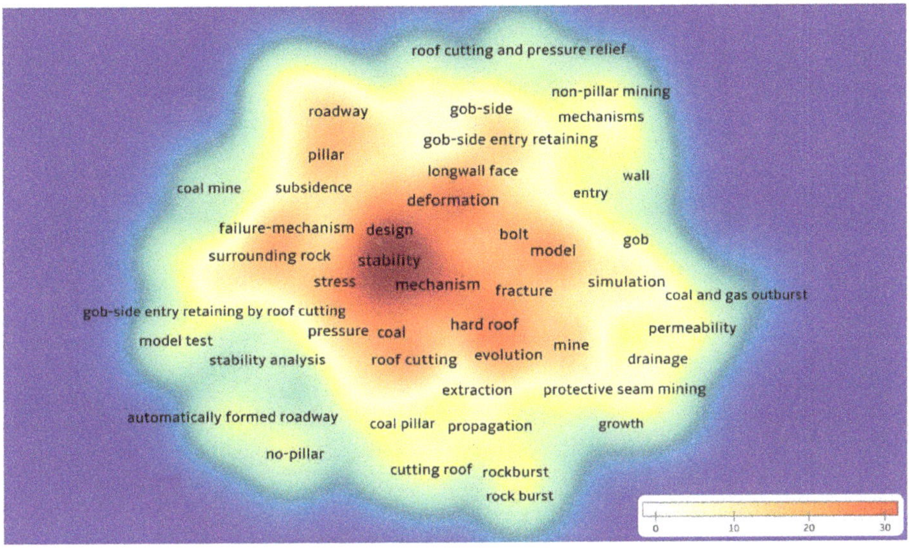

Figure 2. Research hotspot map of pre-splitting and roof-cutting technology.

China is the technological powerhouse country in coal underground mining. Pre-splitting and roof-cutting technology is extensively implemented in Chinese coal mines [2,32]. By searching the keywords "pre-splitting and roof-cutting" in CNKI (China National Knowledge Internet), 180 related papers were obtained in the past 10 years. As shown in Figure 3, the subject terms of these research papers are roof cutting and pressure relief application, gob-side entry retaining technology, pre-split blasting technology, working face pressure relief mining, technical parameter study, non-pillar mining, hard roof cracking, gob-side entry driving technology, and non-pillar self-forming roadway. Among them, the study of gob-side entry retaining through roof-cutting technology is the hot spot of research, which accounts for 56%. The deformation of the roadway along the gob is extensive and challenging to maintain [31,38]. By cutting the roof, the intense stress concentration of the roof can be released, thus significantly improving the stress environment of the roadway along the gob [34]. In addition, in the past four years, research on pre-splitting and roof-cutting has been at the forefront of the industry in the field of gob-side entry retaining and pressure relief mining at the working face. As the geological conditions of applied coal mine working faces become more diverse and complex, scholars have begun to improve the traditional pre-splitting and blasting technology, and the research on the effects of directional energy-gathered blasting has gradually increased.

With the more widespread application of pre-splitting and roof-cutting technology by Chinese and foreign scholars and engineers in coal mining sites, the technology has been comprehensively developed in the coal mining field. Furthermore, four major application scenarios have gradually been derived: roof cutting and pressure releasing for gob-side entry driving and gob-side entry retaining, non-pillar self-forming roadway, cracking of hard and thick layer roof, and pressure relief of working face and the roadway; this paper presents a comprehensive summary of this technology in coal mining in recent years and systematically describes the basic methods, technical principles, mechanical mechanisms, and application scenarios. Currently, underground coal mines advocate the development of safe and green mining technologies with high recovery rates; this technology is widely used to increase recovery rates by reducing or eliminating pillars. Therefore, this paper takes the pre-splitting and roof-cutting technology that helps coal achieve efficient and green mining as the research object of the review and comprehensively analyzes the current situation and development prospect of the application of this technology; it provides reference suggestions for the project experts who need this technology in the mining field to be applied more effectively in green coal mining.

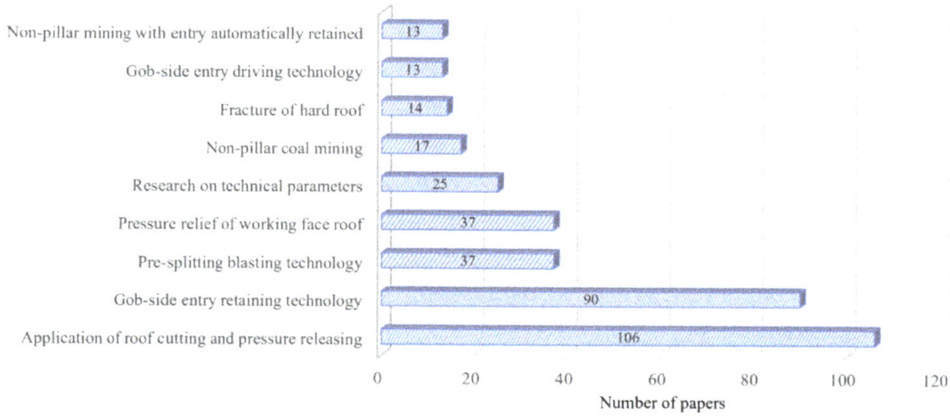

(a)

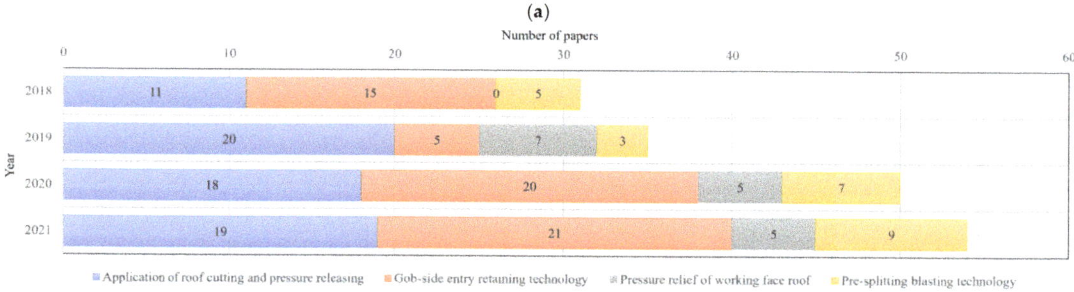

(b)

Figure 3. Category of pre-splitting and roof-cutting technology. (**a**) Nine research hotspots (**b**) Research trends in the past four years.

2. Forms of Pre-Splitting Technology for the Roof

The technology of pre-splitting and roof-cutting is to destroy the roof of the working face and roadway in advance to achieve the effect of releasing its stress concentration, and its core is to fracture the roof successfully [41]. In the field, there are four primary forms of this technology: blasting pre-split technology, hydraulic fracturing technology, liquid CO_2 gas fracturing technology, and mechanical roof-cutting technology.

2.1. Pre-Splitting Roof Technology of Explosive Blasting

2.1.1. Directional Concentrated Blasting Technology

Drill-hole blasting applied in roof-cutting is different from regular blasting. As shown in Figure 4, the blast wave and energy of ordinary blasting spread from the center of the hole to the surrounding area. Due to the anisotropy of the rock mass, the cracks around the hole are randomly extended after blasting. Although this is locally fractured in hard-roof, the blast holes are not effectively connected and do not achieve the "cut-off" effect. Therefore, most field applications of bidirectional energy-gathered tensioning and forming blasting technology. The basic principle of this technology is to control the transmission of blasting shock and stress waves through the energy-gathering pipe, which generates high-pressure gas to stretch the fracture expansion directionally [42–45]. Each blasting hole is blasted simultaneously, and the rock layer is cut off by producing directional fissures.

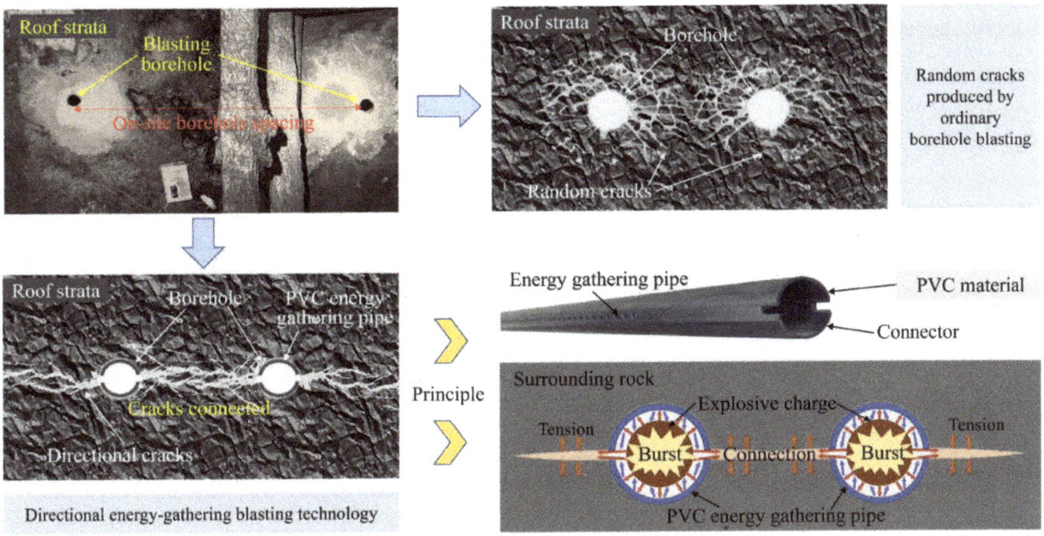

Figure 4. Directional concentrated blasting technology.

In ordinary directional energy-gathered blasting, the explosive is in a dry explosive environment, and its shock wave propagates in the air medium during blasting. Applying this method in underground coal mines generates lots of dust and toxic gases, which is not conducive to the green mining of coal. In order to overcome this drawback, the energy-gathered charge hydrodynamic blasting technology is gradually applied [46]; it is the application of the water medium filled with explosives, the use of water to spread the explosive energy, which reduces the generation of coal dust and toxic gases while enhancing the blasting effect, and it becomes a green and safe method for directional roof-cutting.

2.1.2. Directional Pre-Splitting Technology of Liquid Explosive

Ordinary solid explosives dry blasting drawbacks, improved for water-filled pressure blasting, which is effective in gas and dust control, but still shows the explosive explosion instant power and cracking range is not enough. Therefore, liquid explosives blasting technology gradually developed, the structure of the three as shown in Figure 5.

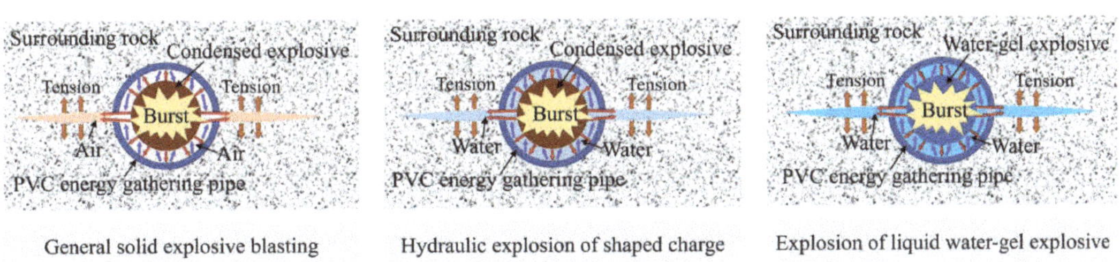

Figure 5. Charge structures of different blasting methods.

Liquid explosives to achieve the pumping method of filling the blast hole, the explosive blast energy in the fracture during the explosion is continuously supplied, the rock breaking load in the seam is robust and long duration, conducive to the full development and expansion of rock fractures. In addition, compared with hydrogel explosives and emulsion explosives, the fracture density of the borehole wall of the surrounding rock after blasting is large, and the fractures are evenly distributed [47–49]. Therefore, liquid explosives in coal mines with blasting the energy gathering pipe can achieve coupled charging in the pipe while playing the flow characteristics; its directional blasting fracturing effect and environmental protection are much better.

2.1.3. Directional Seam-Making Technology of Composite Blasting

In the actual roof-cutting work, the problem of precise and continuous fracturing of the hard and thick roof is often encountered, which requires the composite means of multiple blasting methods. For this reason, the composite blasting directional fracturing technology, which includes energy-gathered injection and high-pressure splitting, is gradually being applied in the field. The core device of this technology is a high-energy perforating gun, which needs to be installed at a certain depth of the blast hole, to be sealed after the excitation of the detonating cord to detonate the perforating bullet; its explosion also stimulates the rapid detonation of the composite propellant to produce high-pressure gas secondary splitting [50]. The result is a radially continuous fracture surface with the fusion of holes and seams to achieve precise control of rock collapse of the roof.

2.2. Hydraulic Fracturing Pre-Splitting Roof Technology

2.2.1. Hydraulic Fracturing Principle

The roadway roof is drilled according to a certain elevation angle, and the radial cutting groove is pre-fabricated on the inner wall of the borehole. After that, the hole is sealed, and the high-pressure water pump is used for water injection fracturing. The roof rock layer cracks and expands, and the crack generated by other boreholes is connected to achieve the effect of cutting off the roof; its core is that the surrounding target rock is pre-cut and fractured by high-pressure pre-water injection, thus weakening the overall strength [51,52].

As shown in Figure 6, for the pre-splitting surrounding rock, the pre-cutting groove is drilled first, and then the two sides of the groove are sealed by the packer, and then the high-pressure water is injected into the sealing fracturing section to realize fracturing of the surrounding rock. Concerning the fracture initiation mechanism, through the force analysis of the coin-shaped crack section, the initial conditions of the coin-shaped fracture in the water-saturated and natural state are obtained:

$$\begin{cases} P_W = 0.506K_{ICW}\sqrt{\frac{\pi}{a}} + \frac{\sigma_v+\sigma_h}{2} - \frac{\sigma_v-\sigma_h}{2}\left(\cos 2\alpha + (-1)^k \frac{2.418}{2-\mu_W}\sin 2\alpha\right) \\ P_N = 0.505K_{ICN}\sqrt{\frac{\pi}{a}} + \frac{\sigma_v+\sigma_h}{2} - \frac{\sigma_v-\sigma_h}{2}\left(\cos 2\alpha + (-1)^k \frac{3.056}{2-\mu_N}\sin 2\alpha\right) \end{cases} \tag{1}$$

where a is the radius of coin-type fracture, mm; σ_V and σ_h are vertical stress and minimum horizontal principal stress, respectively, MPa; α is the elevation angle of the borehole, $\alpha \in (0° \sim 90°)$; K_{ICW} and K_{ICN} are the type-I fracture toughness values under full water and natural state; P_W and P_N are the water pressure of groove initiation, MPa; μ_W and μ_N are Poisson's ratio; for parameter k, $\sigma_V > \sigma_h$ takes an even number, and vice versa takes an odd number. As a result, the trench inclination or borehole elevation angle α, which is most favorable for fracture initiation in hydraulic fracturing, can be calculated under different surrounding rock and stress environments [53,54].

2.2.2. Application of Hydraulic Fracturing Technology

The hydraulic fracturing technology includes fracture hole drilling, hole trenching, hole sealing, roof hydraulic fracturing, and effect monitoring [55,56]. As shown in Figure 7, the 2129 working face has a thick and hard main roof rock layer, designed to be hydraulically fractured to the full extent of the top slab prior to its mining to attenuate the impact of solid rock pressure from the hard roof.

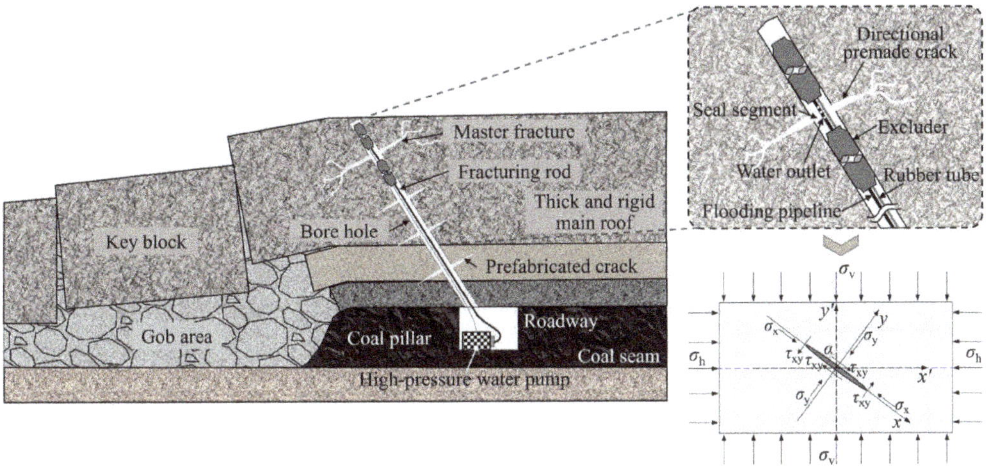

Figure 6. Principle of hydraulic fracturing rock.

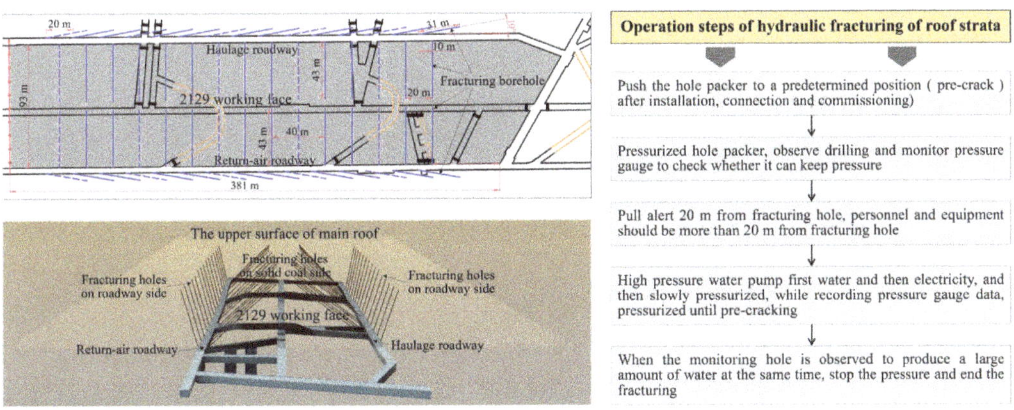

Figure 7. Hydraulic fracturing technology for roof-cutting.

The straightforward steps of the hydraulic fracturing process: firstly, drilling is carried out with a geological drilling rig, and after the drilling is completed, the drilling team will proceed with the next drilling operation. Then a sealer is installed in the completed hole, and the hole is sealed with a manual pump and an energy reservoir. Finally, a high-pressure pump is connected for fracturing. The fracturing of the coal seam to be mined and the hard roof above the roadway is achieved, after which the working face can be advanced and mined [57]; this method is safe, does not produce toxic gas and dust, and is gradually being more widely accepted at coal mine sites.

2.3. Liquid CO_2 Gas Pre-Splitting Roof Technology

CO_2 gas phase pre-fracturing refers to the rapid conversion of CO_2 from liquid to gas by a high-pressure detonation head and its instantaneous volume expansion pressure to fracture rock along the fracture face.

As shown in Figure 8, liquid CO_2 is stored in a storage tank, which requires an environment with $p > 7.35$ MPa and $T < 31$ °C to maintain the liquid form. The volume of liquid CO_2 in the fracture expands 600 times in 20 ms under the action of the heating device, creating an expansion pressure of 80–270 MPa to fracture the rock. Quick steps of liquid CO_2 blasting: After the heating device is activated, the liquid CO_2 in the storage tank expands and destroys the fixed pressure shear sheet. CO_2 is released sharply and generates low temperature and high-pressure gas to act on the pre-fractured part to realize pre-fracture blasting [58]; this method has excellent continuity, does not quickly produce a lot of dust and sparks, has apparent safety and environmental advantages, and is currently being applied in complicated rock fracturing in the excavation and recovery workings.

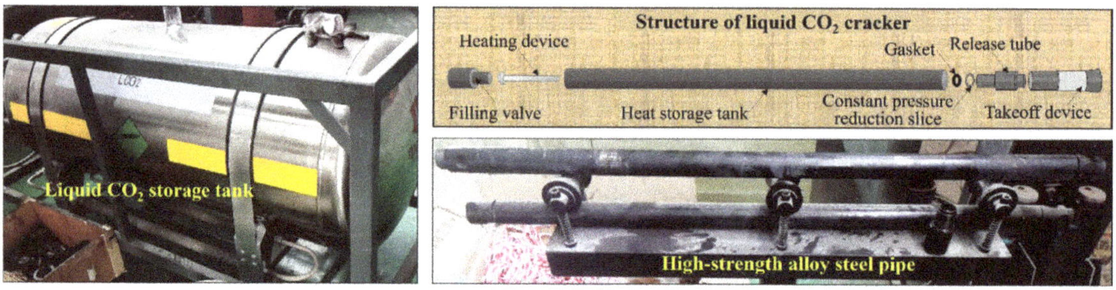

Figure 8. Liquid CO_2 storage tank and cracker.

2.4. Mechanized Directional Roof-Cutting Technology

2.4.1. Continuous Cutting Technology of Chain Arm Saw Machine

The continuous cutting machine with a chain arm saw cuts the rock in a specific direction and inclination with low-speed chipping; it considers both ductile and brittle cutting methods, and the cutting head can make extrusion and shear damage to the rock body, thus forming a continuous through-cutting slit and achieving precise directional cutting of the top slab [59].

This method has a high degree of mechanization and can directly cut the roof by remote directional control, and continuous operation reduces the time of cutting the roof. The low-speed chipping way avoids spark generation and does not consume much water; the mechanical cutting is highly accurate, and the cut surface is smooth and flat [60]. The development of mechanized roof control is an essential step to intelligent mining in coal mines, and the continuous cutting technology of chain arm saw gives an application idea.

2.4.2. Application of Directional Cutting Roof Rig

The directional roof cutting drilling rig is the mechanized roof cutting equipment developed by Academician He Manchao's "N00" construction method [32]; it has the crucial technology of simultaneous drilling in multiple holes and dynamic adjustment in multiple directions, which realizes multi-hole with the same surface and efficient cutting [61]. According to the field test, it effectively cuts off the stress transfer between the goaf's roof and the roadway's roof so that the roof of the roadway is in a specific range to form a short arm beam structure. The method is being gradually improved, popularized, and applied.

3. Pre-Splitting and Roof-Cutting Technology for Non-Pillar Mining

The method of non-pillar mining can significantly improve the coal recovery rate and reduce the waste of coal resources, which is in line with the concept of high-efficiency and green mining and has been extensively applied in the field [62,63].

3.1. Application in Gob-Side Entry Driving

Gob-side entry driving means that a narrow coal pillar (3~8 m) is set on one side of the last working face to drive the roadway of the next working face.

As shown in Figure 9a, the main roof break appears "O-X" shape in the plane when the working face is mined. The roadway and narrow coal pillar are under the "triangle" key block in gob-side entry driving [64,65]. According to the support practice of gob-side entry on-site, the gob-side entry is disturbed strongly by the superimposed stress, and its maintenance is quite tricky. The key influencing factor of mining pressure behavior in gob-side entry driving is the fracture position of the "triangle" key block. The unfavorable fracture position directly leads to large deformation of the roadway and roof collapse. Therefore, it is necessary to actively control the fracture location of critical blocks to alleviate the concentrated stress in the roadway, as shown in Figure 9b.

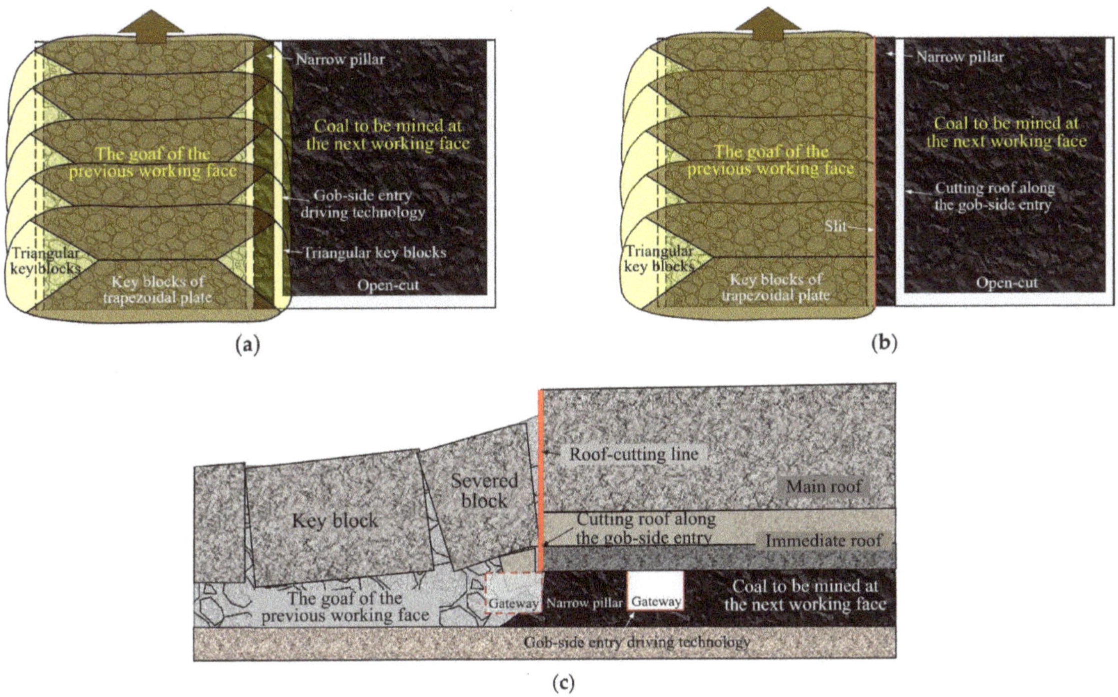

Figure 9. The schematic diagram of roof-cutting for gob-side entry driving. (**a**) Roof structure without roof cutting (**b**) Roof structure after roof cutting (**c**) Cutting roof and unloading the pressure along the gob-side entry.

As shown in Figure 9c, in the mining face, the subsequent mining face side of the roadway driving along the goaf to carry out the pre-splitting operation, the location of roof cutting is in the narrow coal pillar side of the roof. When a section of the roadway roof cutting operation is completed, and the roof collapse in the goaf is stable simultaneously, the roadway digging operation along the goaf. Attention should be paid to the replacement relationship between the mining operation of the previous working face and the tunneling operation of the next working face to ensure that the safety distance is maintained [66,67]. According to the field practice, the stress concentration degree along the empty roadway of this method is small, and the roadway support effect is practical, ensuring the roadway roof's safety.

3.2. Application in Gob-Side Entry Retaining

The gob-side entry retaining technology realizes the succession of the completely non-coal pillar between working faces, an efficient and green mining technology that is booming and applying [68].

3.2.1. Gob-Side Entry Retaining with the Filling Body

This gob-side entry retaining method uses an artificial filling body as a narrow coal pillar in gob-side entry driving. The filling bodies include high water material, paste material, and concrete [69].

(1) Delayed pre-splitting and cutting roof

As shown in Figure 10a, the force on the filling body is similar to that on the narrow coal pillar in the gob-side entry driving. At this moment, the roof structure and fracture state of the gob-side entry is similar to that of gob-side entry driving, which still leads to roadway maintenance difficulties. In order to reduce the pressure of the filling body and make the roof entirely collapse, pressure relief blasting technology with circular shallow holes is used, as shown in Figure 10b. After mining, a row of round holes is drilled towards the goaf next to the filling body about 20 cm to carry out energy-accumulating blasting [70,71]. As a result, the roof above the mining area collapses, falls, and then compacts and continues to support the overlying rock layer, as shown in Figure 10c.

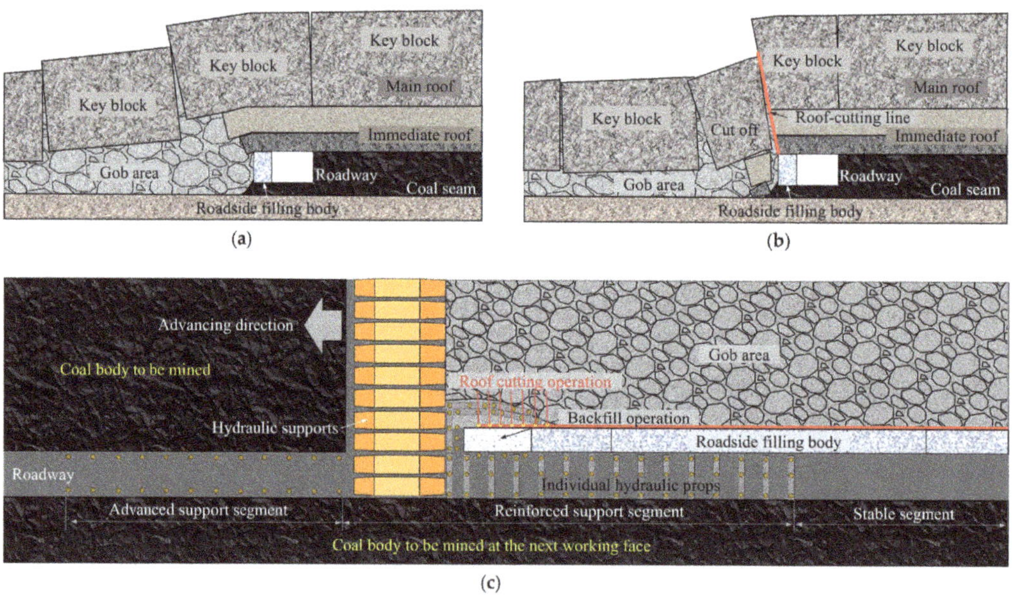

Figure 10. Delayed pre-splitting and cutting roof for gob-side entry retaining. (**a**) Gob-side entry retaining with backfill (**b**) Cutting roof on goaf side (**c**) Working methods of roof-cutting and gob-side entry retaining with backfill.

The disadvantage of this method is that it interferes with the parallel operation of mining and filling of the working face, which cannot realize the fast and high-intensity operation of the comprehensive mining face.

(2) Advance pre-splitting and cutting roof

The method of roof-cutting along the side of the roadway to be mined is the most commonly used in the field before mining the working face. As shown in Figure 11a,b, setting the filling body in the roadway for gob-side entry retaining, the main roof structure above the roadway and the filling body is more integral and stable after cutting the roof in advance [72–74]. According to the size of roadway and backfill, ample ventilation, and laneway equipment size, roadway expansion measures are often taken to ensure that the roadway size is reasonable, as shown in Figure 11c; this method can be applied in thick coal seams, soft rock, and deep high-stress environment.

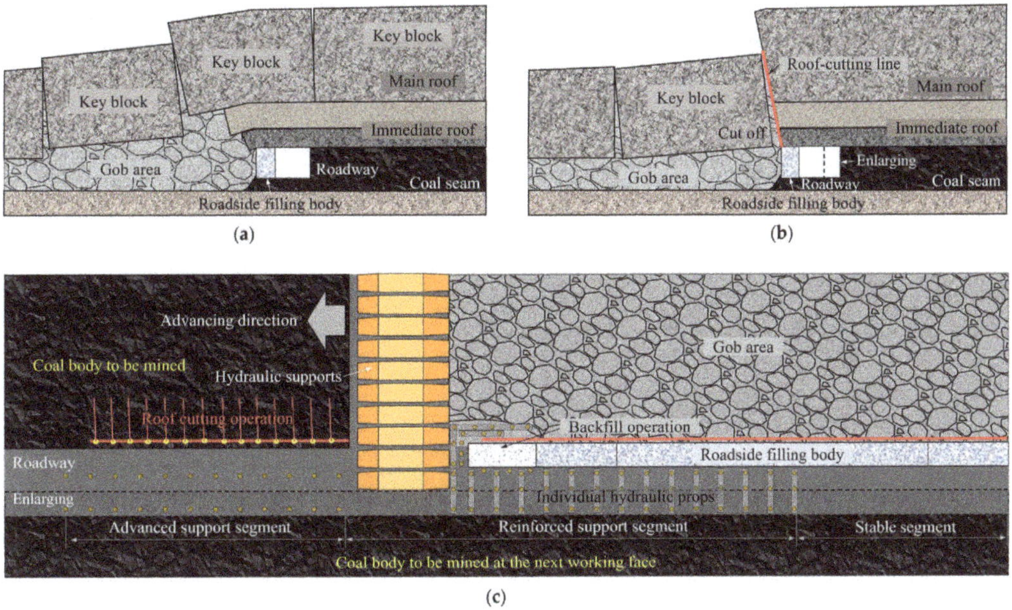

(a) (b) (c)

Figure 11. Advance pre-splitting and cutting roof for gob-side entry retaining. (**a**) Gob-side entry retaining with backfill (**b**) Roof cutting operation in advance (**c**) Working methods of roof cutting in advance and gob-side entry retaining with backfill.

3.2.2. Completely Gob-Side Entry Retaining

The method of fully gob-side entry retaining means not leaving the roadside filling body, using a specific prop or single pillar to support the roof of the goaf side, and setting up gangue protection and Air leakage prevention device [75,76]. Compared with cutting the roof of retaining roadway leaving narrow coal pillar and roadside filling body, this method is more straightforward; it fully uses mining pressure and rock mass fragmentation to realize automatic lane formation.

(1) Cutting roof for retaining roadway with advance pre-splitting

During the working face mining period, pre-fracturing and roof-cutting are overrun along the solid coal side in the roadway to weaken or release the physical-mechanical connection between the roadway's roof and the gob's roof, as shown in Figure 12a,b; it is supported by a temporary prop (portal prop) with a single pillar to provide upward support resistance. The roof of the gob area is cut down as a whole by using mining pressure and artificially created roof-cutting cracks in advance. The roadway is under a complete and stable cantilevered structure, as shown in Figure 12c. According to the mechanism of roof-cutting and roadway-retaining, the site is usually divided into three operational phases: the pre-cracking and roof-cutting stage, the reinforcement support stage, and the stabilization stage of the left roadway [77–80]. The three stages correspond to different locations, so parallel operations are realized between the processes, which will significantly improve the efficiency of the left roadway. Since this method does not require the setting of a filling body, this has little effect on the advancing speed of the working face; this method is being applied under the conditions of different coal seam thickness, burial depth, dip angle, and roof conditions, and it has become a hot spot in the research of no pillar mining methods.

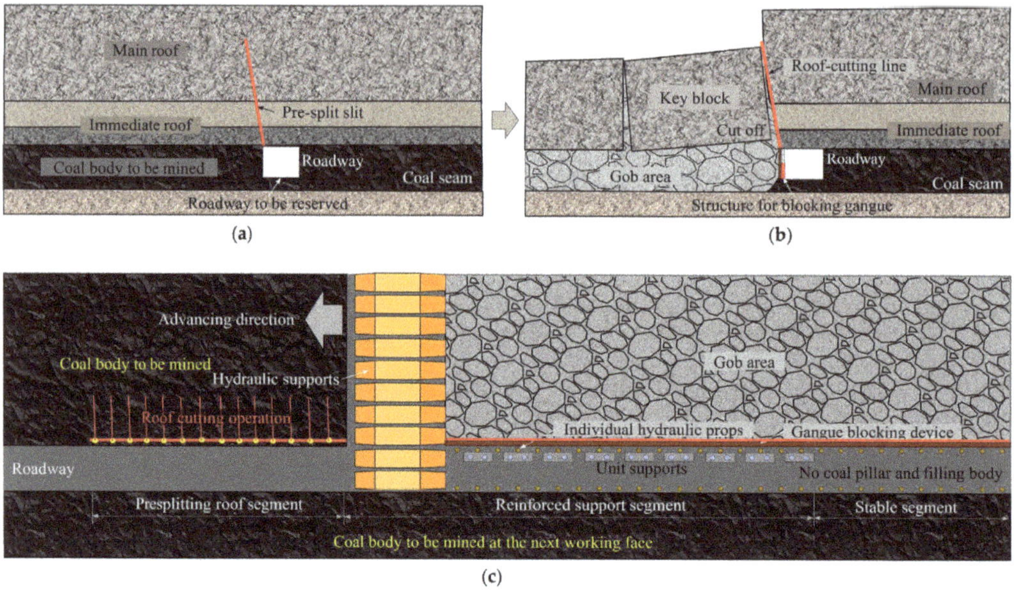

Figure 12. Gob-side entry retaining technology without coal pillar and filling body. (**a**) Pre-splitting roof in advance of the working face (**b**) Cutting roof to form roadway automatically (**c**) Working methods of roof cutting in advance and gob-side entry retaining with no-pillar.

(2) Combined support for cutting roof retaining roadway

For special roof conditions (e.g., thick direct roof), the pre-fracturing and roof-cutting can be re-optimized. For the working face with a thick immediate roof, the roof breaks down and collapses after working face mining. Due to the fragmentation of rocks, the collapsed rock can adequately fill the goaf. At this time, the combined support can be used to achieve roof cutting and roadway retaining, that is, to strengthen the support of the roadway to be retained in advance; it mainly includes the firm anchoring of the deep hole of the high-strength bolt cable on the roof and the support on the goaf side (π-shaped steel, the cross-hinged roof beam, and the single pillar of the gob-side entry retaining); they provide robust support resistance and use the mining pressure of the working face to realize the cutting of the immediate roof of the goaf side and fill the goaf to complete the roadway retaining operation. The overall approach covers four technologies: combined support roof-cutting technology, advanced-area reinforcement support technology, prevention, control technology of gangue in goaf, and beam-column auxiliary support technology of roadway [81,82].

This method directly eliminates the process of over-advance pre-fracturing and roof-cutting; meantime, it avoids the impact on the working face advancement efficiency due to blasting or hydraulic fracturing, which realizes the efficient retention of the roadway with the rapid advancement of the working face. Although its application conditions are still limited at present, its method indicates the future need to develop a mechanized way of cutting the roof and leaving the roadway without blasting.

3.3. "N00" Construction Method

The "N00" coal mining method efficiently combines the working face mining and roadway tunneling to realize the mining of N working faces, the tunneling of 0 roadways, and the retaining of 0 district sublevel coal pillars, as shown in Figure 13 [32]. In this method, the working face is mined to form the next working face's roadway, and the roadway's roof along the goaf is treated by pre-splitting and roof-cutting. The directional cutting method, such as energy-accumulating blasting, was successful at the initial test stage; however, due to the large-scale blasting operation behind the working face, there are hidden dangers to safety. After that, the team researched and developed the supporting equipment for the "N00" method and developed a directional roof-cutting rig, which is located on the side of the left-over roadway and follows the hydraulic support of the working face to push forward the roof-cutting operation [83,84].

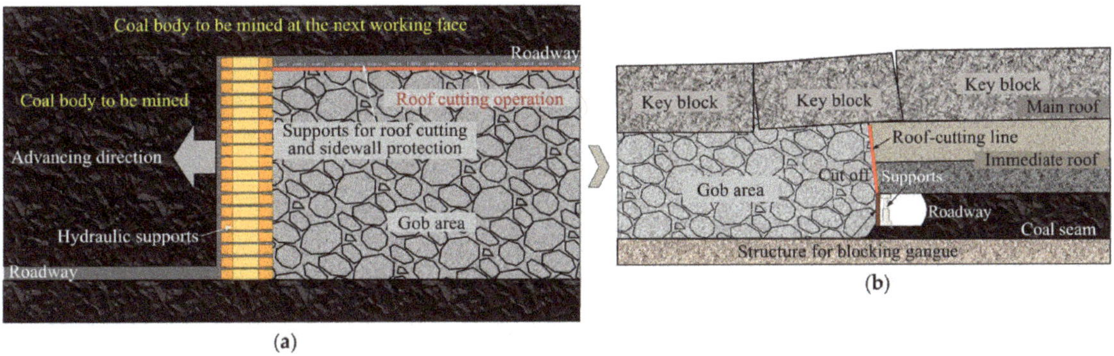

Figure 13. Cutting roof and retaining roadway technology in the "N00" construction method. (**a**) Plane diagram of "N00" construction method (**b**) The diagram of the retained roadway.

In addition, with the roof-cutting prop to support the roof on the side of the mining area of the roadway, to achieve automatic mechanical connection of coal mining, roadway formation, roof-cutting and pressure relief, support, and other processes of the working face; this method is being gradually improved and promoted in the field, providing a complete set of solutions to achieve coal pillar-free automated mining in new mining areas ultimately.

4. Pre-Splitting and Roof-Cutting Technology for Pressure Relief Mining of Working Face

The hard roof is often encountered when the long-wall working face is mined. The hard roof is difficult to collapse, and the support of the working face continues to bear the high load, which requires pre-cracking and unloading the roof [85,86].

4.1. Fracturing Operation of Thick and Hard Main Roof

The physical and mechanical properties of the main roof above the coal seam directly affect the smooth mining of the longwall working face. With the increase of mining depth, the stress level during coal seam mining gradually increases. For the coal seam with a thick and hard main roof, the weighting step and pressure of the main roof increase significantly when the working face is mined. The intense pressure threatens the safety of equipment and workers, such as hydraulic support in the working face [87–89]. Therefore, the pre-splitting operation is usually carried out before the mining face for the thick and hard main roof.

As shown in Figure 14, the main roof rock is pre-cracked by hydraulic fracturing, and the crack location includes the roof of the coal body to be mined, the roof of the roadway on both sides, and the roof of the working face's open-off cut. After the main roof's pre-splitting, the working face's coal wall pressure is slight during the weighting period. The shrinkage of the pillar is minor, and the roof collapse is closely followed by the support, indicating that the main roof activity is not intense during the fracturing [90–92]; this method has been widely used in pressure relief mining of working faces with the thick and hard main roof.

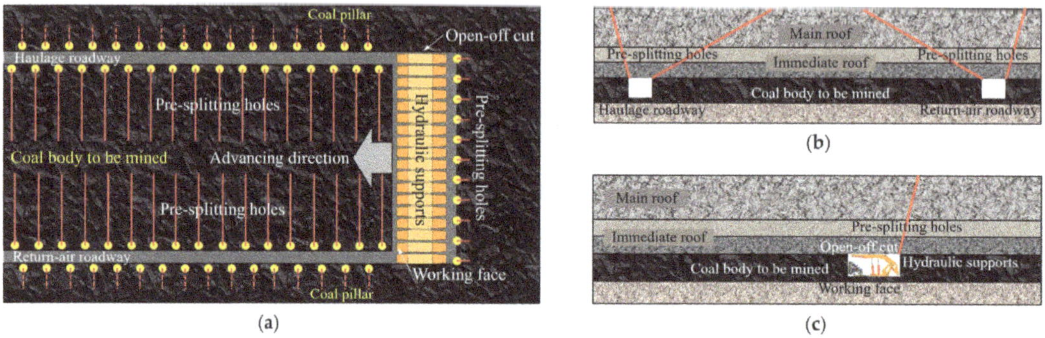

Figure 14. Fracturing operation of the thick and hard main roof. (**a**) Plane diagram of pre-splitting drilling hole (**b**) Side view of boreholes in roadway (**c**) Side view of boreholes in the open-off cut.

4.2. Fracturing Operation of Hard Roof at the End of the Working Face

For the broken roof of the longwall working face, there is a "triangle" key block structure in the end area [93,94]. When the roof of the working face is hard, it is often challenging to collapse on both sides of the roof, resulting in an excessive hanging roof area. Improper treatment will cause air leakage in the working face, gas accumulation in the goaf, and sudden caving of the sagging roof. Therefore, the hard roof at the end side should be cracked in time, and the borehole layout is shown in Figure 15. Due to the goaf behind the end support, there is a risk of explosive blasting, and hydraulic fracturing is generally used to crack the roof [95,96]. According to the field practice, in the subsequent mining process, the end top plate, after cracking, is no longer suspended but gradually collapses with the advance of the end support, which dramatically reduces the pressure of the support and achieves a good pressure relief effect.

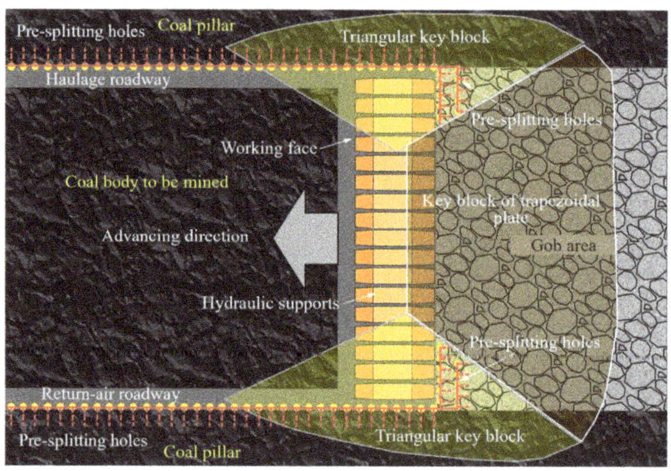

Figure 15. Fracturing operation of a hard roof at the end of the working face.

5. Pre-Splitting and Roof-Cutting Technology for Roadway's Pressure Relief

For gob-side entry, tunneling entry, and deep roadway with a high-stress environment, standard support methods are often tricky to work, and active pressure relief is needed for the position of a continuous deformation [97].

5.1. Large Deformation of Roadway along Goaf

When one side of the working face is being mined, the phenomenon of large deformation occurs in the local area of the roadway along the goaf of the adjacent digging face, especially in the roadway section within the influence of the working face mining, which rate is speedy and challenging to control [98,99]. In order to reduce the influence of mining-induced stress on gob-side entry as much as possible, pre-splitting and roof-cutting operations are designed in the adjacent roadways of this working face. The borehole layout is shown in Figure 16. By cutting the roof in advance, the continuous structure between the working face and the roof of the gob-side entry is severed, hindering mining-induced stress transmission. Thus, the surrounding rock stress environment of the roadway along the goaf is improved, and the stability of the roadway can be successfully maintained by reinforcing the support means [100–103]. Pre-splitting and cutting the roadway's roof in the overhead section of the working face to unload the pressure for the roadway along the goaf is commonly implemented in the field, which has become an important method to control the large deformation of the roadway along the goaf.

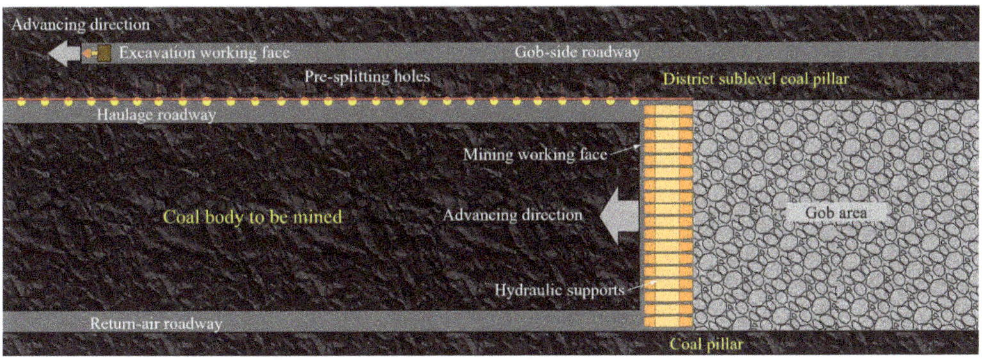

Figure 16. Pressure relief technology of roof-cutting for large deformation of roadway along goaf.

5.2. Blasting Pressure Relief Technology for Roadways

The deformation characteristics have apparent continuity and persistence for the deep high ground stress and robust rheological environment of the roadways. The general reinforcement support method cannot prevent the rheological characteristics of the surrounding rock, and the roadway needs to be expanded and repaired many times. At this point, the key to controlling deformation is to relieve pressure and cut off the transmission path of continuous high stress [104,105]. As shown in Figure 17, illustrates three blasting methods to relieve pressure to cope with substantial deformation of the roof, gangs, and floor. Blasting to relieve pressure means creating cracks by blasting to destroy the integrity and continuity of the rock to release the concentrated stress and cut off the stress transmission path to control the rheology of the surrounding rock in the roadway.

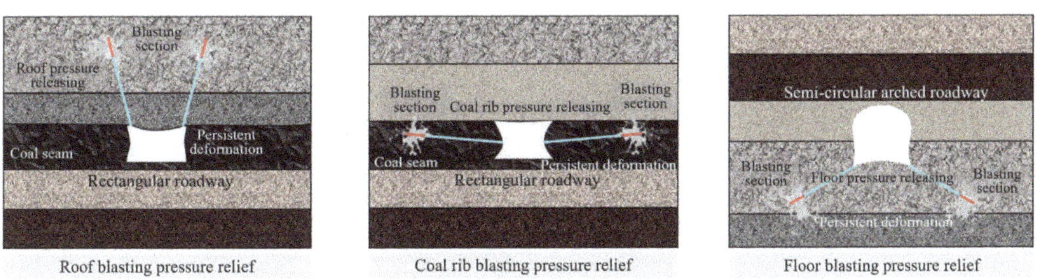

Figure 17. Blasting pressure relief technology for roadways.

For the pressure relief of the roadway under the complex high-stress environment, concentrated drilling is often carried out in the surrounding rock of the complex high-stress area. The small-scale pressure relief and fracture area are generated through the deformation and collapse of the drilling hole. At present, directional pre-splitting and roof-cutting pressure relief technologies such as concentrated energy blasting and hydraulic fracturing have been widely applied [28]. According to different surrounding rock environments and the mechanical properties of rock mass, a reasonable pressure relief method is selected to better realize the stability of the roadway under complex conditions.

With the development of the method, the internal pressure relief method and hydraulic pressure relief method of roadway surrounding rock are gradually evolved [106], which have achieved outstanding field application effect in controlling deep continuous deformation roadway.

6. Discussion

This paper summarizes the four forms of application of pre-splitting and roof cutting technology (explosive blasting, hydraulic fracturing, liquid CO_2 gas fracturing, and mechanized roof cutting) and elaborates its process in three major fields (non-pillar mining, working face's pressure relief, and roadway's pressure relief). The following conclusions and prospects are obtained:

(1) The pre-splitting method of explosive blasting has gradually developed to low pollution, low dust, high precision, and high safety. The methods of water-filled pressure blasting, liquid explosive blasting, and liquid CO_2 gas phase blasting improve environmental protection and safety significantly. The composite blasting form is committed to establishing a continuous fracture surface to achieve precise roof cutting.

(2) Hydraulic fracturing technology is gradually becoming more and more widespread in the application of hard-roof fracturing, especially in roof pre-splitting near the working face and the goaf; this method does not produce toxic gas and dust, which has high application safety for gas mines.

(3) The research hotspots of pre-splitting and roof cutting focus on pressure relief technology and gob-side entry retaining technology. For deep high-stress mining environments, applying roof pre-splitting technology will be more widely, including the safe replacement of deep coal seam working face, deep hard roof cracking, and deep roadway pressure relief.

(4) As an essential step of non-pillar mining, pre-splitting and roof cutting technology is also developing towards mechanized roof cutting, while the mining method is mechanization and automation. The research, development, and application of the chain arm saw machine and directional cutting roof rig are representative; it is anticipated that the future pre-splitting and roof-cutting technology will balance precision, automation, and greening, which will help the development of intelligent and unmanned mining.

Author Contributions: Conceptualization, S.X., D.C. and Y.W.; methodology, D.C., Y.W. and F.G.; software, Y.W. and H.Z.; formal analysis, Y.W., H.Z. and F.G.; investigation, Y.W., H.Z. and X.Z.; data curation, Y.W., H.Z. and X.M.; writing—original draft preparation, Y.W.; writing—review and editing, Y.W.; supervision, R.L. and C.W.; project administration, D.C.; funding acquisition, D.C. All authors have read and agreed to the published version of the manuscript.

Funding: This work was financially supported by the National Natural Science Foundation of China (Grant No. 52074296), the Fundamental Research Funds for the Central Universities (Grant No. 2022YJSNY18), the National Natural Science Foundation of China (Grant No. 52004286), the Fundamental Research Funds for the Central Universities (Grant No. 2022XJNY02), the China Postdoctoral Science Foundation (Grant No. 2020T130701, 2019M650895), all of which were gratefully acknowledged.

Institutional Review Board Statement: Not applicable.

Informed Consent Statement: Informed consent was obtained from all subjects involved in the study.

Data Availability Statement: Not applicable.

Acknowledgments: No conflict of interest exists in the submission of this manuscript, and the manuscript is approved by all authors for publication.

Conflicts of Interest: The authors declare no conflict of interest.

References

1. He, M.; Wang, Q.; Wu, Q. Innovation and future of mining rock mechanics. *J. Rock Mech. Geotech. Eng.* **2021**, *13*, 1–21. [CrossRef]
2. Yu, B.; Tai, Y.; Gao, R.; Yao, Q.; Li, Z.; Xia, H. The sustainable development of coal mines by new cutting roof technology. *R. Soc. Open Sci.* **2020**, *7*, 191913. [CrossRef] [PubMed]
3. Ma, Z.; Wang, J.; He, M.; Gao, Y.; Hu, J.; Wang, Q. Key Technologies and Application Test of an Innovative Noncoal Pillar Mining Approach: A Case Study. *Energies* **2018**, *11*, 2853. [CrossRef]
4. Wang, E.; Xie, S. Determination of coal pillar width for gob-side entry driving in isolated coal face and its control in deep soft-broken coal seam: A case study. *Energy Sci. Eng.* **2022**, *10*, 2305–2316. [CrossRef]
5. Xie, S.; Gao, M.; Chen, D.; Sun, Y.; Pan, H.; Su, H.; Lan, S. Stability influence factors analysis and construction of a deep beam anchorage structure in roadway roof. *Int. J. Min. Sci. Technol.* **2017**, *28*, 445–451.
6. Xie, S.; Pan, H.; Zeng, J.; Wang, E.; Chen, D.; Zhang, T.; Peng, X.; Yang, J.; Chen, F.; Qiao, S. A case study on control technology of surrounding rock of a large section chamber under a 1200-m deep goaf in Xingdong coal mine, China. *Eng. Fail. Anal.* **2019**, *104*, 112–125. [CrossRef]
7. Chen, D.; Wang, E.; Xie, S.; He, F.; Wang, L.; Zhang, Q.; Wu, X.; Jiang, Z.; Li, Y.; Shi, S. Roadway surrounding-rock under multi-coal seam mining: Deviatoric stress evolution and control technology. *Adv. Civil Eng.* **2020**, *2020*, 9891825. [CrossRef]
8. Kang, H.; Zhang, X.; Wang, D.; Tian, J.; Yi, Z.; Jiang, W. Strata control technology and applications of non-pillar coal mining. *J. China Coal Soc.* **2022**, *47*, 16–44.
9. Qian, M.; Miao, X.; Xu, J. Theoretical Study of Key Sratum in Ground Conrol. *J. China Coal Soc.* **1996**, *21*, 2–7.
10. He, S.S.; Xie, S.R.; Song, B.H.; Zhou, D.Q.; Sun, Y.D.; Han, S. Breaking laws and stability analysis of damage main roof in close distance hypogynous seams. *J. China Coal Soc.* **2016**, *41*, 2596–2605.
11. He, F.L.; Chen, D.D.; Xie, S.R. The kDL effect on the first fracture of main roof with elastic foundation boundary. *Chin. J. Rock Mech. Eng.* **2017**, *36*, 1384–1399.
12. Chen, D.D.; Guo, F.F.; Wu, Y.Y. Study on the first fracture of the main roof plate structure with one side goaf and elastic-plastic foundation boundary. *J. China Coal Soc.* **2021**, *46*, 3090–3105.
13. Chen, D.D.; He, F.L.; Xie, S.R.; Zeng, J.C. Time-space relationship between periodic fracture of plate structure of main roof and rebound in whole region with elastic foundation boundary. *Chin. J. Rock Mech. Eng.* **2019**, *38*, 1172–1187.
14. Xie, S.; Cheng, D.; Ceng, J.; Gao, M.; Zhang, T.; Liu, J. First fracturing characteristics of main roof plate structure considering coal pillars and elastic coal foundation on both sides of stope. *J. China Coal Soc.* **2019**, *44*, 115–126.
15. He, F.; He, W.R.; Chen, D.D.; Xie, S.R.; Li, H.; He, C.X. First fracture structure characteristics of main roof plate considering elastic-plastic deformation of coal. *J. China Coal Soc.* **2020**, *45*, 2704–2717.
16. Li, Z.; Changzhi, Z.; Xie, H.; Yang, F. Similar simulation test for breakage law of working face floor in coal mining above aquifer. *Chin. J. Rock Mech. Eng.* **2011**, *30*, 1571–1578. [CrossRef]
17. Zhao, Z.; Ma, N.; Guo, X.; Zhao, X.; Fan, L. Falling principle and support design of butterfly-failure roof in large deformation mining roadways. *J. China Coal Soc.* **2016**, *41*, 2932–2939.
18. Guo, X.F.; Guo, L.F.; Ma, N.J.; Zhao, Z.Q.; Li, C. Applicability analysis of the roadway butterfly failure theory. *J. China Univ. Min. Technol.* **2020**, *49*, 646–653.
19. Tao, Z.; Zheng, X.; Zhu, C.; Zhang, H.; Zhang, X. Framework and engineering application of large data monitoring system for a non-pillar mining method with gob-side entry formed automatically. *J. China Univ. Min. Technol.* **2021**, *50*, 983–991.
20. Xie, S.; Guo, F.; Wu, Y. Control Techniques for Gob-Side Entry Driving in an Extra-Thick Coal Seam with the Inflence of Upper Residual Coal Pillar: A Case Study. *Energies* **2022**, *15*, 3620. [CrossRef]
21. Wang, Q.; Jiang, B.; Xin, Z.X.; He, M.C.; Li, S.C.; Zhang, P.; Wang, Y.; Jiang, Z.H. Development of a 3D geomechanical model test system for non-pillar mining with automatically formed roadway and its engineering application. *Chin. J. Rock Mech. Eng.* **2020**, *39*, 1582–1594.
22. Xie, S.; Wu, Y.; Chen, D.; Liu, R.; Han, X.; Ye, Q. Failure analysis and control technology of intersections of large-scale variable cross-section roadways in deep soft rock. *Int. J. Coal Sci. Technol.* **2022**, *9*, 19. [CrossRef]
23. Liu, X.; He, M.; Wang, J.; Ma, Z. Research on non-pillar coal mining for thick and hard conglomerate roof. *Energies* **2021**, *14*, 299. [CrossRef]
24. Chen, X.; Li, L.; Guo, Z.; Chang, T. Evolution characteristics of spontaneous combustion in three zones of the goaf when using the cutting roof and release pressure technique. *Energy Sci. Eng.* **2019**, *7*, 710–720. [CrossRef]
25. Wang, Q.; Qin, Q.; Jiang, B.; Jiang, Z.; He, M.; Li, S.; Wang, Y. Geomechanics model test research on automatically formed roadway by roof cutting and pressure releasing. *Int. J. Rock Mech. Min. Sci.* **2020**, *135*, 104506. [CrossRef]
26. Chen, X.; Du, Y.; Wang, L.; Zhao, S. Evolution and application of airflow permeability characteristics of gob in roof cutting and pressure releasing mining method. *Energy Sci. Eng.* **2020**, *8*, 2073–2085. [CrossRef]

27. Wu, Y.; Xie, S.; Zhang, Y. Research on stability control of roadway intersections with nested variable cross-section in deep mine. *J. Min. Sci. Technol.* **2022**, *7*, 720–729.
28. Li, G.; Yang, S.; Sun, Y.; Xu, J.; Li, J. Research progress of roadway surrounding strata rock control technologies under complex conditions. *Coal Sci. Technol.* **2022**, *50*, 29–45.
29. Yang, S.; Li, G.; Bi, R.; Yao, B.; Feng, R.; Sun, Y. The Stability of Roadway Groups under Rheology Coupling Mining Disturbance. *Sustainability* **2021**, *13*, 12300. [CrossRef]
30. Sun, Y.; Bi, R.; Sun, J.; Zhang, J.; Taherdangkoo, R.; Huang, J.; Li, G. Stability of roadway along hard roof goaf by stress relief technique in deep mines: A theoretical, numerical and field study. *Geomech. Geophys. Geo-Energy Geo-Resour.* **2022**, *8*, 45. [CrossRef]
31. Wang, Q.; Wang, Y.; He, M.; Jiang, B.; Li, S.; Jiang, Z.; Wang, Y.; Xu, S. Experimental research and application of automatically formed roadway without advance tunneling. *Tunn. Undergr. Space Technol.* **2021**, *114*, 103999. [CrossRef]
32. Zhang, Q.; He, M.; Guo, S.; Qi, J.; Yang, J.; Wang, C.; Xia, M.; Li, L. Investigation on the key techniques and application of the new-generation automatically formed roadway without coal pillars by roof cutting. *Int. J. Rock Mech. Min. Sci.* **2022**, *152*, 105058. [CrossRef]
33. Wang, Y.; He, M.; Yang, J.; Wang, Q.; Liu, J.; Tian, X.; Gao, Y. Case study on pressure-relief mining technology without advance tunneling and coal pillars in longwall mining. *Tunn. Undergr. Space Technol.* **2020**, *97*, 103236. [CrossRef]
34. Li, Z.; Tao, Z.; Meng, Z.; He, M. Longwall mining method with roof-cutting unloading and numerical investigation of ground pressure and roof stability. *Arab. J. Geosci.* **2018**, *11*, 697. [CrossRef]
35. Yang, J.; Liu, B.; Bian, W.; Chen, K.; Wang, H.; Cao, C. Application Cumulative Tensile Explosions for Roof Cutting in Chinese Underground Coal Mines. *Arch. Min. Sci.* **2021**, *66*, 421–435.
36. Xie, S.R.; Pan, H.; Chen, D.D.; Zeng, J.C.; Song, H.Z.; Cheng, Q.; Xiao, H.B.; Yan, Z.Q.; Li, Y.H. Stability analysis of integral load-bearing structure of surrounding rock of gob-side entry retention with flexible concrete formwork. *Tunn. Undergr. Space Technol.* **2020**, *103*, 103492. [CrossRef]
37. Wang, Q.; He, M.; Yang, J.; Gao, H.; Jiang, B.; Yu, H. Study of a no-pillar mining technique with automatically formed gob-side entry retaining for longwall mining in coal mines. *Int. J. Rock Mech. Min. Sci.* **2018**, *110*, 1–8. [CrossRef]
38. Tao, Z.; Song, Z.; He, M.; Meng, Z.; Pang, S. Principles of the roof cut short-arm beam mining method (110 method) and its mining-induced stress distribution. *Int. J. Min. Sci. Technol.* **2018**, *28*, 391–396. [CrossRef]
39. Fan, D.; Liu, X.; Tan, Y.; Yan, L.; Song, S.; Ning, J. An innovative approach for gob-side entry retaining in deep coal mines: A case study. *Energy Sci. Eng.* **2019**, *7*, 2321–2335. [CrossRef]
40. Wang, Q.; He, M.; Li, S.; Jiang, Z.; Wang, Y.; Qin, Q.; Jiang, B. Comparative study of model tests on automatically formed roadway and gob-side entry driving in deep coal mines. *Int. J. Min. Sci. Technol.* **2021**, *31*, 591–601. [CrossRef]
41. Tu, M.; Zhao, G.; Zhang, X.; Bu, Q.; Dang, J. Fracture Evolution between Blasting Roof Cutting Holes in a Mining Stress Environment. *Minerals* **2022**, *12*, 418. [CrossRef]
42. Zhang, Q.; He, M.; Wang, J.; Guo, S.; Guo, Z.; Liu, X.; Hu, J.; Ma, Z.; Fan, L.; Guo, P. Instantaneous expansion with a single fracture: A new directional rock-breaking technology for roof cutting. *Int. J. Rock Mech. Min. Sci.* **2020**, *132*, 104399. [CrossRef]
43. Hua, X.Z.; Liu, X.; Huang, Z.G.; Yang, P.; Ma, Y. Stability mechanism of non-pillar gob-side entry retaining by roof cutting under the coupled static-dynamic loading. *J. China Coal Soc.* **2020**, *45*, 3696–3708.
44. Ma, X.; He, M.; Sun, J.; Wang, H.; Liu, X.; Zhen, E. Neural Network of Roof Cutting Blasting Parameters Based on Mines with Different Roof Conditions. *Energies* **2018**, *11*, 3468. [CrossRef]
45. Fan, D.; Liu, X.; Tan, Y.; Song, S.; Gu, Q.; Yan, L.; Xu, Q. Roof Cutting Parameters Design for Gob-Side Entry in Deep Coal Mine: A Case Study. *Energies* **2019**, *12*, 2032. [CrossRef]
46. Zhang, Q.; Wang, J.; Guo, S.; Gong, W.; Feng, L.; Wang, H.; Ming, C.; Ma, Z. Shaped Charge Hydraulic Blasting: An Environmental, Safe, and Economical Method of Directional Roof Cutting. *Geofluids* **2021**, *2021*, 5511081. [CrossRef]
47. Wang, Q.; Jiang, Z.; Jiang, B.; Gao, H.; Huang, Y.; Zhang, P. Research on an automatic roadway formation method in deep mining areas by roof cutting with high-strength bolt-grouting. *Int. J. Rock Mech. Min. Sci.* **2020**, *128*, 104264. [CrossRef]
48. Zhang, X.; Hu, J.; Xue, H.; Mao, W.; Gao, Y.; Yang, J.; He, M. Innovative approach based on roof cutting by energy-gathering blasting for protecting roadways in coal mines. *Tunn. Undergr. Space Technol.* **2020**, *99*, 103387. [CrossRef]
49. Wang, Y.; Yang, J.; He, M.; Tian, X.; Liu, J.; Xue, H.; Huang, R. Test of a liquid directional roof-cutting technology for pressure-relief entry retaining mining. *J. Geophys. Eng.* **2019**, *16*, 620–638. [CrossRef]
50. Yang, T. *3D Breaking Characteristics of Overlying Strata and Stress Field Evolution Law in Stope with Directional Hard Roof Cutting*; China University of Mining and Technology: Xuzhou, China, 2021.
51. Wang, Q.; Zhang, P.; Jiang, Z.; He, M.; Li, S.; Wang, Y.; Jiang, B. Automatic roadway formation method by roof cutting with high strength bolt-grouting in deep coal mine and its validation. *J. China Coal Soc.* **2021**, *46*, 382–397.
52. Pan, J.F.; Ma, W.T.; Liu, S.H.; Jiaming, G. A prevention technology of rock burst based on directional presplitting of water jet prefabricated slot in hard roof. *Chin. J. Rock Mech. Eng.* **2021**, *40*, 1591–1602.
53. Tang, T.; Liu, D.; Cui, Z.; Han, Z.; Liu, B.; Wang, Z. Initiation pressure evaluation of coal mine roof hydraulic fracturing based on fracture mechanics. *J. China Coal Soc.* **2020**, *45*, 727–735.
54. Lin, J.; Guo, K.; Sun, Z.; Wang, T. Study on fracturing timing of hydraulic fracturing top-cutting and pressure relief in roadway with strong dynamic pressure. *J. China Coal Soc.* **2021**, *46*, 140–148.

55. Huang, B.; Chen, S.; Zhao, X. Hydraulic fracturing stress transfer methods to control the strong strata behaviours in gob-side gateroads of longwall mines. *Arab. J. Geosci.* **2017**, *10*, 236. [CrossRef]
56. Guo, S.; Zhang, Q.; He, M.; Wang, J.; Liu, J.; Ming, C.; Guo, L.; Fan, L. Numerical investigation on rock fracture induced by a new directional rock-breaking technology. *Eng. Fract. Mech.* **2022**, *268*, 108473. [CrossRef]
57. Wu, Y. *Study on De-Stressing Mechanism of Directional Hydraulic Fracturing to Control Deformation of Reused Roadway in Longwall Mining with Two Gateroad Layout and Its Onsite Practices*; China Coal Research Institute: Beijing, China, 2018.
58. Zhang, J.; Gao, Z.; Zhang, H. Parameters determination of coal-rock HJC model and research on blasting characteristics by liquid CO_2. *Chin. J. Rock Mech. Eng.* **2021**, *40* (Suppl. 1), 2633–2642.
59. Kuang, T. *Research on Overlying Rock Structure and Near-Far Roof Control Technology in Extra Thick Coal Seam*; China University of Mining and Technology: Xuzhou, China, 2021.
60. Tai, Y.; Yu, B.; Xia, B.; Li, Z.; Xia, H. Research on stress release for the gob-side roadway using the roof-cutting technology with a chainsaw arm. *R. Soc. Open Sci.* **2020**, *7*, 191663. [CrossRef]
61. Wang, Q.; Feng, H.; Tang, P.; Peng, Y.; Li, C.; Jiang, L.; Mitri, H.S. The structure characteristics and deformation of "short cantilever beam" using a non-pillar mining method with gob-side entry formed automatically. *J. China Univ. Min. Technol.* **2019**, *48*, 718–726.
62. Sun, X.; Liu, Y.; Wang, J.; Li, J.; Sun, S.; Cui, X. Study on three-dimensional stress field of gob-side entry retaining by roof cutting without pillar under near-group coal seam mining. *Processes* **2019**, *7*, 552. [CrossRef]
63. Yang, J.; Wang, H.; Gao, Y.; Wang, J.; Liu, H. Fracture characteristics of the roof in gob-side entry retaining with roof cutting and pressure release. *J. Min. Saf. Eng.* **2019**, *36*, 1137–1144.
64. Tian, C.; Wang, A.; Liu, Y.; Jia, T. Study on the migration law of overlying strata of gob-side entry retaining formed by roof cutting and pressure releasing in the shallow seam. *Shock Vib.* **2020**, *2020*, 8821160. [CrossRef]
65. Chen, D.; Guo, F.; Xie, S.; Wang, E.; Wu, Y.; Jiang, Z.; Wang, L.; Cui, J.; Zhang, X.; Liu, R. Mining-induced failure characteristics and surrounding rock control of gob-side entry driving adjacent to filling working face in the deep coal mine. *Energy Sci. Eng.* **2022**, *10*, 2593–2611. [CrossRef]
66. Zhang, B.S.; Wang, P.F.; Cui, S.Q.; Fan, M.Z.; Qiu, Y.M. Mechanism and surrounding rock control of roadway driving along gob in shallow-buried, large mining height and small coal pillars by roof cutting. *J. China Coal Soc.* **2021**, *46*, 2254–2267.
67. Song, X. *Pressure Relief at Cutting Top of Mining Face Study on the Regularity of Mine Pressure in Retaining Small Coal Pillar*; Taiyuan University of Technology: Taiyuan, China, 2021.
68. Wang, F.; Shang, J.; Zhao, B.; Cao, Q. Surrounding rock structural characteristics and anchor-cable strengthened support technology of the gob-side entry retaining with roof cutting and pressure releasing. *Chin. J. Rock Mech. Eng.* **2021**, *40*, 2296–2305.
69. Zhang, Y.; Xu, H.; Song, P.; Sun, X.; He, M.; Guo, Z. Stress evolution law of surrounding rock with gob-side entry retaining by roof cutting and pressure release in composite roof. *Adv. Mate. Sci. Eng.* **2020**, *2020*, 1961680. [CrossRef]
70. Wang, K. *Study on Surrounding Rock Control of Gob-Side Entry Filling in Lingzhida Hard Top Plate Working Face*; China University of Mining and Technology: Xuzhou, China, 2020.
71. Zhen, E.; Gao, Y.; Wang, Y.; Wang, S. Comparative study on two types of non-pillar mining techniques by roof cutting and by filling artificial materials. *Adv. Civil Eng.* **2019**, *2019*, 5267240. [CrossRef]
72. Li, H.; Zu, H.; Zhang, K.; Qian, J. Study on filling support design and ground pressure monitoring scheme for gob-side entry retention by roof cutting and pressure relief in high-gas thin coal seam. *Int. J. Environ. Res. Public Health* **2022**, *19*, 3913. [CrossRef]
73. Xie, S.; Wang, E.; Chen, D.; Li, H.; Jiang, Z.; Yang, H. Stability analysis and control technology of gob-side entry retaining with double roadways by filling with high-water material in gently inclined coal seam. *Int. J. Coal Sci. Technol.* **2022**, *9*, 52. [CrossRef]
74. Xie, S.; Wang, E.; Chen, D.; Sun, Y.; Cheng, Q.; Ji, C.; Yan, Z.; Xiao, H. Failure analysis and control mechanism of gob-side entry retention with a 1.7-m flexible-formwork concrete wall: A case study. *Eng. Fail. Anal.* **2020**, *117*, 104816. [CrossRef]
75. Liu, X.; Hua, X.; Yang, P.; Huang, Z. A study of the mechanical structure of the direct roof during the whole process of non-pillar gob-side entry retaining by roof cutting. *Energ. Explor. Exploit.* **2020**, *38*, 1706–1724. [CrossRef]
76. He, M.; Ma, X.; Yu, B. Analysis of strata behavior process characteristics of gob-side entry retaining with roof cutting and pressure releasing based on composite roof structure. *Shock Vib.* **2019**, *2019*, 2380342. [CrossRef]
77. Gao, Y.; Wang, J.; Gao, H.; Yang, J.; Zhang, Y.; He, M. Mine pressure distribution and surrounding rock control of gob-side entry formed by roof cutting and pressure release under the influence of faults. *Chin. J. Rock Mech. Eng.* **2019**, *38*, 2182–2193.
78. Liu, H.; Zhao, Y.; Ren, T.; Wang, F.; Li, Z. Characteristics of overburden failure and fracture development in gob of mining with gob-side entry formed by cutting roof. *J. China Univ. Min. Technol.* **2022**, *51*, 77–89.
79. Li, L.; Li, G.; Gong, W.; Wang, J.; Deng, H. Energy evolution pattern and roof control strategy in non-pillar mining method of goaf-side entry retaining by roof cutting—A case study. *Sustainability* **2019**, *11*, 7029. [CrossRef]
80. Wang, Q.; Guo, Z.; Ouyang, Z.; Yin, S.; Peng, R. The influence of crushed rock size on bearing-deformation performance of gangue rib of gob-side entry retaining formed by roof fracturing. *J. China Univ. Min. Technol.* **2022**, *51*, 100–106.
81. Xie, S.; Wu, X.; Chen, D.; Sun, Y.; Zeng, J.; Zhang, Q.; Ji, C.; Cheng, Q.; Wang, E.; Ren, Y. Automatic roadway backfilling of caving gangue for cutting roofs by combined support on gob-side entry retaining with no-pillars: A case study. *Adv. Mater. Sci. Eng.* **2019**, *2019*, 8736103. [CrossRef]
82. Chen, D.D.; He, F.L.; Xie, S.; Gao, M.M.; Song, H.Z. First fracture of the thin plate of main roof with three sides elastic foundation boundary and one side coal pillar. *J. China Coal Soc.* **2017**, *42*, 2528–2536.

83. Wang, Y.; He, M.; Wang, Q.; Yang, J.; Wang, J.; Liu, H.; Chen, F. Design of equipment system and surrounding rock control for N00 mining method without coal pillar left and roadway excavation. *J. China Coal Soc.* **2022**. [CrossRef]
84. Wang, J.; Gao, R.; Yu, G.; Gong, W.; Zhu, D. Experimental study on infrared radiation characteristics of overburden movement of self-forming roadway under pressure relief. *J. China Coal Soc.* **2020**, *45*, 119–127.
85. He, H.; Dou, L.; Fan, J.; Du, T.; Sun, X. Deep-hole directional fracturing of thick hard roof for rockburst prevention. *Tunn. Undergr. Space Technol.* **2012**, *32*, 34–43. [CrossRef]
86. Yan, S.; Liu, T.; Bai, J.; Wu, W. Key Parameters of Gob-Side Entry Retaining in A Gassy and Thin Coal Seam with Hard Roof. *Processes* **2018**, *6*, 51.
87. Liu, H.; Dai, J.; Jiang, J.; Wang, P.; Yang, J. Analysis of Overburden Structure and Pressure-Relief Effect of Hard Roof Blasting and Cutting. *Adv. Civ. Eng.* **2019**, *2019*, 1354652.
88. Zhang, L.; Zhao, J.; Zang, C.; Wang, S. An Innovative Approach for Gob-Side Entry Retaining by Roof Cutting in Steeply Pitching Seam Longwall Mining with Hard Roof: A Case Study. *Min. Metall. Explor.* **2020**, *37*, 1079–1091.
89. Yang, H.; Liu, Y.; Cao, S.; Pan, R.; Wang, H.; Li, Y.; Luo, F. A caving self-stabilization bearing structure of advancing cutting roof for gob-side entry retaining with hard roof stratum. *Geomech. Eng.* **2020**, *21*, 23–33.
90. Gao, K.; Huang, P.; Liu, Z.G.; Liu, J.; Wang, F.; Shu, C.M. Pressure Relief by Blasting Roof Cutting in Close Seam Group Mining under Thick Sandstone to Enhance Gas Extraction for Mining Safety. *Processes* **2021**, *9*, 693.
91. Hu, J.; He, M.; Wang, J.; Ma, Z.; Wang, Y.; Zhang, X. Key Parameters of Roof Cutting of Gob-Side Entry Retaining in a Deep Inclined Thick Coal Seam with Hard Roof. *Energies* **2019**, *12*, 934.
92. Ma, X.; He, M.; Wang, J.; Gao, Y.; Zhu, D.; Liu, Y. Mine Strata Pressure Characteristics and Mechanisms in Gob-Side Entry Retention by Roof Cutting under Medium-Thick Coal Seam and Compound Roof Conditions. *Energies* **2018**, *11*, 2539.
93. Chen, D.D.; Guo, F.F.; Wu, Y.Y. Study on the first fracture of the main roof plate structure with the long side goaf and elastic-plastic softening foundation boundary. *J. China Coal Soc.* **2022**, *47*, 1473–1489.
94. Chen, D.D.; Xie, S.R.; He, F.L.; Zeng, J.C.; Xie, F.X.; Cheng, Q. First fracturing of thin plate of main roof with elastic foundation boundary on both sides of the long side of goaf (coal pillars). *J. China Coal Soc.* **2018**, *43*, 3273–3285.
95. Guo, Z.; Wang, H.; Ma, Z.; Wang, P.; Kuai, X.; Zhang, X. Research on the transmission of stresses by roof cutting near gob rocks. *Energies* **2021**, *14*, 1237.
96. Li, P.; Lai, X.; Gong, P.; Su, C.; Suo, Y. Mechanisms and applications of pressure relief by roof cutting of a deep-buried roadway near Goafs. *Energies* **2020**, *13*, 5732. [CrossRef]
97. Yang, X.; Liu, C.; Sun, H.; Yue, S.; Ji, Y.; Zhang, X.; Hou, L. Research on the deformation mechanism and directional blasting roof cutting control measures of a deep buried high-stress roadway. *Shock Vib.* **2020**, *2020*, 6742504. [CrossRef]
98. Wang, Y.; Wang, H.; He, M.; Wang, Q.; Qiao, Y.; Yang, J. Mine pressure behavior characteristics and control methods of a reused entry that was formed by roof cutting: A case study. *Shock Vib.* **2020**, *2020*, 4276730. [CrossRef]
99. Wang, Q.; Jiang, B.; Wang, L.; Liu, B.; Li, S.; Gao, H.; Wang, Y. Control mechanism of roof fracture in no-pillar roadways automatically formed by roof cutting and pressure releasing. *Arab. J. Geosci.* **2020**, *13*, 274. [CrossRef]
100. Wang, Y.; Gao, Y.; Wang, E.; He, M.; Yang, J. Roof deformation characteristics and preventive techniques using a novel non-pillar mining method of gob-side entry retaining by roof cutting. *Energies* **2018**, *11*, 627. [CrossRef]
101. Ma, Q.; Tan, Y.; Zhao, Z.; Xu, Q.; Wang, J.; Ding, K. Roadside support schemes numerical simulation and field monitoring of gob-side entry retaining in soft floor and hard roof. *Arab. J. Geosci.* **2018**, *11*, 563. [CrossRef]
102. Sun, B.J.; Hua, X.Z.; Zhang, Y.; Yin, J.; He, K.; Zhao, C.; Li, Y. Analysis of roof deformation mechanism and control measures with roof cutting and pressure releasing in gob-side entry retaining. *Shock Vib.* **2021**, *2021*, 6677407. [CrossRef]
103. Guo, J.G.; Li, Y.H.; Shi, S.H.; Jiang, Z.S.; Chen, D.D.; He, F.L.; Xie, S.R. Self-forming roadway of roof cutting and surrounding rock control technology under thick and hard basic roof. *J. China Coal Soc.* **2021**, *46*, 2853–2864.
104. Zhao, S.K.; Deng, Z.G.; Qi, Q.X.; Li, H.Y. Theory and application of deep hole floor-break blasting in floor rock burst coal mine. *J. Min. Saf. Eng.* **2016**, *33*, 636–642.
105. Liu, S.; Feng, M.; Pan, J.; Wang, S. Experimental analysis on vibration of roadway with strong probability of rock burst subjected to pressure relief blasting. *Chin. J. Rock Mech. Eng.* **2016**, *35*, 906–918.
106. Xie, S.R.; Wang, E.; Chen, D.D.; Jiang, Z.S.; Li, H.; Liu, R.P. Collaborative control technology of external anchor-internal unloading of surrounding rock in deep large-section coal roadway under strong mining influence. *J. China Coal Soc.* **2022**, *47*, 1946–1957.

energies

MDPI

Article

Experimental Study on Deformation Behavior and Permeability Evolution of Sandstone Responding to Mining Stress

Yang Liu [1,2,3], Tong Zhang [2,3,*], Jun Wu [1,2], Zhengyang Song [1,4] and Fei Wang [1,2,*]

1 State Key Laboratory of Mining Response and Disaster Prevention and Control in Deep Coal Mines, Anhui University of Science & Technology, Huainan 232001, China
2 Institute of Energy, Hefei Comprehensive National Science Center, Hefei 230031, China
3 School of Mining, Anhui University of Science & Technology, Huainan 232001, China
4 School of Civil and Resource Engineering, University of Science and Technology Beijing, Beijing 100083, China
* Correspondence: zhangt_1990@sina.com (T.Z.); kingfly@mail.ustc.edu.cn (F.W.)

Abstract: Mining-induced hydromechanical behavior of sandstone is critical to mining safety and disaster prevention. To investigate the evolution behavior of the mechanical and permeability properties of sandstone, mining-induced stress was imitated by increasing axial stress and decreasing confining stress, and a set of hydromechanical experiments were further performed, incorporating the effect of in situ stress, pore pressure, and mining stress. The results show the similar variation tendencies of the deformation and permeability of sandstone under different loading paths of in situ stress and pore pressure. Most sandstone samples maintain a compression state for the peak stress condition. The failure mode evolved from shear failure to shear–tension failure with the increase in in situ stress. The stress-relief effect significantly effects the permeability, since the permeability of sandstone increases exponentially with decreasing effective confining stress. The growth rate of permeability in Stage II is significantly greater than that in Stage I. One order of magnitude of permeability was presented at the peak stress situation. A fitting exponential model based on the alteration of effective confining stress was proposed to describe the permeability evolution dominated by the stress-relief effect, and the discovered permeability model can accurately describe the experimental results. The research results provide significant guidance for understanding the hydromechanical behavior and water hazard prevention for underground coal mines.

Keywords: sandstone; deformation behavior; permeability evolution; mining stress path

Citation: Liu, Y.; Zhang, T.; Wu, J.; Song, Z.; Wang, F. Experimental Study on Deformation Behavior and Permeability Evolution of Sandstone Responding to Mining Stress. *Energies* **2022**, *15*, 7030. https://doi.org/10.3390/en15197030

Academic Editor: Manoj Khandelwal

Received: 5 August 2022
Accepted: 13 September 2022
Published: 25 September 2022

1. Introduction

The mechanical properties and seepage characteristics of sandstone, a typical sedimentary rock, have been widely investigated in underground rock mass engineering, such as carbon geological storage, coal mining, water hazard prevention of mines, and so on. As underground activities move towards the deeper crust, the stress–fracture–seepage field behavior in the rock becomes more complex [1–3]. Therefore, the research on rock deformation behavior and permeability evolution under hydromechanical coupling is an urgent basic topic in the field of rock mechanics, which has extreme scientific significance and application prospects for deep underground activities.

Existing related studies mainly concentrated on the mechanical property of the triaxial compression test, and the seepage law in this loading process has been widely and deeply discussed. At present, many researchers have carried out triaxial total stress–strain and cyclic loading–unloading seepage tests on different types of rocks, such as sandstone, limestone, and granite, including intact and fractured rocks [4–9]; analyzed the mechanical and seepage characteristics with rock brittleness–ductility under different confining pressures and osmotic pressures [10–15]; and obtained the variation relationship between permeability and axial stress before and after peak strength [16]. Liu et al. [17] studied the hydromechanical behavior of sandstone with different physical properties, and modeled the damage variables and permeability evolution of the Weibull random distribution. Xiao et al. [18] investigated the hydromechanical properties of red sandstone, collected in Ganzhou, with triaxial compression seepage tests, and established a piecewise model for stress-dependent permeability to discuss the impact of osmotic pressure on the strength and stiffness of red sandstone before the peak strength. Li et al. [19] carried out triaxial compression tests of sandstone to research the effect of different confining pressures and seepage pressures on the failure mechanism and permeability evolution of reservoir sandstone, and established COMSOL numerical model to accurately inverse the laboratory experimental results. Moreover, CT scanning technology, as a hot visualization technology, is widely used in this research field [20,21].

The current rock mechanics theory is mostly based on the total stress–strain test obtained from the triaxial rock mechanics standardization test [22–24]. However, when analyzing engineering disturbance and disaster, it can be found that this standardized test is separated from engineering activities and does not reflect the influence of in situ stress state and mining disturbance path. Therefore, Xie et al. [25,26] took the lead in proposing the mining stress path of coal under different mining layouts, i.e., protective coal-seam mining, top-coal caving mining, and non-pillar mining, which established a bridge between laboratory experiment and field practice. In addition, corresponding mining-induced mechanical behavior, fracture distribution, and coal permeability evolution were analyzed. Zuo et al. [27] investigated the mechanical behavior of limestone under different stress paths, including the complete stress–strain process and mining stress path. Xia et al. [28] conducted loading and unloading triaxial mechanical tests with real-time ultrasonic monitoring for single fissure sandstone and studied the deformation characteristics, the failure form, and the crack evolution of single fissure sandstone with specific angles and lengths under the fully mechanized top-coal caving mining stress path.

As stated above, most of the existing work focused on the mechanical behaviors and permeability evolution for coal in the mining process. However, there is still a lack of quantitative analysis on the mining-induced mechanical behavior and seepage property of sandstone. Moreover, it would be of great significance to investigate the differences in deformation and permeability responding to the change of the mining-induced underground environment. Therefore, in this study, the influence of in situ stress, pore pressure, and mining-induced stress on the deformation and transport behavior of sandstone was investigated. The damage deformation at different mining-induced stress states was dynamically monitored, and the failure mode under the peak stress condition was analyzed. Besides, the effective stress-based permeability evolution was discussed.

2. Experimental Section

2.1. Sample Preparation

A gray sandstone block taken from a coal mine in the Ordos Basin was used in this study. Some 50 mm diameter cylindrical samples were drilled from the sandstone block in the same direction by a core-drilling machine. Then, the cylindrical samples were cut to about 50 mm in diameter and 100 mm in length by a grinding machine with the high-precision standard proposed by the ISRM. The X-ray diffraction (XRD) results show that the sandstone is mainly composed of quartz (59.2%), clinochlore (12.3%), albite (10.1%), and other minerals (18.4%) (Figure 1a). The pore spaces of sandstone are completely dominated by intergranular pore, intercrystalline pore, and microfracture based on the scanning electron microscope (SEM) results (Figure 1b). Moreover, the average density and porosity of the sandstone are 2.44 g/cm^3 and 11.2%, respectively.

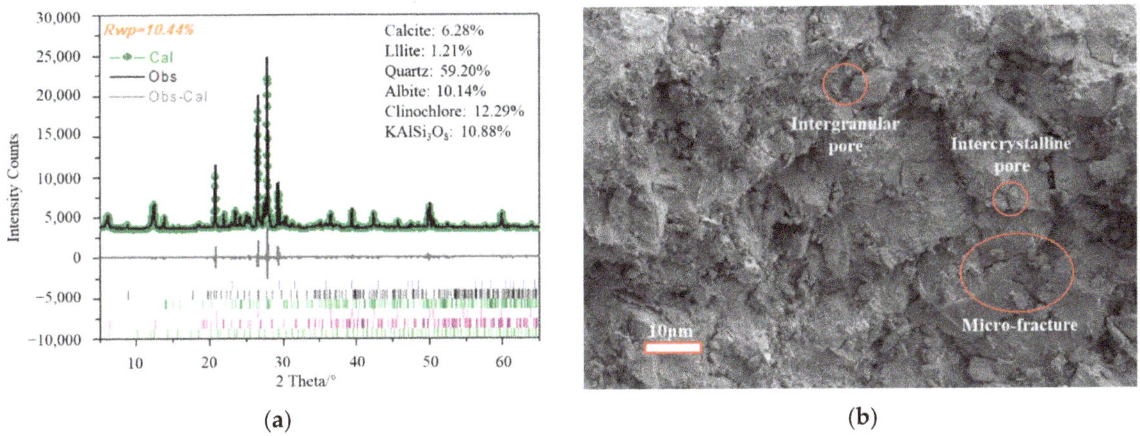

Figure 1. The (**a**) XRD and (**b**) SEM test results.

2.2. Experimental Apparatus

The hydromechanical tests were carried out using a triaxial servo-controlled permeation experimental apparatus. The apparatus, comprising six parts, a servo loading system, a triaxial cell, a pressure control system, a liquid/gas testing system, a data collection system, and an auxiliary system, can supply an axial and confining stress applied to the samples with a stable temperature during the measurement of permeability in this work. The experimental apparatus is shown in Figure 2, and the equipment parameters were described in detail by a previous study [29].

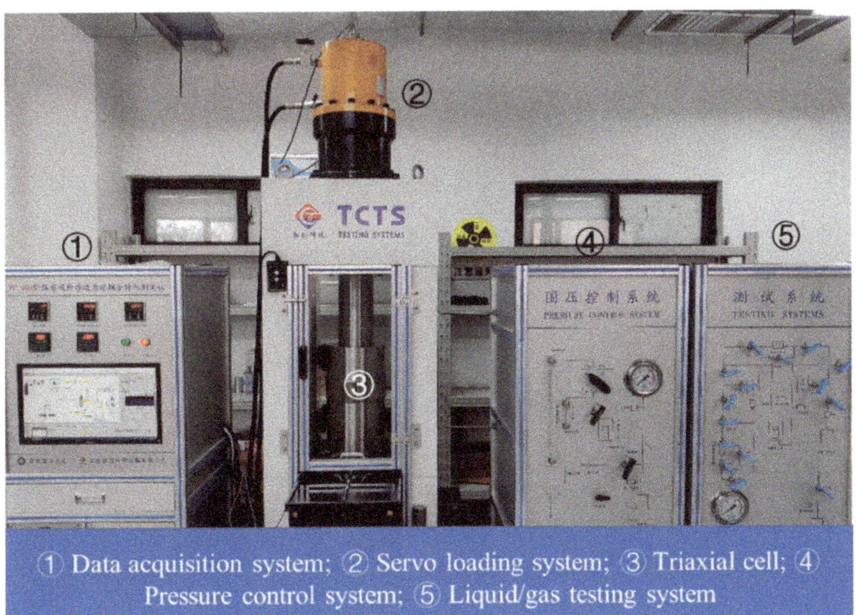

Figure 2. Triaxial servo-controlled mechanical–seepage coupling experimental apparatus [29].

2.3. Experimental Procedure

Nitrogen with a purity of 99.999% was used as the testing gas, and different in situ stresses (15 MPa, 20 MPa, and 25 MPa) and gas pressures (1 MPa, 1.5 MPa, and 2 MPa) were considered in the experiments. Nine sandstone samples were prepared, and each test corresponded to one sample under one stress state. All the experiments were carried out at 20 °C. Moreover, the information on the sandstone samples with designated experimental conditions was summarized in Table 1. The specific mining stress path is described below (Figure 3):

(1) Hydrostatic pressure stage: The sandstone specimens were firstly dried at 100 °C for 48 h. Then, the dried specimens were sealed with a thermal-shrinkable sleeve and placed into the triaxial cell. The specimens were hydrostatically loaded to the target confining pressure (15 MPa, 20 MPa, and 25 MPa) at the rate of 3 MPa/min. Nitrogen with a specified pressure (1 MPa, 1.5 MPa, and 2 MPa) was then injected into the upstream of the specimens.

(2) Stage I: The axial stress was loaded to 1.5 times the initial confining pressure by the stress control mode with a loading rate of 0.625 MPa/min, while the confining stress was unloaded to 0.6 times the initial confining pressure with a loading rate of −0.5 MPa/min.

(3) Stage II: In this study, the concentration coefficient of front abutment pressure was determined as 2.5, and the horizontal stress unloaded to 0.2 times the initial state. So, the axial stress was further loaded to 2.5 times the initial confining pressure by the stress control mode with a loading rate of 1.25 MPa/min, while the confining stress was unloaded to 0.2 times the initial confining pressure with a loading rate of −0.5 MPa/min.

(4) Stage III: The confining pressure remained constant, which ensured that the pore pressure is not larger than the confining pressure and the permeability can continue to be measured while preventing equipment damage. In addition, the load mode of displacement control with a rate of 0.1 mm/min was selected to control the axial stress loading until the specimens were destroyed. Throughout the experiments, the gas flows were continuously recorded at the downstream cylinders for calculating the permeability.

Table 1. Information on sandstone samples.

Sample No.	Diameter (mm)	Length (mm)	Density (g/cm^3)	Initial Confining Pressure (MPa)	Seepage Pressure (MPa)
B1	49.59	99.71	2.45	15	1
B2	49.41	99.72	2.47	15	1.5
B3	49.55	99.87	2.45	15	2
B4	49.99	100.07	2.44	20	1
B5	50.37	99.94	2.40	20	1.5
B6	49.70	99.82	2.44	20	2
B7	49.83	100.01	2.46	25	1
B8	50.01	100.04	2.44	25	1.5
B9	49.88	99.77	2.44	25	2

In this study, the permeability of sandstone was obtained by using Darcy's law:

$$k = \frac{2qP_0\mu L}{\pi\left(\frac{\phi}{2}\right)^2\left(P_1^2 - P_2^2\right)} \quad (1)$$

where k is the permeability of the sample, m^2; q is the gas flow rate, m^3/s; P_0 is the atmospheric pressure, Pa; μ is the gas viscosity, Pa·s; L is the length of the cylindrical sample, m; ϕ is the diameter of the cylindrical sample, m; P_1 is the inlet gas pressure Pa; and P_2 is the outlet gas pressure, Pa.

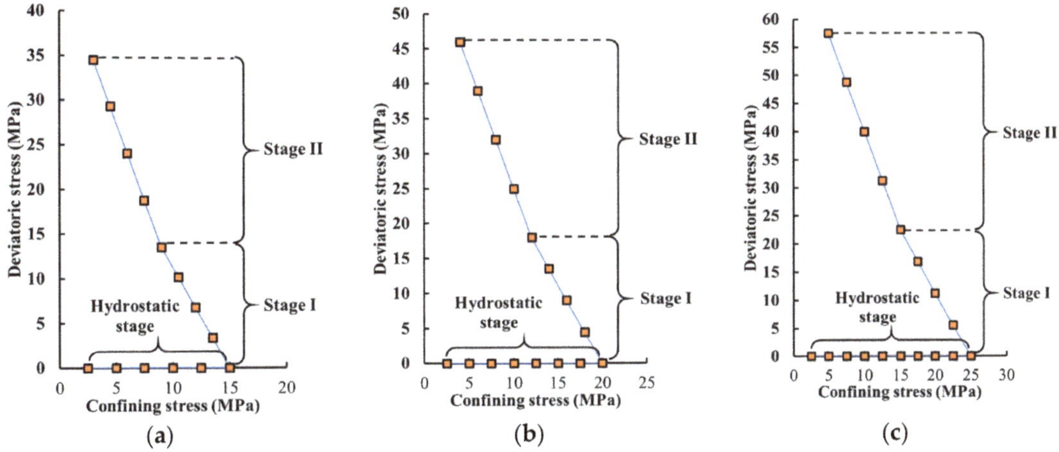

Figure 3. Loading schemes of different sandstone samples ((**a**–**c**) are the initial confining pressure of 15 MPa, 20 MPa, and 25 MPa, respectively).

3. Results and Discussion

3.1. Deformation Behavior of Sandstone under Mining Stress

Taking the Biot coefficient remain constant at 1, the effective stress can be expressed as [30]:

$$\sigma_e = \frac{1}{3}(\sigma_1 + 2\sigma_3) - P \quad (2)$$

where σ_e is the effective stress, MPa; σ_1 is the axial stress, MPa; σ_3 is the confining stress, MPa; and P is the pore pressure, MPa. Besides, ignoring high-order small quantities, the volume strain can be approximately expressed as:

$$\varepsilon_v = \varepsilon_1 + 2\varepsilon_3 \quad (3)$$

where ε_v is the volumetric strain, ε_1 is the axial strain, and ε_3 is the radial strain. The mechanical deformation characteristics of the sandstone samples under the mining stress path at Stages I and II are shown in Figures 4–6. As shown in Figures 4–6, the variation tendency of the axial strain, the radial strain, and the volumetric strain of sandstone under different load conditions of the axial stress, confining stress, and effective stress are approximately similar, respectively. The axial strain of the sandstone shows a linear growth trend with the increase in the axial stress, and the growth rates of Stage I and Stage II remain unchanged (Figure 4). For the radial strain, it presents an exponential function growth trend with the unloading of the confining stress, and the growth rate of Stage II is significantly higher than that of Stage I (Figure 5). As for the volumetric strain of sandstone, its trend with the variation of effective stress roughly depicts a transverse "V" symmetrical distribution, which corresponds to the reduction in the effective stress at Stage I and the increase in the effective stress at Stage II (Figure 6: the arrow in the figure indicates the stress path of the effective stress). In detail, at the end of the loading process in Stage II, the volumetric deformation of sandstone changes from compression to expansion with the increase in initial confining pressure. Under the same initial confining pressure, the influence of mechanical deformation in the sandstone is mainly affected by mining stress compared with pore pressure.

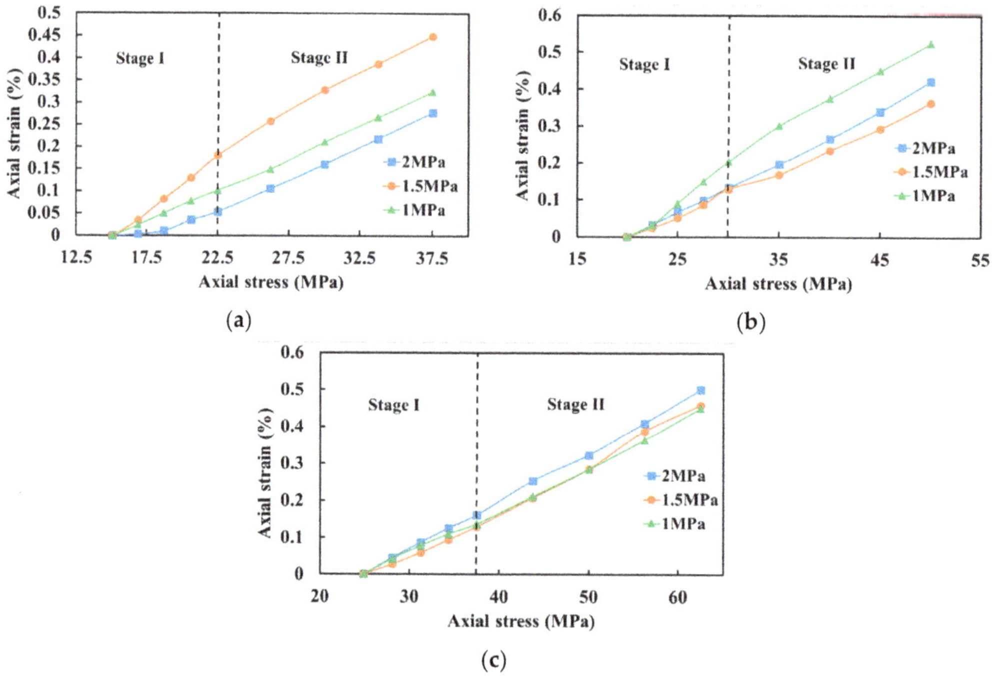

Figure 4. Axial strain variation of sandstone under different initial confining pressures ((**a**–**c**) are the initial confining pressure of 15 MPa, 20 MPa, and 25 MPa, respectively).

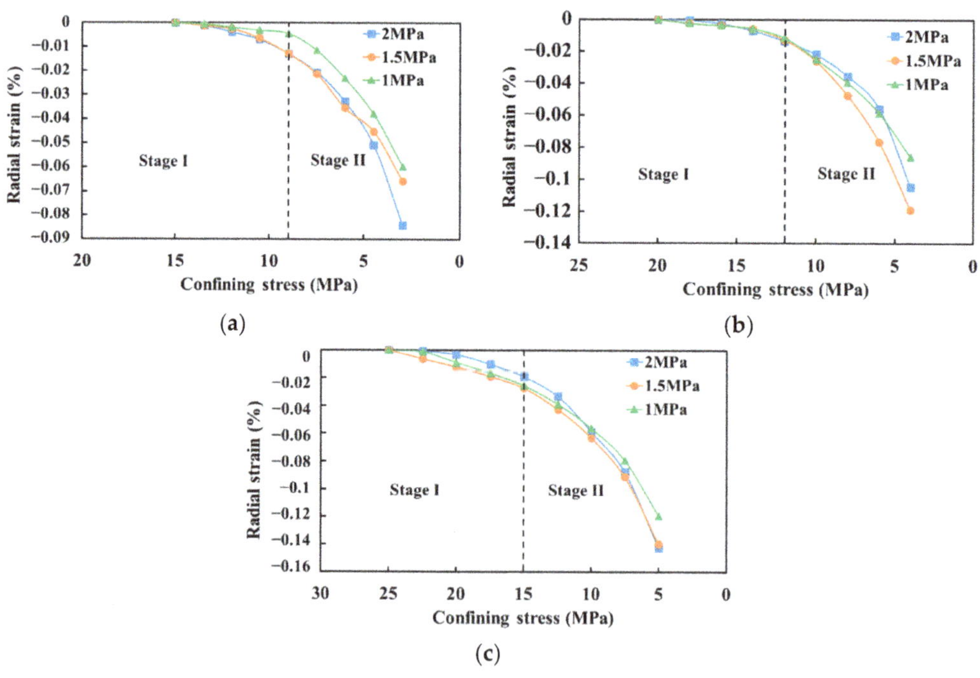

Figure 5. Radial strain variation of sandstone under different initial confining pressures ((**a**–**c**) are the initial confining pressure of 15 MPa, 20 MPa, and 25 MPa, respectively).

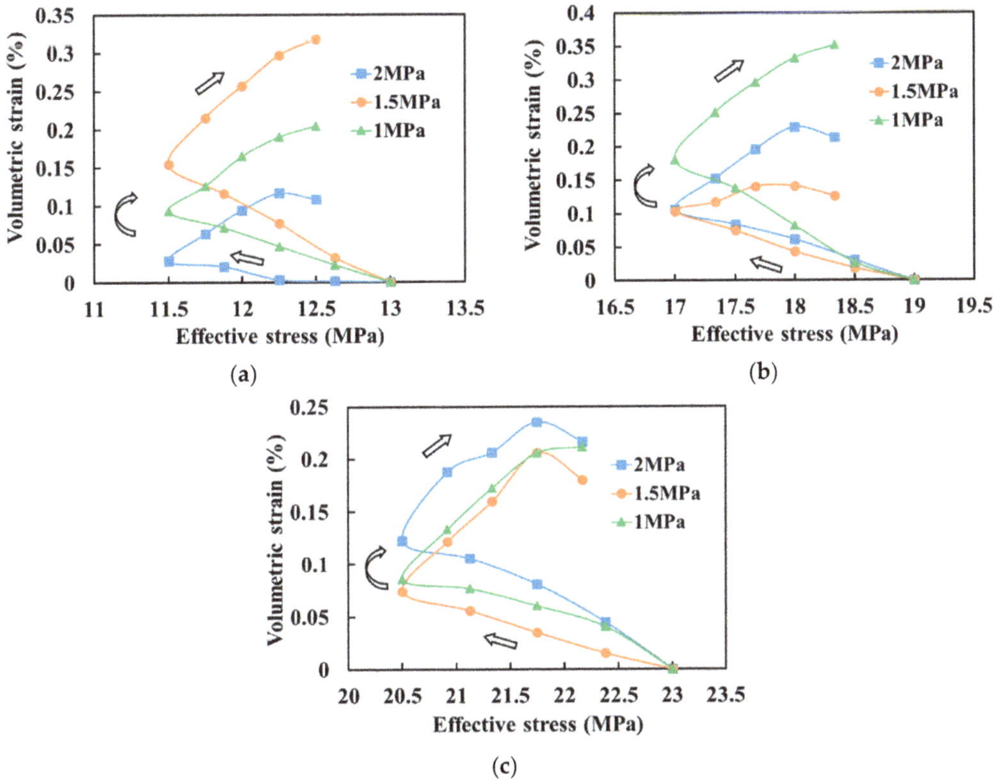

Figure 6. Volumetric strain variation of sandstone under different initial confining pressures ((**a**–**c**) are the initial confining pressure of 15 MPa, 20 MPa, and 25 MPa, respectively).

The peak parameters of sandstone, including deformation and permeability parameters, at the peak stress state in Stage III, are listed in Table 2. To intuitively summarize the corresponding effects of initial confining pressure and pore pressure, Figure 7 is drawn by using the data in Table 2. Under the same initial confining pressure, the peak stress of sandstone reduces with the increase in pore pressure, while the peak stress of sandstone increases with the incremental confining pressure under the same pore pressure. The absolute value of both the axial strain and the radial strain increases with the incremental confining pressure. Under the same confining pressure conditions, the volumetric strain at the peak stress state gradually decreases due to the increase in pore pressure. Most sandstone samples still maintain compression at the peak stress state, except for samples B5 and B8, which was significantly different from the coal samples [26]. The mineral composition and pore–fracture structure are the main reasons for this distinction. The sandstone is mainly composed of brittle minerals and is dense, with fewer pores and microcracks resulting in it being hard to deform. Conversely, the coal is composed of ductile minerals and has more microcracks and cleats, so it becomes more easily deformed at low confining stress. Note that the individual difference in samples leads to a slight deviation between the test results and the overall test law.

Table 2. Information of peak parameters of sandy mudstone sample.

Sample No.	Peak Stress (MPa)	Corresponding Test Value of Peak Stress				Peak Permeability ($\times 10^{-17}$ m^2)
		Axial Strain (%)	Radial Strain (%)	Volumetric Strain (%)	Permeability ($\times 10^{-17}$ m^2)	
B1	89.68	1.251	−0.51	0.23	12.73	52.14
B2	98.71	1.558	−0.698	0.161	13.60	90.77
B3	71.50	0.831	−0.269	0.292	28.76	96.44
B4	113.80	1.695	−0.678	0.339	5.58	61.65
B5	107.30	1.325	−0.746	−0.166	13.89	72.85
B6	87.43	1.065	−0.455	0.154	17.44	88.23
B7	133.11	1.550	−0.672	0.206	7.29	68.21
B8	113.80	1.287	−0.678	−0.069	8.64	89.22
B9	96.93	1.091	−0.506	0.079	12.92	53.98

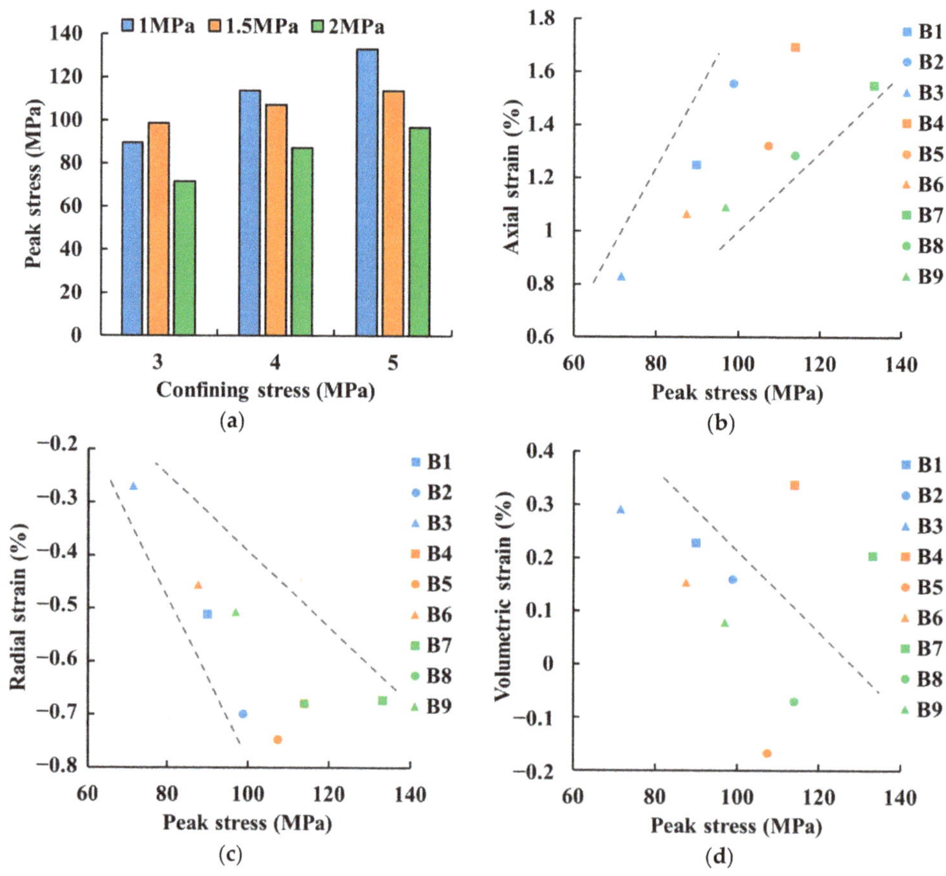

Figure 7. Peak stress and corresponding deformation of sandstone ((**a**–**d**) are the peak strength, corresponding axial, radial, and volumetric strain, respectively).

3.2. Failure Mode of Sandstone under Mining Stress Conditions

The failure characteristics of the sandstone at Stage III are shown in Figure 8. Overall, after the sandstone undergoes a simulated mining stress loading path, its failure mode at Stage III is mainly a shear failure, because the low confining stress makes the sandstone more brittle, which is manifested as follows: the macroscopic shear fracture pierces through the top and bottom end surfaces of the entire sample, so a part of the samples is in shear–tension failure mode, accompanied by some secondary tensile cracks and shear cracks. Under the same pore pressure, the final fracture morphology of the sandstone tended to be complicated with the increase in initial confining pressure. The overall performance is the transition from shear failure to shear–tension failure, which corresponded to the transformation from a single shear fracture to multiple positive and negative shear fractures, which tend to form an approximate "X"-shaped conjugate shear fracture (samples B5 and B8) or to the main shear fracture followed by a secondary shear–tension fracture (samples B7 and B9). At the same time, the number of fractures developed by the final failure form of the sandstone has a good correspondence with the peak permeability under the same initial confining stress, which represents the greater number of fractures, the better connectivity of the pore–fracture system of the sandstone, the increase in the seepage channel of the fluid, and the improved permeation property, which caused the peak permeability to be larger (samples B3, B5, and B8), indicating that the number of fractures has a significant impact on the permeability of the sandstone.

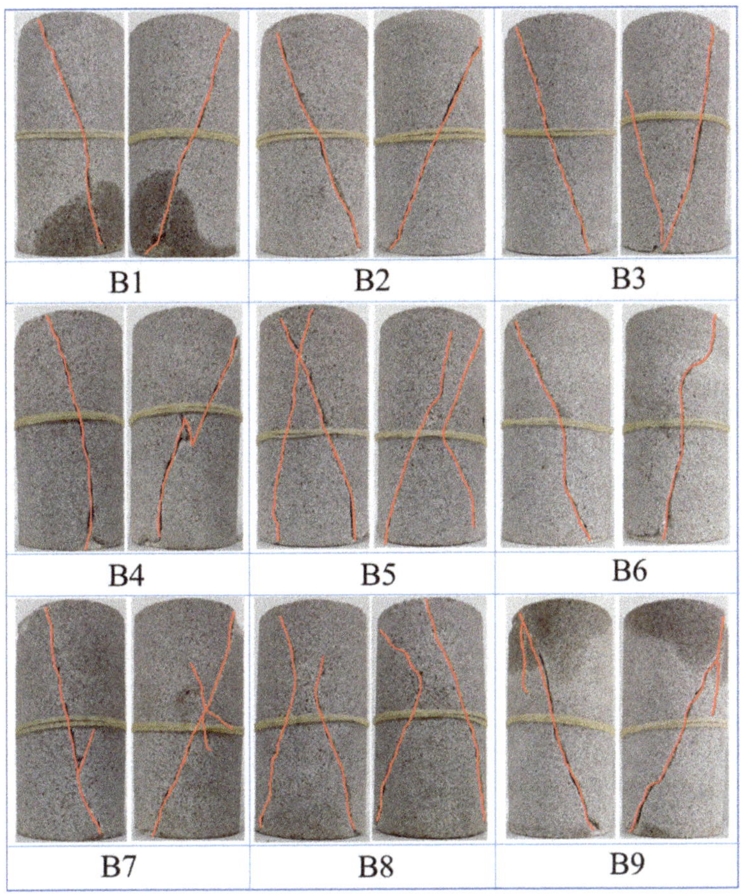

Figure 8. Mining-induced failure characteristics of sandstone.

3.3. *Permeability Evolution of Sandstone under Mining Stress*

Under the different initial confining stress and pore pressure, the variation characteristics of the permeability of sandstone samples in the mining stress path are shown in Figure 9. It can be seen from Figure 9 that the permeability of sandstone samples, under the same initial confining pressure and pore pressure, presents a gradual increasing trend in the form of exponential function with the axial loading and radial unloading, which are mainly induced by the confining stress relief, also called the stress-relief effect. In addition, the growth rate in Stage II is significantly greater than that in Stage I. This phenomenon suggests that under the mining activities, the impact of radial unloading on the permeability of the sandstone (the seepage direction is axial) is dramatically higher than that of axial loading; the main reason for this is that under the action of simulated mining stress, the unloading of the confining stress leads to the aperture broaden of the seepage channels, which are mainly composed of pores and throats, in the sandstone. At the same time, the loading of the axial stress also contributes to the unloading of the confining stress, which can be also proven by the obvious increase in the permeability of the sandstone in Stage II, further enlarging the radii of the pores and the throats and increasing the flow capacity of the seepage path in the sandstone, which results in a gradual increase in the permeability of the sandstone.

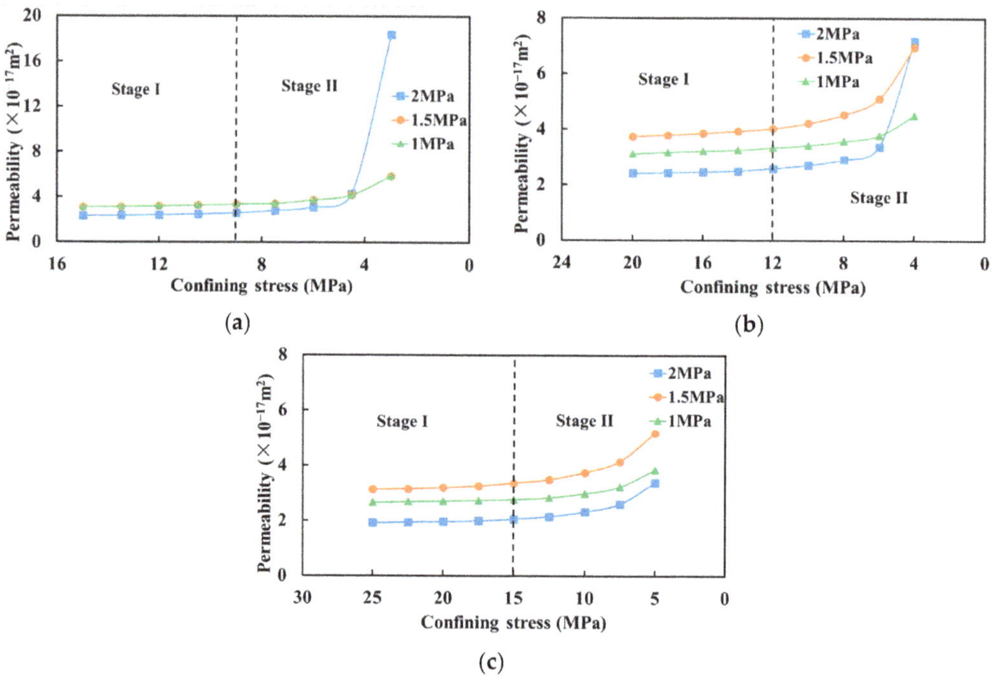

Figure 9. Permeability variation of sandstone under different initial confining pressures ((**a**–**c**) are the initial confining pressure of 15 MPa, 20 MPa, and 25 MPa, respectively).

To represent the variation of the permeability of sandstone samples at different loading stages intuitively, the permeability growth rate η is introduced as follows:

$$\eta = \frac{k_i - k_{ini}}{k_{ini}} \times 100\% = \left(\frac{k_i}{k_{ini}} - 1\right) \times 100\% \quad (4)$$

where η is the permeability growth rate; k_i is the permeability of sandstone, m^2; and k_{ini} is the initial permeability of sandstone, m^2. Under different initial confining pressures and pore pressures, the η of the sandstone samples changed with the mining stress, which is presented in Figure 10. The η of the sandstone samples in Stage I range from 0% to 20%, and the maximum value is 19.92%, indicating that the mining stress in Stage I has a slight impact on the permeability of the sandstone. Once reaching Stage II, the η increases markedly, ranging from 20% to 680%, and the maximum value is 678.81%. With the increasing of the in situ stress, the η of the sandstone samples changes more obviously. When the axial stress was loaded to 2.5 times of the in situ stress, and the confining stress was unloaded to 0.2 times of the in situ stress, the η of the sandstone samples is the largest; this indicates that at the end of Stage II, the sandstone enters the deformation stage of the stable expansion of the cracks, and a large number of new microfractures in the sample germinate and nucleate, connecting and converging with the original seepage channels, so the enlargement of the radii of the original pores and throats finally leads to a sharp increase in the η of the sandstone under this stress state.

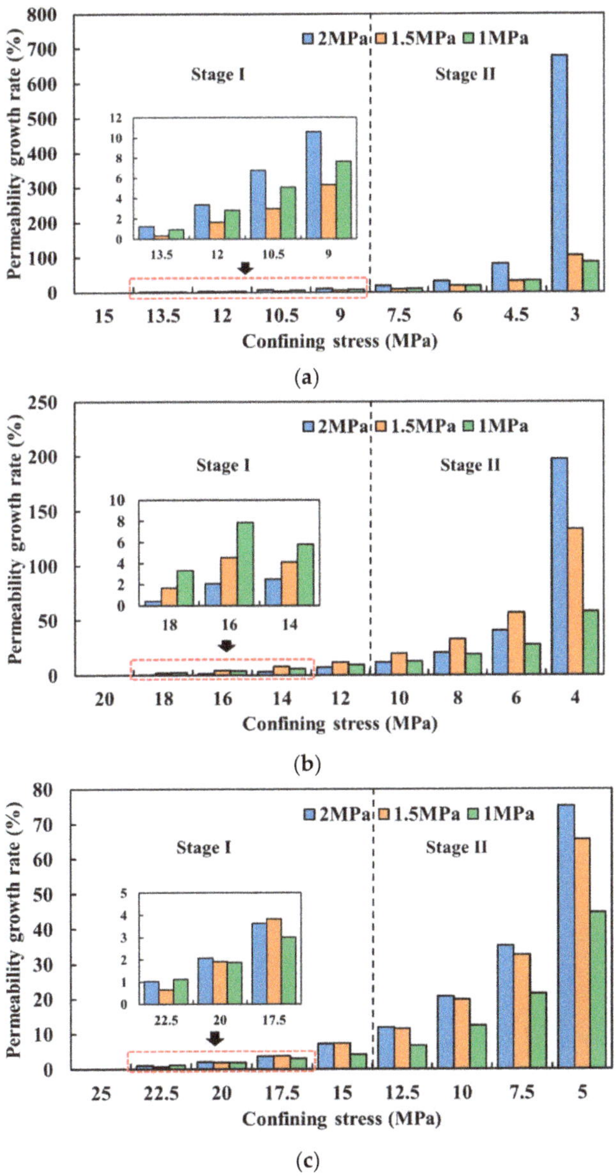

Figure 10. Permeability change ratio variation of sandstone under different initial confining pressures ((**a**–**c**) are the initial confining pressure of 15 MPa, 20 MPa, and 25 MPa, respectively).

The relationships between the permeability and the effective stress under the different initial confining pressure and pore pressure conditions are shown in Figure 11. Under the action of mining stress, the permeability of the sandstone samples increases linearly with the decrease in the effective stress in Stage I and increases exponentially with the increase in the effective stress in Stage II. It can be concluded that selecting the confining stress can better reflect its relationship with the permeability, compared with the axial stress and the effective stress.

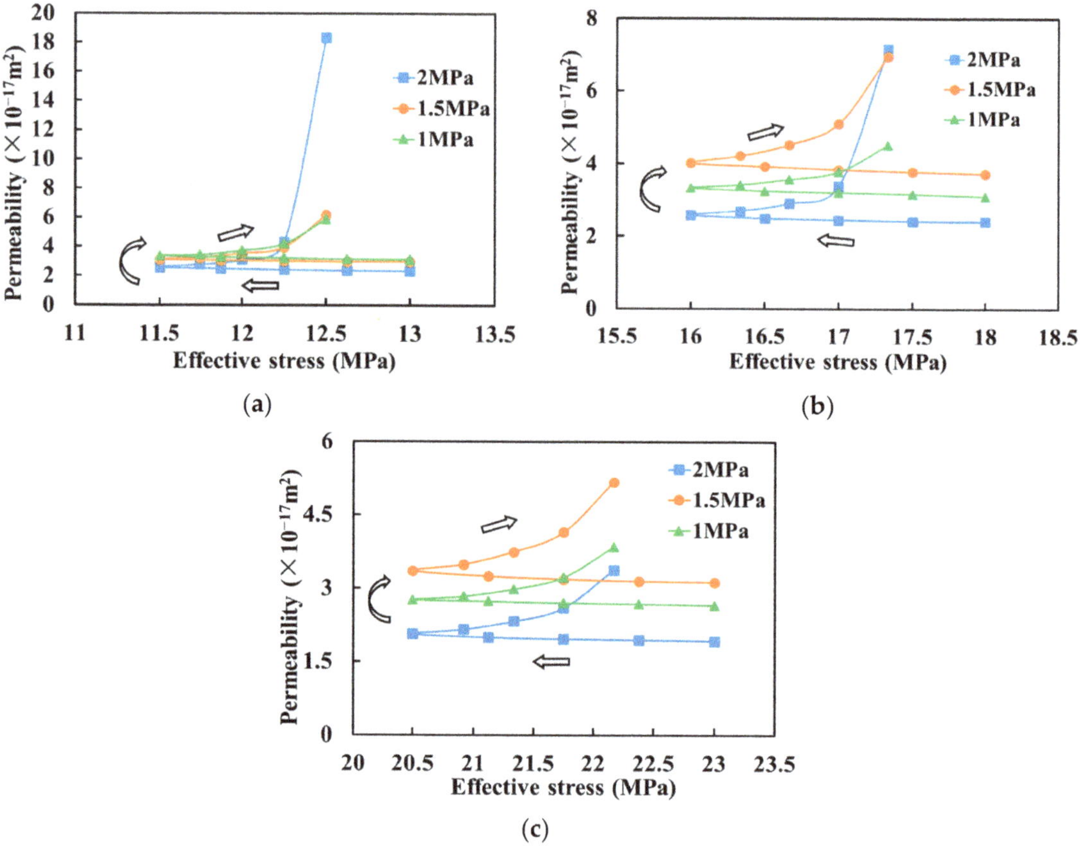

Figure 11. Permeability variation of sandstone under different effective stresses ((**a–c**) are the initial confining pressure of 15 MPa, 20 MPa, and 25 MPa, respectively).

The peak permeability of the sandstone samples in the peak stress state at Stage III is shown in Figure 12. As seen in Figure 12, on the whole, under the same initial confining pressure, the permeability of the sandstone samples at peak stress gradually increased with the increasing pore pressure, while, under the same pore pressure, the permeability of the sandstone samples at peak stress decreased with the increasing initial confining pressure. The permeability of all sandstone samples at peak stress increased by up to one order of magnitude. The influence of the initial confining pressure on the mining-induced seepage behavior of sandstone is stronger than that of the pore pressure.

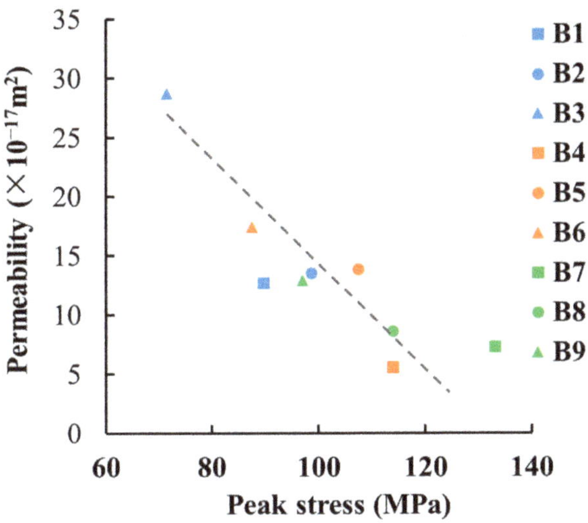

Figure 12. Peak stress and corresponding permeability of sandstone.

3.4. Permeability Fitting Model of Sandstone under Mining Stress

The permeability of the rock is a function of the characteristics of rock media, which describes an average property of the ability to transport fluid. As the unloading confining stress is the dominant factor on the permeability change of sandstone in the mining stress path, the permeability is related to the effective confining stress. In Stages I and II of the mining stress path, the sandstone permeability is mainly contributed to by the variation in pore radius because of the change of the effective confining stress. The exponential relationship between the permeability and the effective confining stress is discovered through regression and analysis using the experimental data, which is expressed as Equation (5).

$$k_i = k_0 + Ae^{-B(\sigma_c - \sigma_{c0})} \tag{5}$$

where k_i is the sandstone permeability, m^2; k_0 is the initial sandstone permeability, m^2; A and B are fitting parameters; σ_c is the effective confining stress, MPa; and σ_{c0} is the initial effective confining stress, MPa.

In this study, furthermore, the effective confining stress σ_c can be calculated by the following equation:

$$\sigma_c = \sigma_3 - \nu(\sigma_1 + \sigma_3) + (2\nu - 1)P \tag{6}$$

where ν is the Poisson's ratio of sandstone, with a value of 0.02 obtained by the mechanical tests.

To verify the mentioned permeability fitting model, the theoretical permeabilities acquired using Equation (5) are compared with the testing permeability data for the sandstone in the simulated mining stress path (see Figure 13). In addition, the corresponding fitting parameters used in the model are listed in Table 3. Figure 13 shows that the permeability of sandstone increases slightly with the decrease in effective confining stress in Stage I of mining stress. In Stage II of mining stress, the permeability grows rapidly with the decrease in effective confining stress. It can be clearly obtained from Figure 13 that the analytical permeability results attained from the exponential function fit well with experimental results based on the original experimental permeability data of sandstone, with the R-square value exceeding 0.98, which indicates that the fitting model can fully capture the permeability evolution for sandstone under the mining stress state. Table 3 illustrates that the permeability model coefficient A has a large variation span, varying from 1.8×10^{-6} to 0.21, while that of B ranges from 0.23 to 1.32. The fitting accuracy increases with increasing in situ stress and pore pressure. Notably, the discovered fitting permeability model is validated with data from sandstone samples under the mining stress state only. Whether this model can be employed on other rocks requires further investigation. Moreover, its physical significance also needs further derivation.

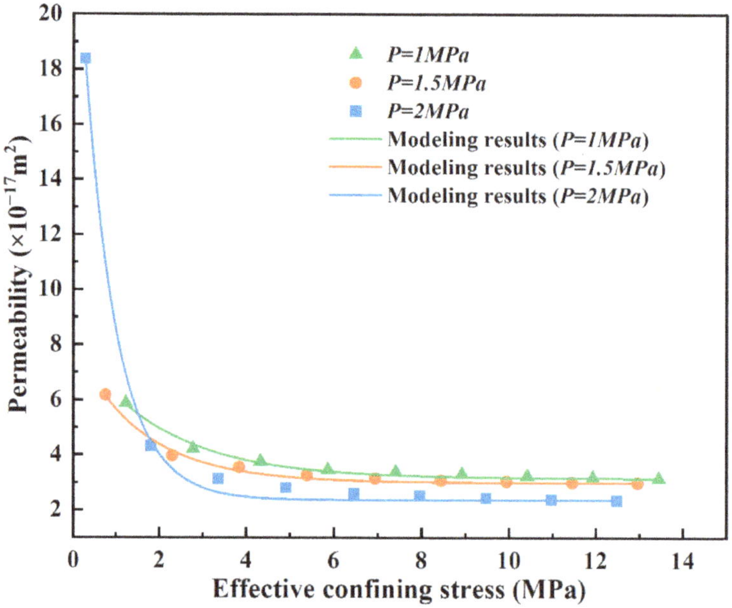

(**a**)

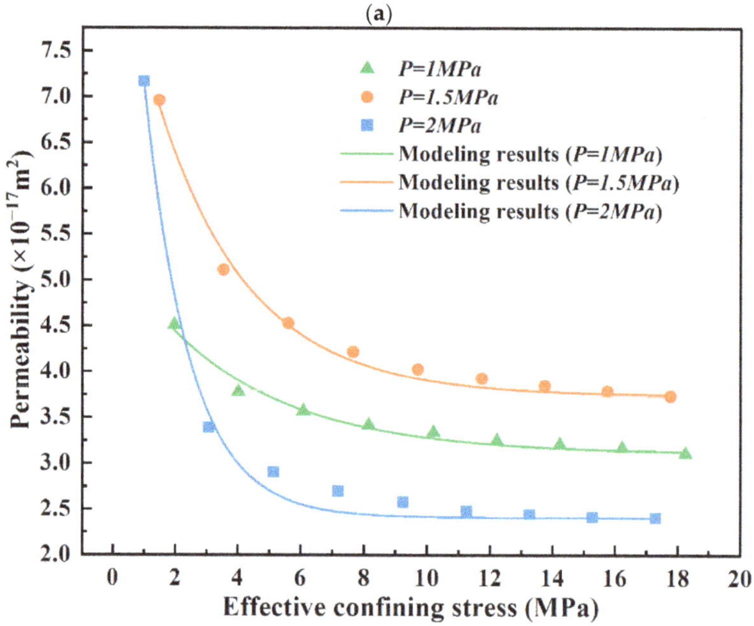

(**b**)

Figure 13. *Cont.*

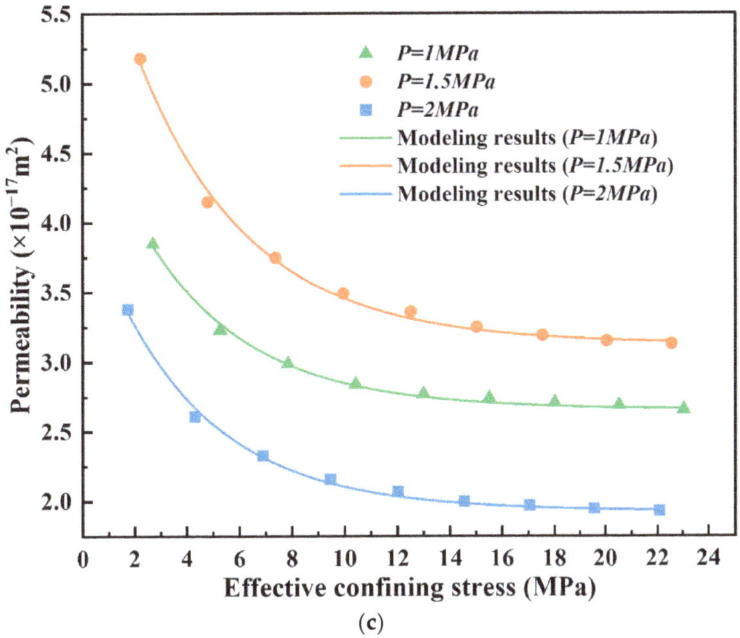

Figure 13. Permeability of the sandstone samples under different effective confining stresses and pore pressures, and modeling results from the proposed model in this study ((**a–c**) are the initial confining pressure of 15 MPa, 20 MPa, and 25 MPa, respectively).

Table 3. Parameters used for data fitting.

Sample No.	In Situ Stress (MPa)	Pore Pressure (MPa)	Parameters		R^2
			A	B	
B1	15	1	5.63×10^{-3}	0.506	0.988
B2	15	1.5	1.12×10^{-3}	0.650	0.989
B3	15	2	1.80×10^{-6}	1.311	0.997
B4	20	1	0.021	0.255	0.981
B5	20	1.5	0.012	0.343	0.991
B6	20	2	5.98×10^{-5}	0.693	0.992
B7	25	1	7.28×10^{-3}	0.250	0.996
B8	25	1.5	0.017	0.235	0.997
B9	25	2	8.23×10^{-3}	0.253	0.996

4. Conclusions

In this study, a series of mechanical–seepage coupling experiments were conducted, considering the mining stress paths. The response of permeability and mechanical properties was observed. A fitting exponential permeability model was proposed for sandstone. The main conclusions are given as follows:

(1) In the imitated mining stress path, the variation tendency of the axial strain, the radial strain, and the volumetric strain for sandstone show a linear growth trend, an exponential function growth trend, and a transverse "V" symmetrical distribution, respectively. Under the same in situ stress, the mechanical deformation in sandstone is more sensitive to mining stress than to pore pressure.

(2) Most sandstone samples maintain compression state at the peak stress condition due to the mineral composition and pore–fracture structure. The absolute value of both the axial strain and the radial strain at the peak stress state increases with the

incremental confining pressure. Under the same in situ stress, the volumetric strain at the peak stress gradually decreases, which is caused by the increased pore pressure. The failure mode of sandstone evolved from shear failure to shear–tension failure with the increase in in situ stress, which corresponds to the transformation from a single shear fracture to form an approximate "X"-shaped conjugate shear fracture or to the main shear fracture followed by a secondary shear–tension fracture.

(3) The sandstone permeability, under the same in situ stress and pore pressure, increases exponentially with the mining stress simulated by axial loading and radial unloading. The permeability under a mining stress state was well-characterized by the stress-relief effect.

Author Contributions: Y.L.: Conceptualization, formal analysis, investigation, data curation, writing—original draft. T.Z.: Conceptualization, writing—review and editing, funding acquisition, supervision. J.W.: writing—original draft, visualization. Z.S.: writing—review and editing, supervision. F.W.: visualization, supervision. The manuscript was written through the contributions of all authors. All authors have read and agreed to the published version of the manuscript.

Funding: This paper was supported by the Institute of Energy, Hefei Comprehensive National Science Center (Grant No. 21KZS216); the National Youth Science Foundation (No. 51904011); the Anhui Provincial Natural Science Foundation (No. 1908085QE183); the Open Fund of State Key Laboratory of Mining Response and Disaster Prevention and Control in Deep Coal Mines (Grant No. SKLMRDPC19ZZ05); and the Collaborative Innovation Project of Colleges and Universities in Anhui Province (Grant No. GXXT-2021-019). The APC was funded by the Institute of Energy, Hefei Comprehensive National Science Center (Grant No. 21KZZ508).

Data Availability Statement: Not applicable.

Conflicts of Interest: The authors declare no conflict of interest.

References

1. He, M.C.; Xie, H.P.; Peng, S.P.; Jiang, Y.D. Study on rock mechanics in deep mining engineering. *Chin. J. Rock Mech. Eng.* **2005**, *16*, 2803–2813.
2. Xie, H.P.; Ju, Y.; Gao, F.; Gao, M.; Zhang, R. Groundbreaking Theoretical and Technical Conceptualization of Fluidized Mining of Deep Underground Solid Mineral Resources. *Tunn. Undergr. Space Technol.* **2017**, *67*, 68–70. [CrossRef]
3. Xie, H.P. Research framework and anticipated results of deep rock mechanics and mining theory. *Adv. Eng. Sci.* **2017**, *49*, 705–713.
4. Fairhurst, C.E.; Hudson, J.A. Draft ISRM Suggested Method for the Complete Stress-Strain Curve for Intact Rock in Uniaxial Compression. *Int. J. Rock Mech. Min. Sci.* **1999**, *36*, 281–289.
5. Ghabezloo, S.; Sulem, J.; Guédon, S.; Martineau, F. Effective Stress Law for the Permeability of a Limestone. *Int. J. Rock Mech. Min. Sci.* **2009**, *46*, 297–306. [CrossRef]
6. Ghanbarzadeh, S.; Hesse, M.A.; Prodanovi, M.; Gardner, J.E. Deformation-Assisted Fluid Percolation in Rock Salt. *Science* **2015**, *350*, 1069–1072. [CrossRef] [PubMed]
7. Liu, L.; Xu, W.Y.; Wang, H.L.; Wang, R.B.; Wang, W. Experimental Studies on Hydro-Mechanical Properties of Metamorphic Rock under Hydraulic Pressures. *Eur. J. Environ. Civ. Eng.* **2016**, *20*, 45–59. [CrossRef]
8. Liu, J.; Yang, H.; Xiao, Y.; Zhou, X. Macro-Mesoscopic Fracture and Strength Character of Pre-Cracked Granite Under Stress Relaxation Condition. *Rock Mech.* **2018**, *51*, 1401–1412. [CrossRef]
9. Zhang, T.; Liu, Y.; Yang, K.; Tang, M.; Yu, X.; Yu, F. Hydromechanical Coupling Characteristics of the Fractured Sandstone under Cyclic Loading-Unloading. *Geofluids* **2020**, *10*, 1–12. [CrossRef]
10. Zhu, Z.D.; Zhang, A.J.; Xu, W.Y. Experimental research on complete stress-strain process seepage characteristics of brittle rock. *Rock Soil Mech.* **2002**, *23*, 55–559.
11. Wang, H.L.; Xu, W.Y.; Yang, S.Q. Experimental investigation on permeability evolution law during course of deformation and failure of rock specimen. *Rock Soil Mech.* **2006**, *10*, 1703–1708.
12. Yin, G.; Jiang, C.; Wang, J.G.; Xu, J. Combined Effect of Stress, Pore Pressure and Temperature on Methane Permeability in Anthracite Coal: An Experimental Study. *Transp. Porous Media* **2013**, *100*, 1–16. [CrossRef]
13. Wang, W.; Xu, W.Y.; Wang, R.B.; Cao, Y.J.; Wang, H.L.; Feng, S.R. Permeability of dense rock under triaxial compression. *Chin. J. Rock Mech. Eng.* **2015**, *34*, 40–47.
14. Jiang, Z.Q.; Ji, L.J.; Zuo, R.S.; Cao, L.W. Correlativity among rock permeability and strain, stress under servo-control condition. *Chin. J. Rock Mech. Eng.* **2002**, *10*, 1442–1446.
15. Yu, J.; Li, H.; Chen, X.; Cai, Y.Y.; Wu, N.; Mu, K. Triaxial experimental study of associated permeability-deformation of sandstone under hydro-mechanical coupling. *Chin. J. Rock Mech. Eng.* **2013**, 32, 1203–1213.

16. Wang, J.-A.; Park, H.D. Fluid Permeability of Sedimentary Rocks in a Complete Stress–Strain Process. *Eng. Geol.* **2002**, *63*, 291–300. [CrossRef]
17. Liu, X.; Xu, M.; Wang, K. Mechanism of Permeability Evolution for Reservoir Sandstone with Different Physical Properties. *Geofluids* **2018**, *2018*, 5327895. [CrossRef]
18. Xiao, W.; Zhang, D.; Wang, X. Experimental Study on Progressive Failure Process and Permeability Characteristics of Red Sandstone under Seepage Pressure. *Eng. Geol.* **2020**, *265*, 105406. [CrossRef]
19. Li, M.; Liu, X. Experimental and Numerical Investigation of the Failure Mechanism and Permeability Evolution of Sandstone Based on Hydro-Mechanical Coupling. *J. Nat. Gas Sci. Eng.* **2021**, *95*, 104240. [CrossRef]
20. Kamel, K.E.M.; Gerard, P.; Colliat, J.-B.; Massart, T.J. Modelling Stress-Induced Permeability Alterations in Sandstones Using CT Scan-Based Representations of the Pore Space Morphology. *Int. J. Rock Mech. Min. Sci.* **2022**, *150*, 104998. [CrossRef]
21. Hu, C.; Jia, Y.; Duan, Z. Pore and Permeability Properties of Reservoir Sandstone under a Uniaxial Compression CT Test. *J. Nat. Gas Sci. Eng.* **2022**, *104*, 104666. [CrossRef]
22. Yang, Y.J.; Song, Y.; Chen, S.J. Test study of coal's strength and deformation characteristics under triaxial compression. *J. China Coal Soc.* **2006**, *31*, 150–153.
23. Li, X.S.; Yi, G.Z.; Zhao, H.B.; Wang, W.Z.; Jing, X.F. Experimental study of mechanical properties of outburst coal containing gas under triaxial compression. *Chin. J. Rock Mech. Eng.* **2010**, *29*, 3350–3358.
24. Zhang, Z.; Zhang, R.; Xie, H.; Gao, M. The Relationships among Stress, Effective Porosity and Permeability of Coal Considering the Distribution of Natural Fractures: Theoretical and Experimental Analyses. *Environ. Earth Sci.* **2015**, *73*, 5997–6007. [CrossRef]
25. Xie, H.P.; Zhou, H.W.; Liu, J.F.; Gao, F.; Zhang, R.; Xue, D.J.; Zhang, Y. Mining-induced mechanical behavior in coal seams under different mining layouts. *J. China Coal Soc.* **2011**, *36*, 1067–1074.
26. Xie, H.P.; Zhang, Z.T.; Gao, F.; Zhang, R.; Gao, M.Z.; Liu, J.F. Stress-fracture-seepage field behavior of coal under different mining layouts. *J. China Coal Soc.* **2016**, *41*, 2405–2417.
27. Zuo, J.P.; Liu, L.F.; Zhou, H.W.; Huang, Y.M. Deformation failure mechanism and analysis of rock under different mining condition. *J. China Coal Soc.* **2013**, *38*, 1319–1324.
28. Xia, B.W.; Liu, S.W.; Ou, C.N.; Gao, Y.G. Experimental study on mechanical properties of sandstone with single fracture under fully-mechanized top-coal caving mining stress path. *Coal Sci. Tech.* **2022**, *50*, 95–105.
29. Liu, Y.; Zhang, T.; Ma, Y.K.; Song, S.B.; Tang, M.; Li, Y.F. Deformation behavior and damage-induced permeability evolution of sandy mudstone under triaxial stress. *Nat. Hazards* **2022**, 1–21. [CrossRef]
30. Chen, D.; Pan, Z.; Shi, J.Q.; Si, G.; Ye, Z.; Zhang, J. A novel approach for modelling coal permeability during transition from elastic to post-failure state using a modified logistic growth function. *Int. J. Coal Geol.* **2016**, *163*, 132–139. [CrossRef]

energies

Article

Experimental Study on Damage Fracture Law of Coal from Solid-Propellant Blasting

Huaibao Chu [1,2], Mengfei Yu [1,*], Bo Sun [1,3,*], Shaoyang Yan [1], Haixia Wei [1], Guangran Zhang [1], Donghui Wang [1] and Jie Xu [1]

1 School of Civil Engineering, Henan Polytechnic University, Jiaozuo 454000, China
2 Henan Key Laboratory Underground Engineering and Disaster Prevention, Jiaozuo 454000, China
3 School of Civil Engineering, Xinyu University, Xinyu 338000, China
* Correspondence: ymf19980922@163.com (M.Y.); bosun0419@163.com (B.S.); Tel.: +86-188-3965-6466 (M.Y.); +86-157-0790-6611 (B.S.)

Abstract: The low permeability of coal seams has always been the main bottleneck restricting coalbed methane drainage. In this paper, a coal seam anti-reflection technology with solid-propellant blasting was proposed, and the composition and proportion of the solid propellants were determined based on the principle of oxygen balance. The authors designed a solid-propellant blasting damage fracture experiment of simulation coal, tested the impact pressure on a blast hole wall, measured the ultrasonic wave velocity, explosive strain and crack propagation velocity, and then revealed the blasting damage fracture process and mechanism of coal based on the experimental results and damage fracture mechanics theory. The history curve of impact pressure time can be divided into three processes including the slow pressurization process, dramatic increase process, and nonlinear pressure relief process. The pressure distribution along the whole blasting hole was uneven, and the peak pressure was relatively small, but the pressure action time was long. The damage and fracture process of coal solid-propellant blasting can be divided into two stages including the rapid damage fracture development stage and the stable slow damage fracture development stage. Firstly, the explosion stress wave produced and rapidly accelerated the radial cracks extension; secondly, the cracks slowly expanded over a large area by the combined effects of the high-pressure gases, the gas, and the original rock stress.

Keywords: coal; solid propellant blasting; borehole wall pressure; damage fracture law; experimental study

Citation: Chu, H.; Yu, M.; Sun, B.; Yan, S.; Wei, H.; Zhang, G.; Wang, D.; Xu, J. Experimental Study on Damage Fracture Law of Coal from Solid-Propellant Blasting. *Energies* **2022**, *15*, 8104. https://doi.org/10.3390/en15218104

Academic Editors: Shengrong Xie and Dongdong Chen

Received: 22 September 2022
Accepted: 26 October 2022
Published: 31 October 2022

1. Introduction

China's energy situation of 'rich coal, less oil, and natural gas' and the stage of economic and social development determine the main energy status of coal, and it will continue to play a supporting role in energy supply for a longer period in the future [1]. At the same time, the energy strategic direction of 'Reducing Coal, Stabilizing Oil and Increasing Gas, Developing New Energy' was proposed by the Chinese government in the 'Energy Development Strategic Plan for 2020–2050'. The proportion of natural gas in energy consumption will increase to about 15% in 2050. The 'increasing reserves and production' of coal-measure gas is one of the important ways to make up for the gap in the natural gas supply. Large-scale firedamp extraction and utilization is an important part of the coal-measure gas industry. Therefore, the effective development and utilization of coal-measure gas resources is an important way to ensure the safety of the national natural gas supply, help the safe production of coal, and achieve the goal of 'peak carbon emissions by 2030 and reach carbon neutrality by 2060' in China.

The reserve of only coalbed methane is about 36.8 trillion m^3 (shallow coalbed methane buried depth of 2000 m) in China, ranking third in the world, and the recoverable quantity of coalbed methane is about 10 trillion m^3. However, the occurrence conditions of coalbed methane resources in China are complex, the technical requirements are high, the regional adaptability is poor, and the proportions of high-stress tectonic coal and low permeability coal-measure gas resources are high [2,3]. The low permeability of coal seams has become the key to restricting coalbed methane extraction in China. Therefore, it is one of the major scientific and technological needs to promote the construction of a large coal-measure gas industry to develop and improve the key technology of high-efficiency and low-cost coalbed methane extraction, improve the fracture development level of coalbed methane reservoirs, dredge the seepage channel, and improve the permeability of coal reservoirs and achieve the efficient extraction of coalbed methane.

Researchers at domestic and foreign levels have carried out multi-path research to enhance the permeability of coal seams with low permeability and high-gas content and have achieved fruitful research results. The anti-reflection measure of interlayer pressure relief seam making is adopted for low permeability and high gas coal seams with protective layer mining conditions, having good anti-reflection effects, and is a mature technology [4,5]. For most high-gas and low-permeability coal seams without protective layer mining conditions, hydraulic anti-reflection technologies including hydraulic fracturing, hydraulic punching, and hydraulic slotting are the main measures for anti-reflection transformation of low-permeability coal seams [6–9]. However, hydraulic measures require a lot of water resources, and the invasion and retention of water will also cause water lock and water sensitivity damage to the coal reservoir [10]. At the same time, a large amount of water-based fracturing fluid occupies the gas flow channel, thus affecting gas production. In addition, the penetration of hydraulic fracturing fluid containing chemical additives into the surface and underground drinking water layer will also produce a certain degree of pollution [11]. Therefore, researchers in various countries are also actively seeking alternative methods for hydraulic anti-reflection measures. The non-hydration anti-reflection technologies including deep-hole presplit blasting fracturing technology [12], high-energy and high-pressure gas impact [13,14], controllable shock wave impact [15,16], acoustic shock and ultrasonic disturbance [17,18], and liquid nitrogen injection fracturing technology [19] have been widely used. Among them, the fracturing technology with high-energy and high-pressure gas is considered to be a fracturing anti-reflection technology that can be comparable to hydraulic fracturing.

High-energy, high-pressure gas fracturing refers to the use of rapid discharge of high-energy, high-pressure gas impact on the borehole wall to achieve the purpose of fracturing and increasing permeability. Aerospace solid propellants can be treated as weak explosives, they can be excited to produce a large number of high-energy, high-pressure gas products [20]. Therefore, it is feasible to use solid propellants to fracture and increase the permeability of the coal body. Studies have shown that propellant blasting technology can easily fracture coal seams and create multiple cracks that are not controlled by ground stress due to the advantages of the pressure strength and pressure transfer rate [21]. It also generates oscillating high-pressure pulses to enlarge the fracture extension, and it does not require large equipment and is highly adaptable to the environment [22]. At present, solid propellants have been widely used in the field of conventional oil and gas extraction in highly brittle rock formations [23], but its application in the field of coal reservoir permeability enhancement is still in the initial stage, especially the damage fracture mechanism of the coal body from solid-propellant blasting is still unclear.

Aiming at the characteristics of high gas and low permeability of coal reservoirs in China, this paper proposes a solid-propellant blasting, permeability-increasing technology. Firstly, the composition and proportion of solid propellants were determined according to the principle of propellant gas production and oxygen balance, and then the solid-propellant blasting test of simulation coal bodies was carried out. Based on the results and the theory of damage fracture mechanics, the blasting damage fracture process and mechanism of coal from solid-propellant blasting were analyzed. It provides a reference for the practical application of solid-propellant blasting technology in low-permeability coal seams.

2. Development of Solid Propellant

Solid propellants are mainly composed of a heating agent, gas supply agent, oxygen supply agent, and other components. The oxygen supply agent provides oxygen to the heating agent, causes the oxidation reaction to occur, and releases a lot of heat, and then the gas generating agent is thermally decomposed to produce a large amount of gas. Solid propellants utilized as a coal seam anti-reflection technology should first meet the requirements of gas produced to be non-toxic, harmless, and non-corrosive; secondly, the coal body after solid-propellant blasting does not produce a small crushing zone, and the energy utilization rate is improved as much as possible. Thirdly, the gas production of a solid propellant is large, which can steadily drive the crack to expand within the range. Finally, safety must be ensured during the production process and usage of solid propellants.

A solid propellant is essentially a weak explosive. Therefore, based on the principle of zero oxygen balance of explosives, the authors chose aluminum powder as the heating agent, ammonium perchlorate as the oxygen donor and the main gas generating agent to make a solid propellant. The solid propellant's composition and ratios are shown in Table 1. The main role of HTPB is as an adhesive to bond various materials together, and can give the propellant moisture-proof and antistatic properties. The role of ammonium oxalate is to decompose after heating to produce ammonia to inhibit the decomposition rate of ammonium perchlorate, thereby controlling the gas production rate. The oxygen balance values of three solid propellants with high, medium, and low burning rates were 1.6%, −6.2%, and −18.6%, respectively. The positive oxygen-balance explosives have better detonation performance, but the micro-negative oxygen-balance explosives are safer.

Table 1. Composition and ratio of solid propellant.

Raw Material	Specification	Percentage (%)		
		High Burning Rate	Secondary Burning Rate	Low Burning Rate
HTPB TDI	Type I Industrial grade	13.39	17.37	24.19
KZ	Industrial grade	3.50	4.50	4.23
AP	Type I	0.00	68.00	0.00
AP	Type III	73.00	0.00	61.00
AP	Type IV	5.00	5.00	5.20
Ammonium oxalate	Industrial grade	0.00	0.00	2.50
AL	Industrial grade	5.00	5.00	0.00
Others		0.11	0.13	2.88

Note: (1) HTPB (HO- [CH2CH = CHCH2] n-OH) is hydroxyl-terminated polybutadiene; (2) TDI($C_9H_6N_2O_2$) is toluene diisocyanate; (3) KZ($C_{26}H_{50}O_4$) is dioctyl sebacate; (4) AP(NH_4ClO_4) is ammonium perchlorate; and (5) AL is aluminum.

Under high-temperature conditions, ammonium perchlorate decomposes to produce oxygen, and aluminum powder is oxidized by oxygen to release a lot of heat, which in turn promotes the thermal decomposition of the mixture dominated by ammonium perchlorate to produce a large number of high-temperature and-high pressure gas products. The main reaction equations are shown in Equations (1)–(3).

$$4Al + 3O_2 = 2Al_2O_3 \tag{1}$$

$$2NH_4CLO_4 = N2 + 4H_2O + CL_2 + 2O_2 \tag{2}$$

$$(NH_4)_2C_2O_4 = 2NH_3 + CO + CO_2 + H_2O \tag{3}$$

In order to understand the basic performance parameters of solid propellants, the burning rate, pressure resistance, and temperature resistance of three solid propellants were tested according to conventional explosives testing methods. The basic performance parameters of the solid propellants are shown in Table 2.

Table 2. Performance parameters of solid propellant.

Name	Recipe Code	m/kg	v/(mm/s)	V_m(L/kg)	Q/(kJ/kg)	ρ/(g/cm^3)	σ/(MPa)	T/(°C/48 h)
Sy-high	YLD	0.3	14	>1400	≥3500	1.4~1.7	120	163
Sy-mid	YLDA	0.3	10	>1200	≥3200	1.4~1.7	120	163
Sy-low	E3	0.3	5	>1091	≥3000	1.4~1.7	120	163

Note: (1) m is explosive quantity; (2) v is burning rat; (3) V_m is gas production; (4) Q is explosion heat; (5) ρ is density; (6) σ is pressure resistance; and (7) T is Temperature resistance.

After a solid propellant is excited, the high-temperature and high-pressure gas is produced in the hole and rapidly expands and impacts the hole wall, thus forming a large range of crack zone around the blast hole, which improves the crack development level of the coal reservoir to a certain extent, dredges the seepage channel, and improves the permeability of coal seam. At the same time, the blasting of a solid propellant will also disturb the dynamic equilibrium state of adsorption and desorption of coalbed methane in the coal reservoir, and finally form disturbance desorption, thermal effect desorption, and replacement desorption of coalbed methane, improving the extraction efficiency of coalbed methane.

3. Experimental Study on Blasting Damage Fracture Process of Coal Solid Propellant

3.1. Test Method

During the field collection and post-processing of the coal samples, secondary disturbance damage occurred inside the coal samples. In addition, the corresponding sensors needed to be placed inside the specimen to obtain the strain waves induced by high-energy gas. This was difficult to achieve in the actual coal samples. Therefore, simulated coal samples were selected as the research sample to carry out the related experimental research work. According to the test results of existing literature [24], the mixing ratio and basic physical and mechanical properties of the simulation coal body are shown in Table 3. Cement, sand, and water were used as the main materials to control the structural strength of the simulated coal body; gypsum, broken coal, perlite, a foaming agent, and mica were used as additional materials to control the micro-cracks, micro-voids, structural planes, and gas of the simulated coal body.

Table 3. Mixing ratio and physical and mechanical properties parameters of simulated coal body.

Materials and Proportioning	Density ρ (g/cm^3)	Compressive Strength σ_e/MPa	Longitudinal Wave Velocity c_p/(m/s)	Young's Modulus E/GPa
Sand: Water: Cement: Gypsum: Crushed Mica: Perlite: Foaming Agent				
1.4:0.6:1.4:0.2:0.028:0.027:0.048	1.71	12.23	1660	2.32

Note: (1) ρ is density, (2) σ_e is compressive strength, (3) c_p is longitudinal wave velocity, and (4) E is Young's modulus.

Three simulated coal test blocks were made to carry out the blasting test of solid propellants with a high burning rate. The size of the test block was 1000 × 1000 × 600 mm, as shown in Figure 1a. The block was made using manual stirring, tamping, and forming with a small vibrating rod in the template, and manually cured for 28 days. A charge hole with a diameter of 50 mm and a depth of 400 mm was reserved in the middle of the test block. After the solid propellant was placed, it was filled with planting glue and detonated by the ignition powder head. The solid propellant samples and ignition powder head are shown in Figure 1b,c.

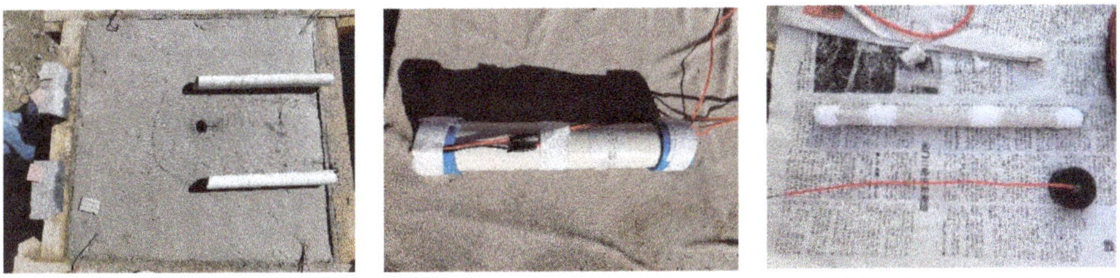

(**a**) Simulated coal test blocks (**b**) Propellant charge reel (**c**) Excitation device

Figure 1. The test block and cartridge of solid propellant blasting.

To explore the load characteristics and pressure distribution law in the borehole well from propellant blasting, a pressure sensor and signal acquisition system were used to test the borehole wall pressure. The pressure sensor used was a PVDF piezoelectric film sensor; the pressure signals were collected by the DH5922 N dynamic strain gauge with a sampling frequency of 200 kHz. The test instrument and pressure sensor were arranged as shown in Figure 2.

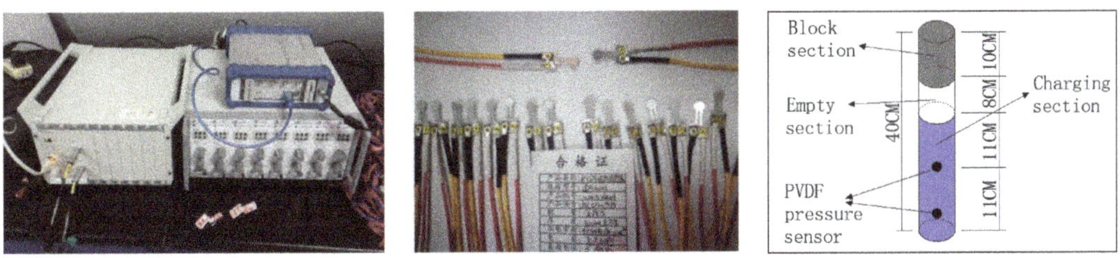

(**a**) DH5922N dynamic strain meter (**b**) PVDF pressure sensor (**c**) Point diagram

Figure 2. The instruments and measuring point arrangement of hole wall pressure testing.

To explore the degree of damage to coal from solid propellant blasting, the ZT-801 non-metallic ultrasonic detector (Figure a) was used to test the ultrasonic wave velocity before and after blasting. The measuring points were arranged on the horizontal line passing through the center of the charge, the first measurement point was 50 mm from the borehole and each subsequent measuring point was arranged at an interval of 100 mm, as shown in Figure 3. The damage value was calculated with the formula [25] $D = 1 - (v/v_0)^2$ (v_0 and v are the ultrasonic velocities of the test block before and after blasting, respectively).

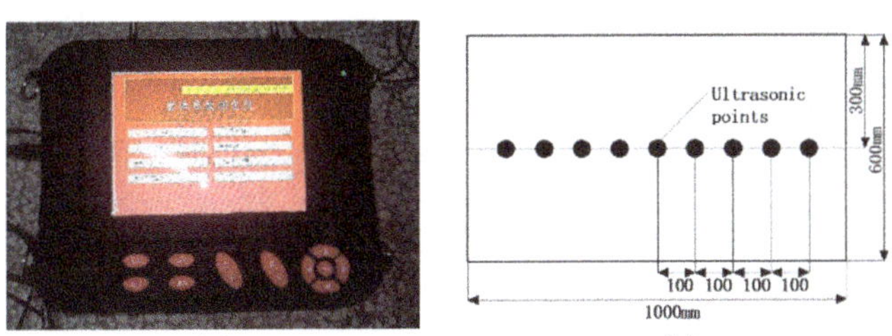

(**a**) (**b**)

Figure 3. Ultrasonic test: (**a**) ZT-801 inorganic non-metallic ultrasonic detector and (**b**) ultrasonic measurement point schematic diagram.

In order to explore the propagation and attenuation law of solid-propellant blasting stress wave in coal, the stress wave was collected by embedding strain bricks in the test block. The strain bricks were embedded in the four positions of 100, 200, 300, and 400 mm from the center height of the cartridge to the center of the reserved hole. The size of the strain brick was 20 mm × 20 mm × 20 mm, and the proportion of the strain brick was consistent with the large test block. The strain gauge on the strain brick was a BF120-3AA strain gauge. The DH5922N dynamic signal acquisition system was used to collect the strain waveform.

In order to test the crack propagation velocity of coal during the blasting process of the solid propellants, an enameled copper wire with a diameter of 0.14 mm was embedded as a metal probe sensor at a distance of 100 mm between the center of the cartridge and from 50 mm the edge of the blast hole, and the enameled wire was fixed with a U-shaped iron wire. The BSW-3A intelligent five-stage detonation velocity meter was used to test the crack propagation speed in the test block.

The position of the stress wave and crack propagation velocity test sensor is shown in Figure 4.

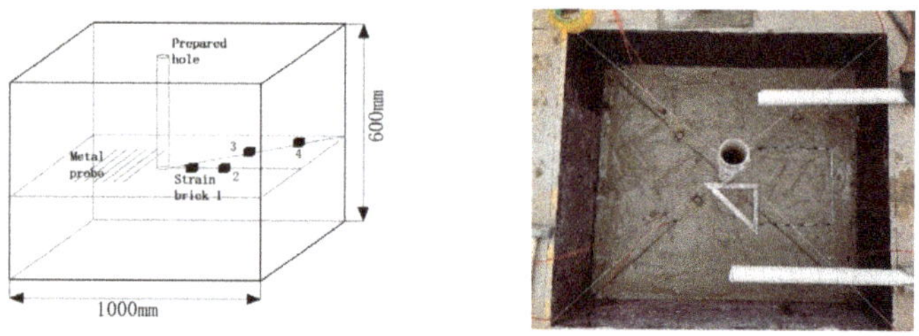

Figure 4. Arrangement of strain brick and metal probes.

3.2. *Results and Analysis of Load on Solid-Propellant Blasting Hole Wall*

A penetrating crack was formed in all three test blocks after blasting. A half-hole mark of the blast hole was complete, and no crushing zone appeared. The test block after blasting is shown in Figure 5. The representative history curves of the impact pressure time for the middle and bottom hole walls of the blast hole are shown in Figure 6.

Figure 5. The test block after blasting.

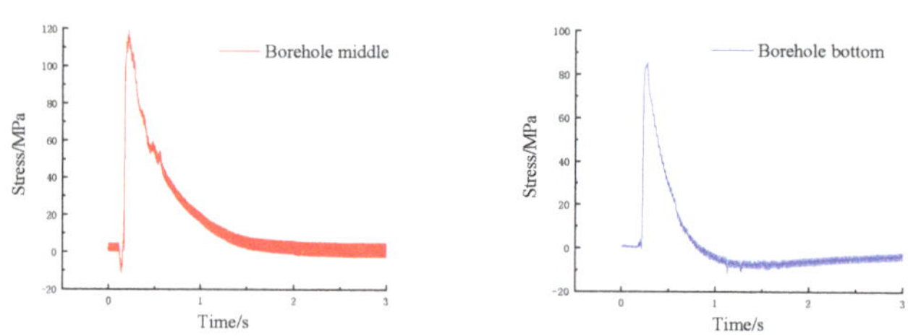

Figure 6. The pressure time history curve of blast hole walls.

It can be seen from Figure 6 that the hole wall pressure after solid-propellant blasting exhibited three stages: transient accumulation pressurization, rapid pressurization, and nonlinear depressurization. The pressure distribution along the whole length of the hole was uneven, the peak pressure was relatively small, and the pressure action time was relatively long.

After a solid propellant is excited, it decomposes and produces a large number of high-temperature and high-pressure gas products. After a short period of accumulation (0.15 s in the middle of the borehole, and 0.21 s at the bottom of the borehole), the gas products rapidly expanded, impacted the hole wall, and generated a stress wave in the test block, and the hole wall pressure rapidly increased. With the propagation of a stress wave, cracks began to appear in the specimen, high-temperature and high-pressure gas quickly wedged into the crack, and the hole pressure began to nonlinearly decline.

The high-temperature and high-pressure gas products produced by the excited solid-propellant impacted the hole wall and induced the stress wave in the coal near the borehole after a short period of accumulation. When the stress wave intensity was higher than the dynamic compressive (tensile) strength of the coal, the initial crack appears near the borehole, and then the high-temperature and high-pressure gas products began to wedge into the initial crack and drove the crack propagation. Because the ignition end of the solid propellant sample in this test was at the upper part, the initial crack appeared at the upper part of the borehole. When the gas flowed from the upper part of the borehole to the bottom of the borehole, some of the gas leaked from the initial crack in the upper part of the borehole, which caused the peak pressure on the borehole wall to gradually decrease from the upper part of the borehole to the bottom of the borehole. This was consistent with the test results, the peak pressure in the hole (118.1 MPa) was greater than the bottom of the hole (85.4 MPa).

The detonation pressure of an explosion was generally 10~40 GPa, and the detonation velocity of an explosive was generally 1000~8500 m/s [26], so the pressure in the hole was basically uniform in distribution and the loading time was very short. However, the reaction rate of the solid propellants was slow (Table 2) and the impact action was small. The chemical reaction process of the solid propellants was parallel to the impact cracking process of the high-pressure gas product generated by the previous reaction, that is, the formation and diffusion of the high-pressure gas product were also simultaneously carried out. As a result, the pressure in the hole after being blasted with the fracturing agent was relatively small and the pressure maintenance time in the hole was relatively long.

3.3. Results and Analysis of Solid-Propellant Blasting Damage Fracture

The test results of peak strain, damage value, and crack propagation velocity of the coal body from solid-propellant blasting and explosive blasting are shown in Table 4. Among them, the results of coal damage fracture tests from explosive blasting refer to the references [27,28], and the solid-propellant blasting damage fracture results are the average of the valid values of the three test block results.

Table 4. Results of Coal body damage and fracture test.

Results \ Grouping and Points	Grouping	1	2	3	4	5
Peak strain	Solid-propellant blasting	4845	2088	1005	560	
	Explosive blasting	8946	3691	1834	875	
Damage value	Solid-propellant blasting	0.655	0.430	0.304	0.263	0.221
	Explosive blasting	0.865	0551	0.275	0.216	0.161
Crack propagation velocity	Solid-propellant blasting	280.8	162.4	137.4	120.6	109.2
	Explosive blasting	559.5	264.8	187.4	132.6	97.5

According to the table data, the variation curves of stress peak strain, damage, and crack propagation velocity of coal solid-propellant blasting, along with the distance from the center of the blast hole are shown in Figure 7. From Figure 7a, it can be seen that the stress wave intensity in the coal body from propellant blasting is lower, but the decay rate is slower than that of explosive blasting; from Figure 7b, it can be seen that in the adjacent area, the damage value of coal from solid-propellant blasting is less than that of explosive blasting, but with the increase of distance, the damage value of coal from propellant blasting gradually approaches or even exceeds that of explosive blasting; from Figure 7c, it can be seen that the cracking rate of the coal from propellant blasting is much smaller than that of explosive blasting in the adjacent area, but the two are gradually approaching parity with the increase of distance.

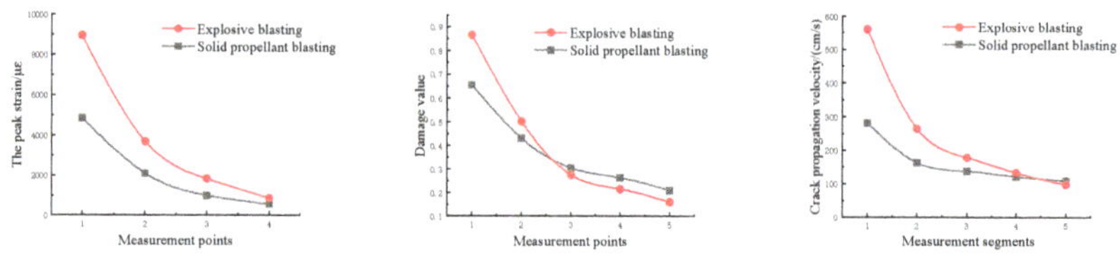

(**a**)Peak strain (**b**) Damage value (**c**) Crack propagation velocity

Figure 7. Peak strain, damage value, and crack propagation velocity of coal body from blasting.

In order to further analyze the process and law of blasting damage and fracturing of coal, the damage test results were expanded by 1000 times, the strain peak was reduced by 10 times, and the crack propagation speed was in the same order of magnitude. The curves of damage, strain peak, and crack propagation speed with distance from the center of the blast hole from propellant blasting and explosive blasting are shown in Figure 8.

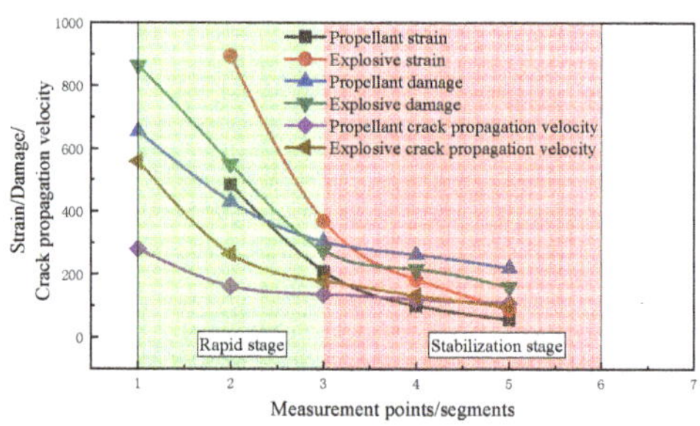

Figure 8. The damage and fracture process curves of coal.

According to Figures 7 and 8, the damage fracture process of the coal body from solid-propellant blasting can be divided into two stages: rapid-damage fracture and slow steady-state damage fracture.

The stress waves generated near the borehole wall by solid-propellant blasting were of low intensity, so no crush zone was generated near the borehole. However, the tensile waves derived from the compressive stress waves can contribute to the initial radial cracking of the coal body that is:

$$\sigma_\theta > (1 - D_0)\sigma_{\text{dtension}}, \sigma_\theta = \sigma_r \cdot \mu / 1 - \mu \quad (1)$$

where σ_θ is the tangential stress generated by the propagation of explosive stress wave in coal, MPa; and μ is Poisson's ratio of coal.

The initial radial crack propagation velocity of coal from solid-propellant blasting was only 280.8 m/s, and the stress wave in the coal caused by propellant blasting rapidly decayed in the adjacent area with the increase of distance, so the crack expansion velocity in the adjacent area of the shell hole, although fast, could expand in a small area. As the distance increased, the attenuated stress wave was unable to continue driving the initial crack propagation in the coal body, as a result, the quasi-static effect of the high-temperature and high-pressure gas products became the main contributor to driving the continued crack propagation. High-temperature and high-pressure gas products wedged into the initial radial crack and drove the crack steady-state expansion from the combined effects of high-temperature and high-pressure gas internal pressure, gas, and original rock stress [29].

Cracks then started to steadily expand when the stress intensity value generated at the crack tip was greater than the fracture toughness of the coal body, that is:

$$2\sqrt{\frac{C}{\pi}}\int_0^a \frac{p(r)}{\sqrt{a^2-r^2}}dr - \sigma_\infty\sqrt{\pi C} \geq K_{IC} \tag{5}$$

where a is the initial radial crack length, m; σ_∞ is the original rock stress, MPa; K_{IC} is the dynamic fracture toughness of coal; $p(r)$ is the gas pressure on the crack surface, MPa; and $P(r) = P_0 \cdot (a-r)/a$, where P_0 is the initial impact compressive stress on the hole wall, MPa.

As the crack continues to expand, the gas pressure in the crack decreases, but the free gas desorbed is involved in driving the further expansion of the crack for the pressure reduction in the crack and the dynamic disturbance of the solid-propellant blasting. At the same time, under the quasi-static stress field of high-pressure gas products and gas, the brittleness of coal increases with the increase of gas pressure. Therefore, when the tangential stress at the crack tip is greater than the critical stress of crack propagation, the crack continues to slowly propagate in a steady state.

$$\frac{a}{r^2}\left[r_0 p_0 + \int_{r_0}^a p(r)dr + \int_{r_0}^a p'(r)dr\right] - \sigma_\infty\left(1+\frac{a^2}{r^2}\right) \geq \sqrt{\pi/4a}K_{IC} \tag{6}$$

where r_0 is the radius of the blast hole, m; $p'(r)$ is the gas pressure at the crack surface, MPa; and because the gas pressure is smaller than the detonation gas pressure, the gas pressure in the crack is usually taken as a constant, namely $p'(r) = p_g$ (p_g is the gas pressure in coal).

At the same time, solid-propellant blasting is a sustainable and slow reaction. The solid propellant that reacts in the middle and late stages of the crack-formation process of the high-pressure gas product produced by the earlier reaction will continue to produce high-temperature and high-pressure gas products. Therefore, the high-pressure gas that drives crack propagation will not decrease too quickly, and is similar to the uniform internal pressure that drives crack propagation. This is the main reason for the steady-state slow propagation process of crack formation in coal from solid-propellant blasting. Moreover, it can be observed from Formula (6) that with the slow propagation of crack length, the critical stress required for the continuous propagation of brittle fracture cracks in coal will continue to decrease, and the lower tangential stress can drive the crack to continue to slowly propagate. Therefore, the crack propagation range is large, although the crack propagation speed in this stage is only 109.2~137.4 m/s.

4. Conclusions

(1) The impact pressure of the hole wall for solid-propellant blasting exhibits three stages: temporary accumulation pressurization, rapid pressurization, and nonlinear depressurization. The pressure is unevenly distributed, and the peak pressure is smaller than that of explosive blasting, but the time of pressure exertion is relatively long, which is more conducive to improving its energy utilization.

(2) The crack propagation velocity and stress wave intensity of coal are small from propellant blasting, but the crack propagation velocity and stress wave slowly attenuate with the increase of distance, which can cause a wide range of crack area and effectively improve the permeability of the coal reservoir.

(3) The damage and cracking of coal from propellant blasting is due to the combined effect of blast stress waves, high temperature and pressure gases, gas, and original rock stress. The damage fracture process can be divided into two stages: the rapid development stage dominated by stress wave action and the slow development stage dominated by the quasi-static action of gas products.

Author Contributions: Conceptualization, H.C. and M.Y.; methodology, H.C. and G.Z.; software, B.S.; validation, H.W.; formal analysis, H.C.; investigation, M.Y.; resources, H.C.; data curation,

D.W.; writing-original draft preparation, H.C.; writing-review and editing, M.Y.; visualization, S.Y.; supervision, D.W.; project administration, J.X. All authors have read and agreed to the published version of the manuscript.

Funding: This research was funded by the National Natural Science Foundation of China (No. 51874123, 11542019), and the Henan Research Program of Foundation and Advanced Technology (No. 162300410032).

Data Availability Statement: The data in this manuscript have been presented in Tables 1–4.

Acknowledgments: The authors thank Siyuan Zhu, Zhen Chen, and Zhiqiang Ren for their guidance on the experiment in this paper. The authors also thank the anonymous reviewers for their remarks which improved the paper.

Conflicts of Interest: The authors declare no conflict of interest. The authors declare that they have no known competing financial interests or personal relationships that could appear to have influenced the work reported in this paper.

References

1. Zou, Q.M. Under The Dual Carbon Target, Coal Should Not Only Practice Decarbonization Transformation, but Also Pay Attention to the Bottom Guarantee. Available online: http://nyj.guizhou.gov.cn (accessed on 31 May 2021). (In Chinese)
2. Wang, X.L. In Proceedings of the Speech at the Opening Ceremony of the 2016 National Coal Fai, Taiyuan, China, 19–20 July 2016. Available online: https://www.meitanwang.com/meitan/b1/851876.html (accessed on 31 May 2021). (In Chinese).
3. Yuan, L. Study on development and utilization strategy of Coalbed methane in China. In Proceedings of the Chinese Academy of Engineering Major Consulting Project Research Achievements, Beijing, China, 29 July 2013. (In Chinese).
4. Hu, J.; Feng, K.W.; Sun, C.; Chen, Y. Research on pressure-relief gas drainage mode of adjacent coal seams in upper protective layer mining of close thin coal seam group. *J. Saf. Sci. Technol.* **2021**, *11*, 65–71. (In Chinese)
5. Liu, S.; Yang, K.; Zhang, T. Rib spalling 3D model for soft coal seam faces with large mining height in protective seam mining: Theoretical and numerical analyses. *Geofluids* **2020**, *2020*. [CrossRef]
6. Zhao, X.; Huang, B.; Wang, Z. Experimental Investigation on the Basic Law of Directional Hydraulic Fracturing Controlled by Dense Linear Multi-Hole Drilling. *Rock Mech. Rock Eng.* **2018**, *51*, 1739–1754. [CrossRef]
7. Hu, S.Y.; Guan, S.W.; Feng, G.R.; Han, D.D.; Chen, Y.B. Effect of hydraulic fracture closure on the void ratio of proppant particles in coalbed methane reservoir. *J. Nat. Gas Sci. Eng.* **2020**, *75*, 103111. [CrossRef]
8. Gao, S.H.; Wang, M.Y. Study on hydraulic fracturing-deep hole pre-splitting blasting composite permeability enhancement technology. *Coal Sci. Technol.* **2018**, *48*, 1–8.
9. Zheng, P.; Xia, Y.; Yao, T.; Jiang, X.; Xiao, P.; He, Z. Formation mechanisms of hydraulic fracture network based on fracture interaction. *Energy* **2022**, *243*, 123057. [CrossRef]
10. Jew, A.D.; Druhan, J.L.; Ihme, M.; Kovscek, A.R.; Battiato, I.; Kaszuba, J.P.; Bargar, J.R.; Brown, G.E., Jr. Chemical and Reactive Transport Processes Associated with Hydraulic Fracturing of Unconventional Oil/Gas Shales. *Chem. Rev.* **2022**, *122*, 9198–9263. [CrossRef]
11. Du, J.; Liu, J.; Zhao, L.; Liu, P.; Chen, X.; Wang, Q.; Yu, M. Water-soluble polymers for high-temperature resistant hydraulic fracturing: A review. *J. Nat. Gas Sci. Eng.* **2022**, 104673. [CrossRef]
12. Li, P.; Zhang, X.; Li, H. Technology of coupled permeability enhancement of hydraulic punching and deep-hole pre-splitting blasting in a "three-soft" coal seam. *Mater. Technol.* **2021**, *55*, 89–96.
13. Xia, J.; Dou, B.; Tian, H.; Zheng, J.; Cui, G.; Kashif, M. Research on initiation of carbon dioxide fracturing pipe using the liquid carbon dioxide phase-transition blasting technology. *Energies* **2021**, *14*, 521. [CrossRef]
14. Yan, S.Y.; Yang, X.L.; Chu, H.B.; Wang, C. Experimental Study on the Dynamic Response and Pore Structure Evolution of Coal under High-Pressure Air Blasting. *ACS Omega* **2022**, *7*, 24475–24484. [CrossRef]
15. Zhang, Y.M.; Meng, Z.Z.; Qin, Y.; Zhang, Z.F.; Zhao, Y.Z.; Qiu, A.C. Innovative engineering practice of soft coal seam permeability enhancement by controllable shock wave for mine gas extraction: A case of Zhongjing Mine, Shuicheng, Guizhou Province, China. *J. China Coal Soc.* **2019**, *44*, 2388–2400. (In Chinese)
16. Lin, B.Q.; Wang, Y.H.; Yan, F.Z.; Zhang, X.L.; Yang, W.; Zhu, C.J. Experimental study of the effect of NaCl solution on the pore structure of coal body with high-voltage electrical pulse treatments. *J. China Coal Soc.* **2018**, *43*, 1328–1334. (In Chinese)
17. Jiang, Y.D.; Xian, X.F.; Liu, Z.F. Experiment and mechanism for enhancing coalbed penetrating coefficient with ultrasonic vibration. *J. Liaoning Tech. Univ.* **2009**, *28*, 236–239. (In Chinese)
18. Shi, Q.M.; Qin, Y.; Li, J.Q.; Wang, Z.W.; Zhang, M.J.; Song, X.J. Simulation of the crack development in coal without confining stress under ultrasonic wave treatment. *Fuel* **2017**, *205*, 222–231. [CrossRef]
19. Qin, L.; Zhai, C.; Liu, S.; Xu, J. Factors controlling the mechanical properties degradation and permeability of coal subjected to liquid nitrogen freeze-thaw. *Sci. Rep.* **2017**, *7*, 3675. [CrossRef] [PubMed]
20. Ao, W.; Fan, Z.; Liu, L.; An, Y.X.; Ren, J.R.; Zhao, M.T.; Liu, P.J.; Li, L.K.B. Agglomeration and combustion characteristics of solid composite propellants containing aluminum-based alloys. *Combust. Flame* **2020**, *220*, 288–297. [CrossRef]

21. Zhang, T.; Zhang, X.; Li, N.; Li, K. High-energy perforation and fracturing (HEPF): Great Revolution of Perforation for 21st Century. In Proceedings of the International Oil and Gas Conference and Exhibition in China, Beijing, China, November 2000. [CrossRef]
22. Jaimes, M.G.; Castillo, R.D.; Mendoza, S.A. High energy gas fracturing: A technique of hydraulic pre-fracturing to reduce the pressure losses by friction in the near wellbore: A Colombian field application. In Proceedings of the SPE Latin America and Caribbean Petroleum Engineering Conference, Mexico City, Mexico, 16–18 April 2012.
23. Li, J.; Cao, L.; Guo, B.; Zhang, X. Prediction of productivity of high energy gas-fractured oil wells. *J. Pet. Sci. Eng.* **2018**, *160*, 510–518 [CrossRef]
24. Chu, H.B.; Yang, X.L.; Yu, Y.Q.; Liang, W.M. Experimental research of the choice for coal blasting simulation material. *Coal Sci. Technol.* **2010**, *8*, 31–33. (In Chinese)
25. Zhao, B.; Wang, H. Different technologies of permeability enhancement of single coal seam in china and new technique of high pressure gas shock. *Blasting* **2014**, *31*, 32–41. (In Chinese)
26. Luo, L.M.; Xie, J.P. Effects of detonation velocity on medium deep hole blasting. *Copp. Eng.* **2021**, *1*, 27–30. (In Chinese)
27. Chu, H.B.; Yang, X.L.; Liang, W.M.; Yu, Y.Q. Study on the damage-fracture process and mechanism of coal blasting. *J. Min. Saf. Eng.* **2018**, *35*, 410–414. (In Chinese)
28. Chu, H.B.; Wang, J.X.; Yang, X.L.; Liang, W.M. Action mechanism of methane gas in the process of coal blasting damage and fracture. *J. Min. Saf. Eng.* **2014**, *31*, 494–498. (In Chinese)
29. Yang, X.L.; Wang, S.R. Meso-Mechanism of damage and fracture on rock blasting. *Explos. Shock. Waves* **2000**, *20*, 247252. (In Chinese)

energies

MDPI

Article

Study on the Influence and Control of Stress Direction Deflection and Partial-Stress Boosting of Main Roadways Surrounding Rock and under the Influence of Multi-Seam Mining

Dongdong Chen, Fangfang Guo, Zijian Li, Xiang Ma, Shengrong Xie *, Yiyi Wu and Zhiqiang Wang

School of Energy and Mining Engineering, China University of Mining & Technology-Beijing, Beijing 100083, China
* Correspondence: xsrxcq@cumtb.edu.cn

Abstract: A large coal pillar (usually more than 90 m) is generally left in place to ensure the stability of main roadway groups, due to its long service lifespan, which commonly also causes a significant loss of coal resources. The design of the width of the protective coal pillar and the control system of the surrounding rock are directly determined by the characteristics of the stress field and the damage mechanism under the influence of the mining activities. However, there are few studies on the effects of the partial-stress boosting and the direction deflection of the stress field on the failure evolution of the surrounding rock (especially in multi-seam mining). In this paper, theoretical analysis and numerical simulation are used to investigate the direction evolution of the maximum principal stress in front of the working face with malposition distances between the upper and lower working faces during the influence of double coal seams mining. Furthermore, a large-scale numerical model is used to study the deviatoric stress evolution of the surrounding rock and the propagation process of the plastic zone in the main roadway group with different widths of protective coal pillars. Then, an asymmetric cooperative anchoring classification method is proposed to strengthen the roadway support, depending on the critical area of the deviatoric stress in the roadway surrounding rock. The peak zone deflection of the deviatoric stress determines the evolution direction of the plastic area, and the peak value of the deviatoric stress presents a typical asymmetric stress boosting on both sides of the roadway. These findings are validated by the on-site ground pressure monitoring results and the practical failure modes of the surrounding rock.

Keywords: main roadway group; mining-induced stress; surrounding rock stability; ground control; deflection of principal stress

Citation: Chen, D.; Guo, F.; Li, Z.; Ma, X.; Xie, S.; Wu, Y.; Wang, Z. Study on the Influence and Control of Stress Direction Deflection and Partial-Stress Boosting of Main Roadways Surrounding Rock and under the Influence of Multi-Seam Mining. *Energies* **2022**, *15*, 8257. https://doi.org/10.3390/en15218257

Academic Editor: Nikolaos Koukouzas

Received: 10 October 2022
Accepted: 2 November 2022
Published: 4 November 2022

1. Introduction

Energy and mineral resources serve as the economic development drivers for countries around the world [1]. China's proven coal resource reserves rank second in the world, and the total consumption of coal in China's primary energy consumption in 2020 still accounts for more than 50% of the proportion [2,3]. Ensuring the safe production of coal means a stable economic development. Long-wall underground mining arose in Europe and is widely used in China because of its high productivity. About 85 percent of China's coal production comes from underground, and long-wall mining is used for most of it, about 20 percent of the total; and about 5 percent of India's production comes from underground [4–7]. Although long-wall mining has a high productivity, because of the characteristics of roadway layout, large coal pillars should be set in terminal stage of long-wall face to ensure the safety of the roadway group [8]. Due to the long service life of the main roadway group, the mine tends to set up excess coal pillars to maintain the main roadway safety, so the study on the width of the roadway protective coal pillar is important.

The main roadways are the lifeblood of mine transportation, ventilation, pedestrians, etc., and they generally need to serve the whole production process of the mine. Therefore, most of the main roadways are built on the roof or floor of the coal seam in underground mining. Furthermore, the coal seam roadways are easy to excavate, and coal resources can be recovered as soon as possible, reducing the financial pressure on mine production. In recent years, the main roadways are increasingly inclined to be placed in coal seams in some mine areas with relatively good geological conditions, such as in Yulin and Erdos, China.

For the roadway problems of a coal mine, scholars around the world have made many efforts to study the properties of the roadway surrounding rock, the state of the deposit, the influence of mining stress, the mechanism of the surrounding rock deformation, and the role of support, etc., and have accumulated rich theoretical and practical engineering experience. Gao et al. [9–11] conducted a series of numerical simulations to study the roadway's squeezing failure and shear failure, using the UDEC Trigon method and analyzed the effects of rock bolts on the roadway support. Wang et al. [12] tested the properties of surrounding rock considering the influence of the lower coal seam on the stress state by a numerical simulation. The creep characteristics of the surrounding rock of the mining roadway and its effects on the deformation were systematically performed. Pan et al. [13] focused on the deformation and failure process and the mechanism of the surrounding rock induced by the stress adjustment in a deeply arched roadway. A self-developed large-scale simulation test was carried out. Mohammadi et al. [14] developed a geometric computational model to calculate the extension of the excavation-damaged zone above the gate roadways, induced by longwall mining and proposed a new mathematical formulation to determine the face influence coefficient.

Huang et al. [15] clarified the connotation of roadway mining and large deformation under the consideration of the deviatoric stress. They proposed a theoretical framework of a large deformation of surrounding rock rheology and structural instability in deep mining roadways. Y et al. [16] investigated the problems of roadway stability with repeated mining-induced disturbance and focused on the direction and emphasis of the stability control of roadway surrounding rock. Shan et al. [17] summarized the current research status of coal tunnel support technology at home and abroad, in recent years.

Moreover, they proposed that roadway support adopted a more diversified integrated active support technology and showed the trend of gradually moving into intelligent support. Some researchers [18–24] concentrated on the significant deformation of the surrounding rock in deep roadways and targeted control technologies were proposed to solve this problem. These papers are dedicated to discovering the failure mechanism of the roadway or efficient roadway support technology to prevent the instability of the roadway. The Coal pillar width does contribute an essential part to the stability of the roadway. The non-pillar technologies (gob-side entry driving or gob-side entry retaining) have been widely used in China, due to the high coal resource recovery rate [25–32]. Currently, for a coal seam main roadway group in China, the protective coal pillar is usually larger than 90 m for the stability of the roadway, which causes a tremendous waste of coal resources. Some experts have also studied the width of coal pillars to protect the main roadways. For example, some scholars [33,34] studied the stability control of the main deep roadway groups, considering the characteristic rheology coupled with multiple disturbances of mining activities of the adjacent working faces. It determined that it was rational to leave a protective coal pillar for main roadway groups with a width of 100 m.

However, at present, for the close multi-seam mining, when the width of the end-mining coal pillar of each coal seam is different, the stress superposition and direction deflection law in the area in front of the working face, caused by the influence of mining and the evolution law of the damage to the main roadway surrounding rock, is rarely studied. Therefore, it is impossible to determine the location of the high deviatoric stress area and the critical control orientation of the main roadway surrounding rock. This also leads to the roadway surrounding rock support design that is not a targeted basis.

The study background of this paper is to examine the specific engineering geological conditions of close two-seam coal mining and the different widths of end-mining coal pillars. On this basis, the following studies are conducted: the law of the principal stress direction deflection and plastic damage evolution in the main roadway surrounding rock, the stress superposition and deflection of the peak deviatoric stress area. Through the study, the location of critical high deviatoric stress zones and support orientation of the roadway surrounding rock are obtained, and the asymmetric reinforcement support technology of the channel steel-anchor cable truss with a bi-directional reinforcement function is proposed. The experimental results show that the technique is effective.

2. Project Overview

2.1. Geological Conditions

The No.3 and No.4 coal seams are mainly exploited in the mine, and the No.4 coal seam is located 25 m above the No.3 coal seam. The relationship between the two coal seams is shown in Figure 1. Both are near horizontal coal seams adopting the comprehensive mechanized top coal caving mining method. The thickness of the No.4 coal seam is about 7.5 m, which has been recovered, and a 90 m coal pillar was left to ensure the stability of the main roadway group. The geological histogram of the long-wall working face (LWF) 8216 is shown in Figure 1. The long-wall working face (LWF) 8216 in the No.3 coal seam under the residual coal pillar is the main mining face, with an average coal seam thickness of about 5.5 m and a burial depth of about 400 m. The spacing between the two adjacent roadways is 30 m, and there is slight deformation in the surrounding rock of the roadway group during the upper coal seam recovery. Therefore, during the retreating of the lower coal seam, the mine intended to further reduce the width of the protective coal pillar for the roadway group to maximize the recovery of coal resources. The location of the project is shown in Figure 2.

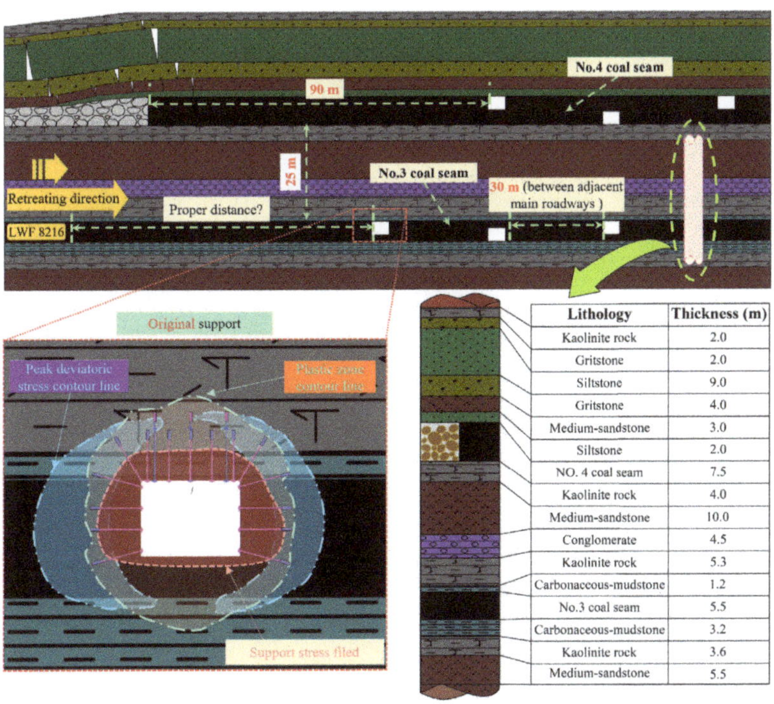

Figure 1. Main roadway group layout of the multiple coal seam and generalized stratigraphic column in the test site.

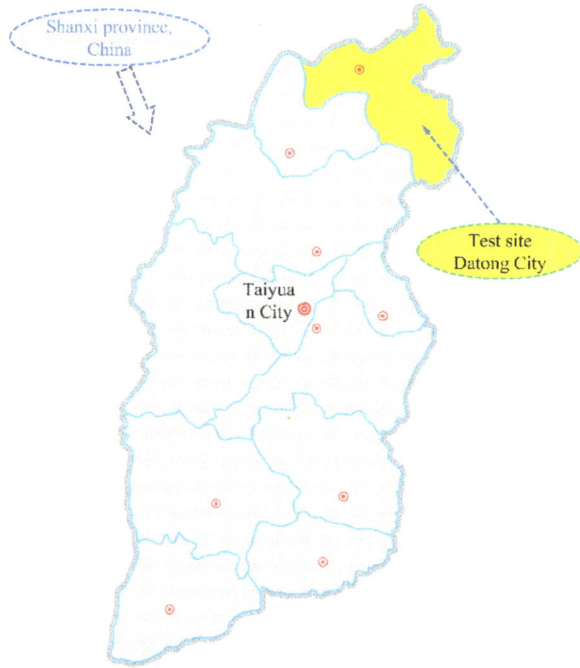

Figure 2. Location of the Project.

2.2. Description of the Mining Conditions

The No.4 coal seam above the longwall working face 8216 of the No.3 coal seam in the mine has been mined out, and the width of the protective coal pillar of the main roadway group is 90 m. During the mining process of the No.3 coal seam, it usually chooses a giant protective coal pillar (90 m~140 m) to avoid the damaging effect with multi-seam mining stress disturbance to the main roadway. The main roadway is supported by a symmetrical anchor bolting and cable combination using a large coal pillar width. The anchor cables are Φ17.8 × 7000 mm with an inter-row distance of 1500 × 3000 mm and pallet size of 300 × 300 × 16 mm (length × width × height); the anchor bolts are left hand non-longitudinal rebar threaded steel anchor bolts with Φ20 × 2400 mm and inter-row distance of 900 × 1000 mm, and the pallet is a dished pallet with 150 × 150 × 10 mm (length × width × height). The roof, floor, and two ribs of the roadway use a shotcrete support. The roof and two ribs' concrete thickness is 100 mm, and the concrete grade is C20. The support parameters of the main return-air roadway and main haulage roadway are shown in Figure 3.

Although the giant protective coal pillar prevented the main roadway group from destruction and instability, it also caused the loss of coal resources. In addition, the asymmetrical damage to the main roadway can occur under symmetrical anchor rope support conditions after the suspension of mining at specific working faces of the No.3 coal seam. In the ribs of the main return-air roadway and the main haulage roadway, the support structure showed apparent damage, as shown in Figure 4. This is inconsistent with our traditional understanding of the location of the roadway damage. The roadway failure location under the influence of mining stress should first appear in the area near the working face. Therefore, in the study of multi-seam mining conditions, the direction of the evolution of the roadway damage is conducive to the better design of targeted support programs to improve support efficiency.

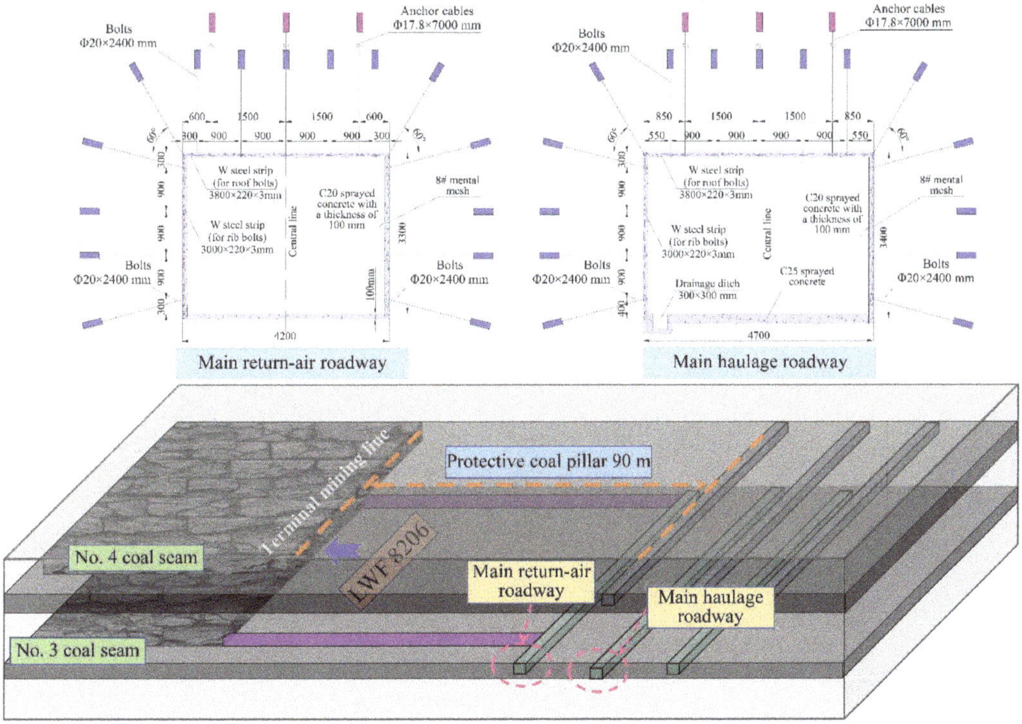

Figure 3. Support parameters of the main roadway group.

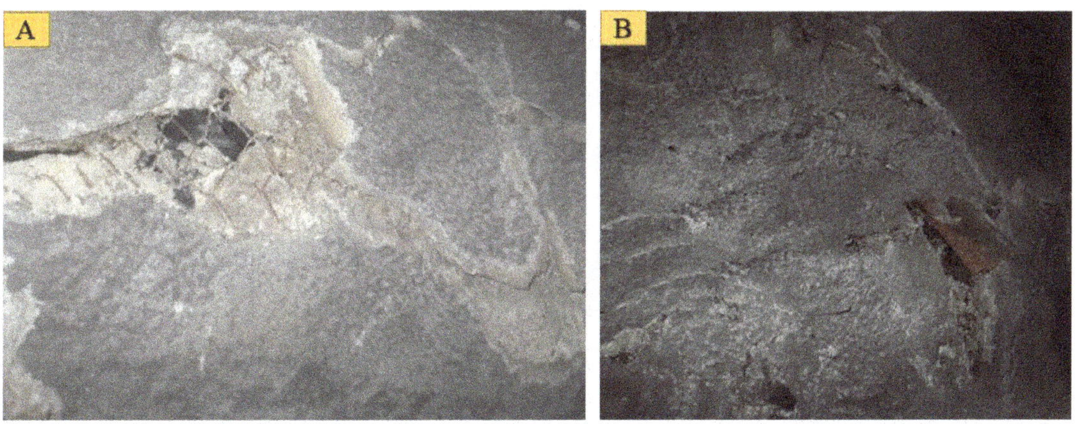

Figure 4. Asymmetric damage of the main roadway after stopping mining. (**A**) main return-air roadway, and (**B**) main haulage roadway.

3. Theoretical Analysis of the Direction Deflection of the Stress and Partial-Stress Boosting in the Surrounding Rock

3.1. Solution for the Direction of Stress Deflection and Partial-Stress Boosting in the Mining Superposition Conditions

The problems of stress distribution and plastic damage depth of a coal mass have been of great concern. Especially the superimposed loading law of the stress field with multi-seam mining, which is an essential guideline for engineering practice.

The traditional view mainly focuses on the change of lead vertical pressure (vertical direction). However, vertical stress only represents one of the nine stress components. The direction is all vertical, so the simple vertical stress cannot scientifically and comprehensively characterize the change of stress in the over-front area, caused by the mining of the working face, and it cannot effectively propose the damage characteristics of the advanced abutment pressure on the main roadway and the critical direction of control.

Deviatoric stress synthesizes three principal stresses, making the analysis more comprehensive and scientific than the single vertical stress index. According to the elastic-plasticity theory [35], the deviatoric stress plays a dominant role in the plastic deformation of the surrounding rock, so the analysis of the mining superposition main roadway surrounding stress increase law by the deviatoric stress index, can more comprehensively reflect the essence of the surrounding rock deformation damage development.

The S_1 is the principal deviatoric stress, which plays a dominant role in the stress tensor, and the commonly referred deviatoric stress refers to the maximum principle of the deviatoric stress, which is also the analytical index of the deviatoric stress in this section, and its formula is as follows.

$$S_1 = \sigma_1 - \frac{\sigma_1 + \sigma_2 + \sigma_3}{3} \tag{1}$$

In the three-dimensional stress field, the magnitude of the principal stress and its direction can be solved by the magnitude of nine (six independent) stress components (σ_{xx}, σ_{yy}, σ_{zz}, τ_{xz}, τ_{xy}, τ_{yz}, τ_{zx}, τ_{zy}, τ_{yx}). Take any unitary body for an oblique section ABC, as in Figure 5, and let the outer normal of the plane ABC be N, whose direction cosine is

$$cos(N,x) = l,\ cos(N,y) = m,\ cos(N,z) = n \tag{2}$$

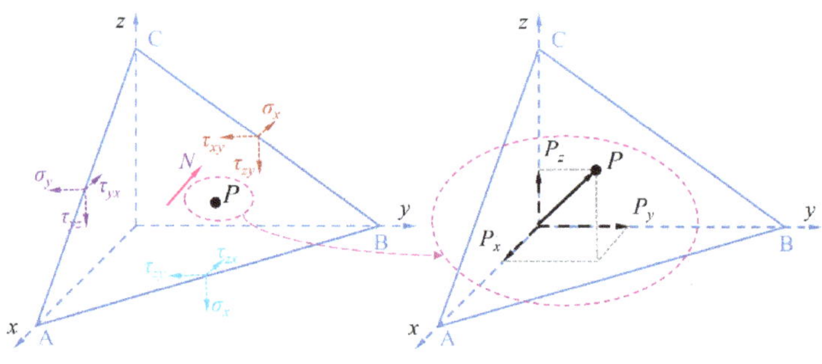

Figure 5. Stress components of the oblique section of the element body.

The full stress P of the triangle ABC can be decomposed into p_x, p_y, p_z, and the following formula is satisfied:

$$\left.\begin{aligned} p_x &= l\sigma_x + m\tau_{yx} + n\tau_{zx} \\ p_y &= m\sigma_y + n\tau_{zy} + l\tau_{xy} \\ p_z &= n\sigma_z + l\tau_{xz} + m\tau_{yz} \end{aligned}\right\} \tag{3}$$

Let the positive stress on triangle ABC be σ_N. Then from the projection, we have

$$\sigma_N = lp_x + mp_y + np_z \tag{4}$$

It follows that

$$\sigma_N = l^2\sigma_x + m^2\sigma_y + n^2\sigma_z + 2mn\tau_z + 2nl\tau_{zx} + 2lm\tau_{xy} \tag{5}$$

$$p^2 = \sigma_N^2 + \tau_N^2 = p_x^2 + p_y^2 + p_z^2 \tag{6}$$

Suppose there is a principal stress plane at P, and the projection of the full stress (principal stress σ) on this plane on the coordinate axis is:

$$p_x = l\sigma,\ p_y = m\sigma,\ p_z = n\sigma \tag{7}$$

Satisfies Equation (9)

$$l^2 + m^2 + n^2 = 1 \tag{8}$$

Equation (9) can thus be obtained

$$\left.\begin{aligned} (\sigma_x - \sigma)l + \tau_{yx}m + \tau_{zx}n = 0 \\ \tau_{xy}l + (\sigma_y - \sigma)m + \tau_{zy}n = 0 \\ \tau_{xz}l + \tau_{yz}m + (\sigma_z - \sigma)n = 0 \end{aligned}\right\} \tag{9}$$

Following a simplification, Equation (10) can be derived

$$\begin{cases} \tau_{yx}\frac{m_1}{l_1} + \tau_{zx}\frac{n_1}{l_1} + (\sigma_x - \sigma_1) = 0 \\ (\sigma_y - \sigma_1)\frac{m_1}{l_1} + \tau_{zy}\frac{n_1}{l_1} + \tau_{xy} = 0 \end{cases} \tag{10}$$

The two equations solved for the two unknowns (m_1/l_1) and (n_1/l_1) so that l_1, as well as m_1 and n_1, can be obtained by bringing in the following equation:

$$l_1 = \frac{1}{\sqrt{1 + \left(\frac{m_1}{l_1}\right)^2 + \left(\frac{n_1}{l_1}\right)^2}} \tag{11}$$

l_1, m_1, and n_1 are the cosine of the angle between the principal stress of the principal plane and the three axes. Therefore, based on the nine stress components at any point, the three principal stresses and the direction of the principal stresses can be solved.

3.2. Solution for the Degree of the Stress Direction Deflection and the Partial-Stress Boosting and the Damage Analysis with the Disturbance of Multi-Seam Mining

As shown in Figure 6, the stress components and the changes of the principal stress during the mining of each coal seam can be extracted through a numerical calculation. Then, the changes in the magnitude and direction of the principal stress before and after being affected by the mining superposition at any location, can be obtained. At the same time, as the magnitude of the principal stress and its direction change, the peak area of the deviatoric stress of the roadway surrounding rock is bound to change, which will inevitably lead to asymmetric damage to the surrounding rock. That is, through the size of the principal stress field of the roadway surrounding rock and its direction, changes to determine the critical damage area of the roadway, to guide the determination of the vital reinforcement direction of the roadway.

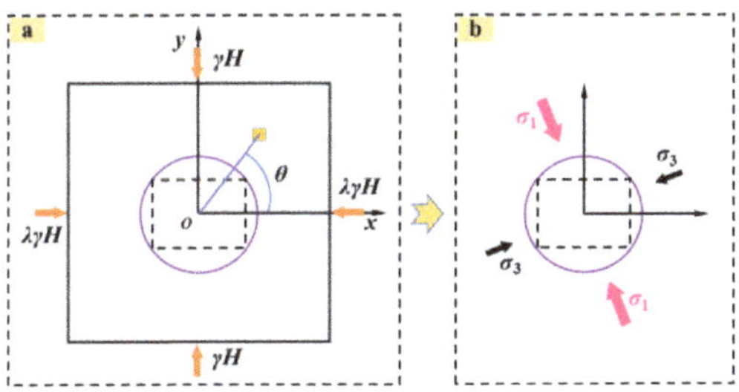

Figure 6. Expression of the principal stresses for the stress components. (**a**) Arbitrary stress field, (**b**) The direction deflection of the principal stress field and the partial-stress boosting.

According to the principal stress field, to solve the damage range of the roadway, using the stress distribution formula around the hole in the elastoplastic mechanics, the stress expression at any point of the surrounding rock of a circular roadway in polar coordinates [36] can be found as follows.

$$\begin{cases} \sigma_r = \left[3(\lambda-1)\cos 2\theta \frac{a^4}{R_\theta^4} + (1+\lambda)\left(1-\frac{a^2}{R_\theta^2}\right) + (\lambda-1)\cos 2\theta - 4(\lambda-1)\cos 2\theta \frac{a^2}{R_\theta^2}\right]\frac{\gamma H}{2} \\ \sigma_\theta = \left[(1+\lambda)\left(1+\frac{a^2}{R_\theta^2}\right) - \cos 2\theta(\lambda-1) - 3\cos 2\theta(\lambda-1)\frac{a^4}{R_\theta^4}\right]\frac{\gamma H}{2} \\ \tau_{r\theta} = \left[\sin 2\theta(\lambda-1) + 2\sin 2\theta(\lambda-1)\frac{a^2}{R_\theta^2} - 3\sin 2\theta(\lambda-1)\frac{a^4}{R_\theta^4}\right]\frac{\gamma H}{2} \end{cases} \tag{12}$$

where: σ_r is the radial stress at any point; σ_θ is the circumferential stress at any point; $\tau_{r\theta}$ is the shear stress at any point; γ is the weight density of rock; H is the burial depth of the roadway; λ is the lateral pressure coefficient; a is the radius of the circular roadway; the polar coordinates of any point are set as (θ, R_θ). In the polar coordinate system, the formula for calculating the principal stress determined by the stress component [36] is:

$$\begin{cases} \sigma_1 = \frac{\sigma_r+\sigma_\theta}{2} + \frac{1}{2}\sqrt{(\sigma_r-\sigma_\theta)^2 + 4\tau_{r\theta}^2} \\ \sigma_3 = \frac{\sigma_r+\sigma_\theta}{2} - \frac{1}{2}\sqrt{(\sigma_r-\sigma_\theta)^2 + 4\tau_{r\theta}^2} \end{cases} \tag{13}$$

The stress state of a point can be decomposed into spherical stress and deviatoric stress, where the deviatoric stress is the dominant factor leading to the plastic deformation of the surrounding rock, so use the deviatoric stress for the theoretical analysis.

When conducting the study of the stress in the roadway enclosure, the problem can be simplified to a plane problem, which results in the principal deviatoric stress calculation formula:

$$S_1 = \frac{\gamma H}{6}(1+\lambda) + \frac{\gamma H a^2(1-\lambda)\cos 2\theta}{3R_\theta^2} + \frac{\gamma H}{2}\sqrt{\left[(1+\lambda)(-\frac{a^2}{R_\theta^2}) + (\lambda-1)(1+3\frac{a^4}{R_\theta^4} - 2\frac{a^2}{R_\theta^2})\cos 2\theta\right]^2 + \left[(\lambda-1)\left(1-3\frac{a^4}{R_\theta^4}+2\frac{a^2}{R_\theta^2}\right)\sin 2\theta} \tag{14}$$

The surrounding rock plasticity criterion satisfies the following relation [37,38].

$$\frac{{\sigma_c}^2(\sigma_1+\sigma_3)}{2(\sigma_c+2\sigma_t)} + \frac{\sigma_t{\sigma_c}^2}{\sigma_c+2\sigma_t} - \frac{{\sigma_c}^4}{8(\sigma_c+2\sigma_t)^2} - \frac{(\sigma_1-\sigma_3)^2}{2} = 0 \tag{15}$$

By combining the deviatoric stress formula and its plasticity criterion, we can analyze the location of the high deviatoric stress zone in the roadway surrounding rock. We can also make it possible to quantify the deflection of the high deviatoric stress zone when the principal stress direction changes. Combined with the numerical simulation tests, the evolution of the peak stress zone of the surrounding rock in the roadway, from no stress deflection and partial-stress boosting to stress deflection and partial-stress boosting conditions, is shown in Figure 7.

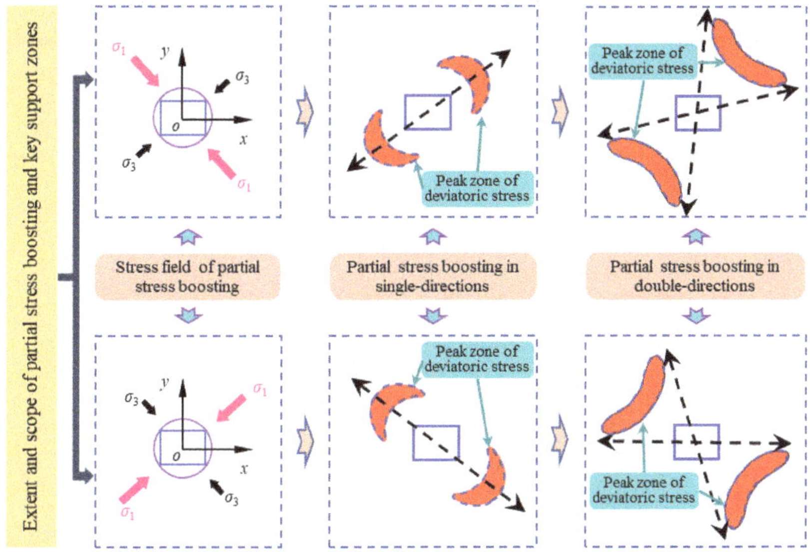

Figure 7. Evolutionary pattern of surrounding rock damage under different stress direction deflection and partial-stress boosting conditions.

With different degrees and directions of stress deflection and partial-stress boosting, there exist different deviatoric stress critical zones in the roadway surrounding rock, and this zone highlights the roadway surrounding rock's key damage development area and direction: When the degree of stress deflection and partial-stress boosting is small, the roadway surrounding the rock's deviatoric stress critical zone mainly develops symmetrically in two directions of the roadway; when the degree continues to increase, the deviatoric stress critical zone is symmetrically distributed on both sides of the roadway surrounding rock, and the damaged area tends to develop from four directions to the deep part of the surrounding rock. Therefore, under stress deflection and partial-stress boosting conditions, the surrounding rock is not gradually and uniformly damaged to the deep part of the surrounding rock, but it is a non-uniform process. The different stress direction deflection and partial-stress boosting conditions lead to two directions of damage development or four directions of the damage development pattern of the surrounding rock, which has an essential guidance significance for the design of the surrounding rock control scheme.

4. Numerical Simulation

The influence with multi-seam mining will cause a stress increase effect on the main roadway. Under different principal stress field conditions, the roadway's stress peak area and damage development area will be deflected, forming the phenomenon of stress deflection and partial-stress boosting. There is an essential difference between the development of the surrounding rock deformation and the critical area of its control. The numerical simulation can more intuitively study the law of the deflected loading of the main roadway surrounding rock, caused by mining multiple coal seams. This is essential for analyzing the plasticization law of the main roadway surrounding rock and determining the critical reinforcement location of the main roadway surrounding rock.

The model selects the LWF 8216 as the test working face. The *X*-axis length is 400 m in the length direction, the *Y*-axis length is 100 m and parallel to the main roadway group, and the *Z*-axis length is 90 m in the vertical direction, as shown in Figure 8. The boundary displacement of the model is constrained in the horizontal and bottom. The upper of the model is subjected to a stress of 7.75 MPa, equivalent to the weight of the overburdened rock, and the lateral stress coefficient is 1.2. The model is calculated using the Mohr–

Coulomb model. In contrast, the double-yield model is used for the job. The mechanical parameters of the coal and rock mass were very important to the accuracy of the numerical simulation results. Cai et al. [39] suggested that the tensile strength, cohesion and elastic modulus of the coal and rock mass could be estimated in the range of 0.10–0.25 times of the experimental results in the laboratory. Mohammad et al. [40] suggested that the average uniaxial compressive strength of the numerical model should be 0.284 of the laboratory strength and the average stiffness should be 0.469 of the laboratory stiffness. Combined with the generalized Hoke–Brown failure criterion [41–43], the simplified physical parameters of the coal seam are shown in Table 1.

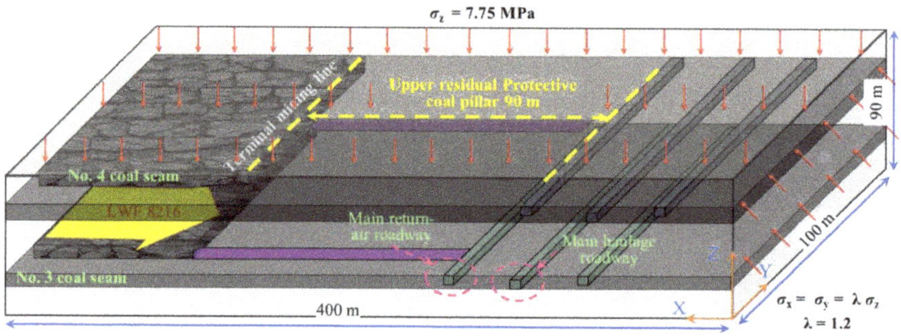

Figure 8. Numerical model overview.

Table 1. Properties of the rock mass in LWF 8216.

Lithology	Density ($kg \cdot m^{-3}$)	Elastic Modulus (GPa)	Friction (°)	Cohesion (MPa)	Tensile Strength (MPa)
Gritstone	2650	17.43	32	3.16	2.42
Siltstone	2602	14.59	35	3.07	2.17
No.4 coal seam	1400	3.79	18	1.49	1.21
Conglomerate	2660	18.08	34	3.5	3.13
No.3 coal seam	1400	3.80	18	1.49	1.21
Carbonaceous-mudstone	2200	14.77	29	2.94	2.22
Kaolinite rock	2570	17.00	38	5.18	2.78
Medium-sandstone	2557	12.69	38	6.26	3.63

The numerical model is used to study two aspects: (1) The change of the abutment pressure in front of the working face when the upper and lower coal seams are in different positions (outer fault, overlap, internal fault). (2) To investigate the stress distribution and the mechanism of the surrounding rock damage in the surrounding rock of the roadway when the width of the coal pillar is different in the working face of the lower coal seam, when the protective pillar of the upper coal seam is 90 m.

4.1. Stress Field Advance of the Working Face, Caused by Multi-Seam Mining

4.1.1. Deflection of the Principal Stress

Multi seam mining generally adopts a downward mining practice, so the relative position relationship between the underlying coal seam and the overlying coal seam mainly includes three types. Namely: external staggering (the end-mining line of the lower coal seam is under the gob of the upper coal seam), overlapping (the end-mining line of the upper and lower coal seams overlap), and internal staggering (the end-mining line of the lower coal seam is under the entity coal of the upper coal seam). Suppose the width of the final end-mining coal pillar of the lower coal seam is smaller than that of the upper coal seam. In that case, the working face of the lower coal seam will experience the whole process of the above three types of positional relationships.

The stress component of each point before each working face, is extracted through a numerical calculation. The maximum principal stress direction of the leading working face under three positional relations, is calculated using the stress component. It is specified that the angle of the maximum principal stress line rotating counterclockwise to the vertical direction line is "+" and vice versa is "−". The absolute value of the principal stress direction angle is less than 90°. When the upper and lower coal seams are at different off-set distances, the deflection characteristics of the principal stress of the element in the center of the lower coal seam with variable distances from the working face are shown in Figure 9.

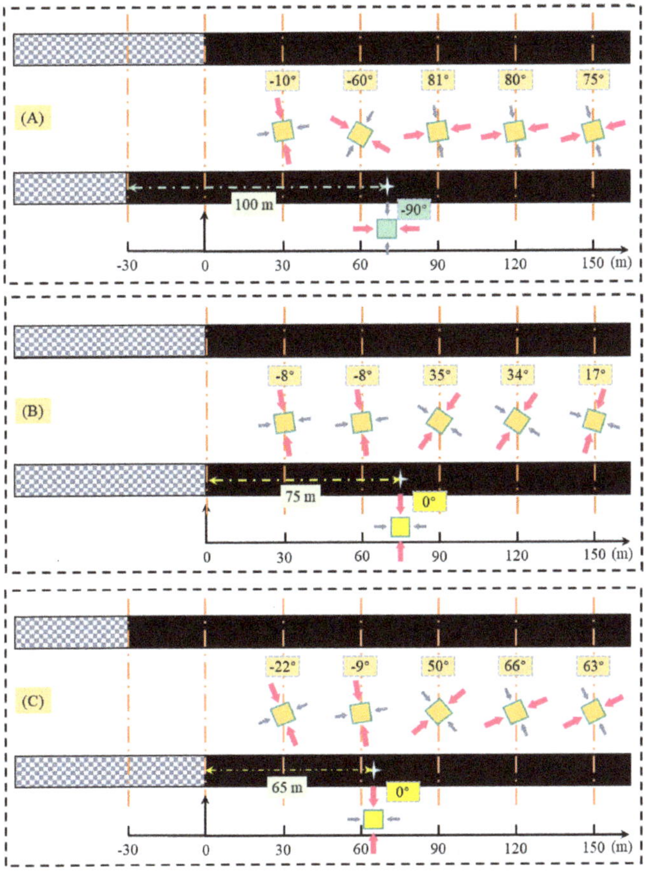

Figure 9. Deflection law of the principal stress in front of the different layout modes of the upper and lower coal seams. (**A**) External-offset of 30 m. (**B**) Overlapping arrangement. (**C**) Internal-offset of 30 m.

The primary conclusions are as follows.

(1) The external staggered arrangement differs from the overlapping and internal staggered arrangement. The evolution trajectory of the maximum principal stress direction is approximately vertical → horizontal → deflection to the vertical direction in the external staggered arrangement; while that in the overlapping and internal staggered arrangement is approximately vertical → vertical → inclined to the two ribs.

(2) The distance between the turning point of the maximum main stress direction (deflection angle of 0° or 90°) and the mining area, gradually decreases as the arrangement of the upper and lower coal seams changes from outer staggered type → overlap-

ping type → internal staggered type. It is reasonable to adopt symmetrical support measures for the roadway at the turning point of the principal stress.

(3) Under the condition of multi-seam mining, the maximum principal stress direction within 60 m in front of the lower coal seam is approximately vertical and deflected to the mining void. If the roadway is arranged here, the roadway surrounding rock will be damaged first along the direction of the minimum principal stress, namely the entity coal top angle of the roadway.

4.1.2. Direction Deflection of the Principal Stress

Under the influence of the repeated mining of multiple coal seams, the bearing stress in front of the workings will be subjected to the cyclic loading-unloading effect of mining stress. The geological rock body shows a typical non-homogeneity in the vertical stress direction, and the nature of the different rock seams varies. Therefore, the different relative positions of the upper and lower coal seams will significantly affect the peak and distribution characteristics of the stress field in front of the workings. The vertical stress of the model is selected as the index, and the distribution of the bearing stress field of the working face of the lower coal seam, when the upper and lower coal seams are at different staggered distances, is shown in Figure 10.

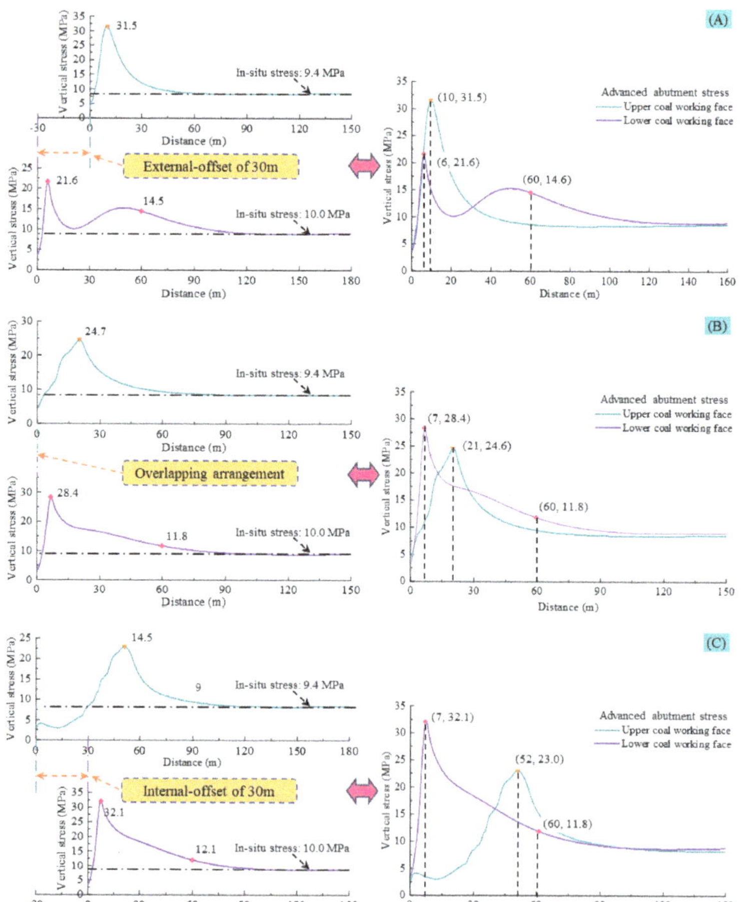

Figure 10. Abutment stress field distribution in the lower coal seam with different offset distances. (**A**) External-offset of 30 m. (**B**) Overlapping arrangement. (**C**) Internal-offset of 30 m.

As shown in Figure 10:

(1) When the working face is staggered outward, the upper and lower coal seam abutment stress shows a trend of increasing to the peak first, then gradually stabilizing. The peak stress ratio between the lower and upper coal seam is $R_s = 0.69$, and the depth ratio is $R_d = 0.6$. When the working face of the lower coal seam is ahead by 60 m, the loading factor of the peak stress, compared with the original rock stress is $L = 1.46$. Furthermore, the abutment stress curve of the working face of the lower coal seam shows double peak characteristics, due to the position of the working face of the upper coal seam.

(2) When the working face overlaps, the abutment stress trend in the upper and lower coal seams increases first and then decreases. The ratio of the peak stress between the lower coal seam and upper coal seam is $R_s = 1.15$, and the depth ratio is $R_d = 0.33$; when the working face of the lower coal seam is ahead by 60 m, the stress peak, compared to the original rock stress has a loading factor of $L = 1.18$.

(3) When the working face is staggered internally, the trend of the abutment stress in the upper and lower coal seams still increases first and then decreases. The ratio of the peak stress between the lower and upper coal seams is $R_s = 1.40$, and the depth ratio is $R_d = 0.13$; when the working face of the lower coal seam is ahead by 60 m, the loading factor of the peak stress, compared with the original rock stress is $L = 1.18$.

(4) In the process of the upper and lower working face from the outer fault $\rightarrow$ overlap $\rightarrow$ inner fault arrangement, the abutment stress peak of the lower coal seam working face increases continuously, and the value is gradually larger than that of the upper coal seam. The peak stress in the upper coal seam is the opposite of the peak stress in the upper coal seam. The peak stress depth of the lower coal seam does not change much and remains at a 6 to 7 m depth. In contrast, the peak stress depth in the upper coal seam is significantly affected by the working face layout, and the depth increases from 10 m to 52 m, an increase of 520%.

4.2. Roadway Group Stress Field Characteristics under Various Protective Coal Pillars

The relative positional relationship of the upper and lower coal seams working face significantly affects the abutment stress distribution condition in front of the working face. Due to the superimposed influence of the mining stress in multiple coal seams, the stress loading law of the main roadway group in the process of the upper and lower coal seams working face going from an outer staggered arrangement $\rightarrow$ overlapping arrangement $\rightarrow$ inner staggered arrangement, is significant for the selection of the width of the protection coal pillar. This section explores the stress evolution process of this seam's main roadway group affected by mining when the end-mining coal pillar of the upper coal seam is 90 m, with the gradual mining of the lower coal seam, as shown in Figure 11.

As shown in Figure 11:

(1) When the end-mining coal pillar of the upper coal seam is 90 m, the protective coal pillar's width of the lower coal seam main roadway reduces from ∞ m (which means that the lower coal seam is not mined) to 30 m, the abutment stress' peak of the lower coal seam main roadway group is gradually increasing. The maximum loading increase factor is $L = 2.03$. Furthermore, when the width of the coal pillar is less than 60 m (i.e., the overlapping arrangement of the upper and lower coal seams), the stress of the main roadway surrounding rock increases sharply.

(2) The surrounding rock of the main roadway group shows a typical asymmetric effect of the direction deflection of the stress and partial-stress boosting. The peak stress of the mining side is more extensive than its entity coal side of the main return-air roadway and main haulage roadway, and the degree of stress direction deflection and partial-stress boosting increases with the shortening of the end-mining coal pillar width. The direction deflection of stress and the partial-stress boosting phenomenon of the main roadway group surrounding rock shows that the stress distribution of the

surrounding rock is asymmetric, so it is necessary to use targeted support solutions in some places where the surrounding rock is weak.

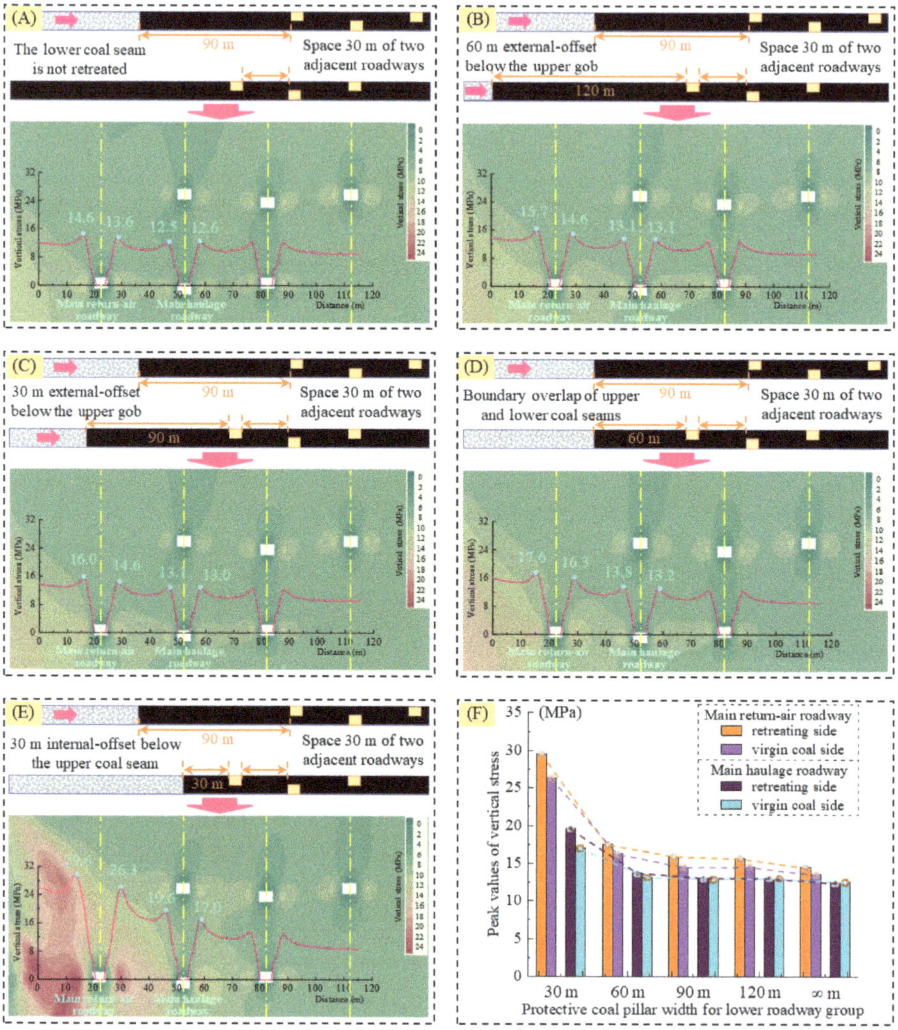

Figure 11. Stress evolution process of the main roadway group during the lower coal seam recovery. (**A–E**) with the protective coal pillar widths of ∞ m, 120 m, 90 m, 60 m, and 30 m, respectively, and (**F**) distribution of peak stress value in both ribs.

4.3. Law of Stress Direction Deflection and Partial-Stress Boosting of the Deviatoric Stress Field of the Surrounding Rock with Different End-Mining Coal Pillar Widths

Deviatoric stress is the stress that deviates from hydrostatic stress and causes deformation, reflecting the essence of plastic deformation at a point in the surrounding rock. The high deviatoric stress area in the surrounding rock indicates that the area is in a state of damage development. Mining activities, especially the disturbance of multi-seam mining, will promote the direction deflection of stress and partial-stress boosting effect of the main roadway surrounding rock, further leading to the formation and transfer of the high deviatoric stress zone in the surrounding rock. It is of great significance to investigate the development and evolution of the high deviatoric stress zone in the surrounding rock to develop the damage mechanism and key reinforcement technology of the roadway surrounding rock. This section explores the evolution of the peak deviatoric stress zone and plastic zone of this seam's main roadway group affected by mining when the end-mining coal pillar width of the upper coal seam is 90 m, with the gradual mining of the lower coal seam, as shown in Figure 12.

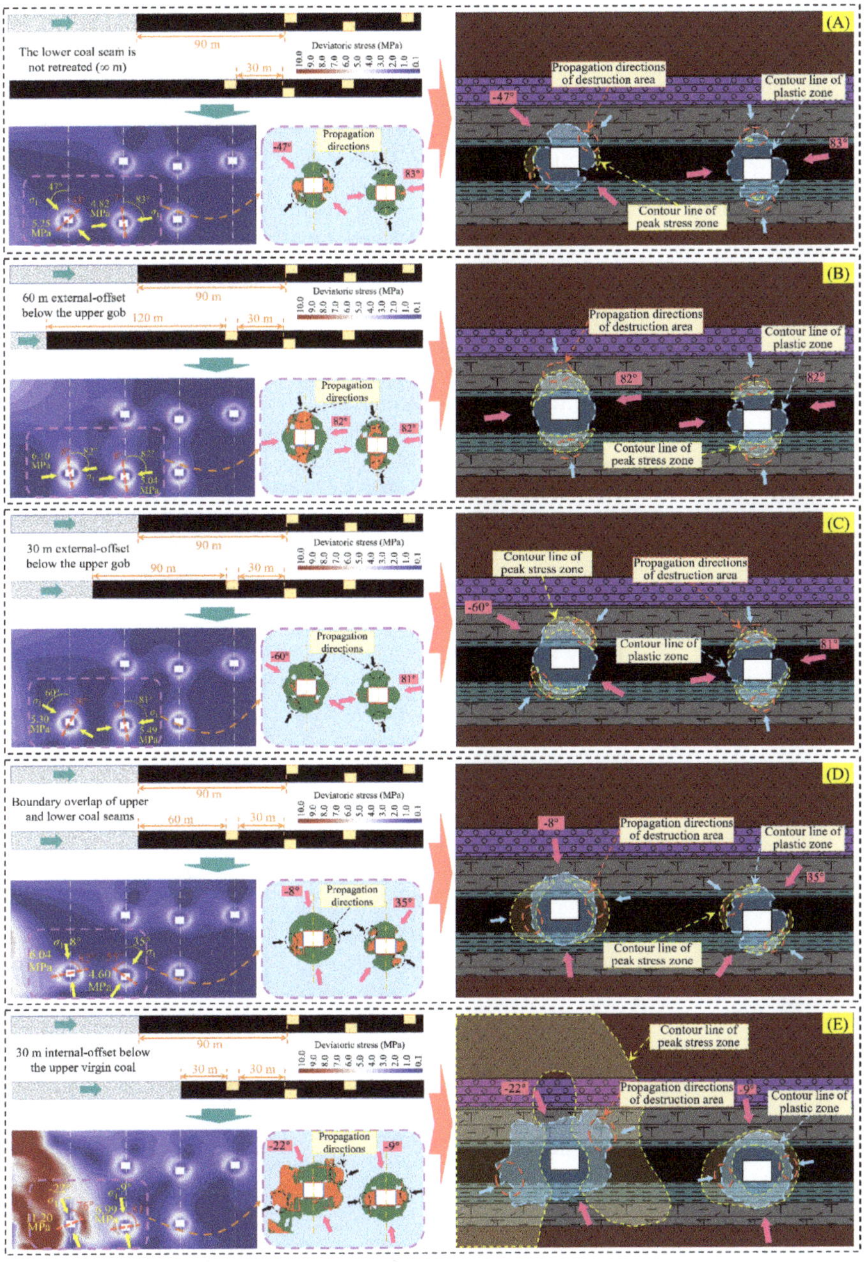

Figure 12. Development characteristics of the deviatoric stress peak zone and plastic zone. (**A–E**) with the protective coal pillar widths of ∞ m, 120 m, 90 m, 60 m, and 30 m, respectively.

As shown in Figures 12 and 13:

(1) The peak value of deviatoric stress shows an overall increasing trend with advancing the lower coal seam workings. When the width of the end-mining coal pillar is from ∞ m $\rightarrow$ 60 m $\rightarrow$ 30 m, the peak value of deviatoric stress in the main return-air roadway is from 5.25 MPa $\rightarrow$ 6.04 MPa $\rightarrow$ 11.20 MPa. The maximum stress loading factor is L = 2.1, and the peak value of the deviatoric stress in the main haulage roadway is from 4.82 MPa $\rightarrow$ 4.60 MPa $\rightarrow$ 6.99 MPa, and the maximum stress loading factor is L = 1.45. When the width of the lower end-mining coal pillar is more than 60 m, the peak value of the deviatoric stress in the main roadway group is more fluctuating. When the width of the lower end-mining coal pillar is bigger than 60 m, the peak value of the deviatoric stress in the main roadway group is more fluctuating, indicating that the peak distribution is not significantly affected by the larger width of the end-mining coal pillar.

(2) The peak deviatoric stress area gradually expands during the gradual decrease of the lower coal seam end-mining coal pillar width. As the influence of mining intensifies, the peak deviatoric stress area gradually surrounds the roadway surrounding rock and shows an apparent directional deflection. The location of the peak zone gradually transitions from the roof and floor to both ribs of the roadway. The plastic zone's contour line of the roadway passes through the core area of the peak of the deviatoric stress. Furthermore, the plastic zone evolution direction also has obvious directionality; with the minimum principal stress direction deflection, the plastic zone development position occurs with a corresponding shift.

(3) The maximum principal stress direction shows a significant deflection with the advancing lower coal seam workings (main return-air roadway). With the increase of the distance of the main roadway from the gob area, the direction of the maximum principal stress gradually transitions from the direction of the deflected gob area to the direction of the deflected entity coal. The direction of the minimum principal stress passes through the center of the peak deviatoric stress area of the two ribs of the roadway. The location of the peak of the deviatoric stress shows a significant asymmetry, and this type is to be avoided in the actual project.

(4) When the width of the end-mining coal pillar in the lower seam is 60 m, the direction of the maximum principal stress of the main return-air roadway is approximately perpendicular to the roof and floor of the roadway (deflected 8° toward the side of the gob area). That means the direction of the minimum principal stress in the roadway points to the entity coal top angle of the roadway, which also corresponds to the direction of the development of the plastic zone in the roadway. The failure results of the main roadway by a numerical simulation are consistent with the failure position of the main roadway under field working conditions, and the failure mechanism of the surrounding rock of the roadway is well revealed. This is conducive to applying an early warning and asymmetric control technology for the deformation orientation of the main roadway surrounding rock.

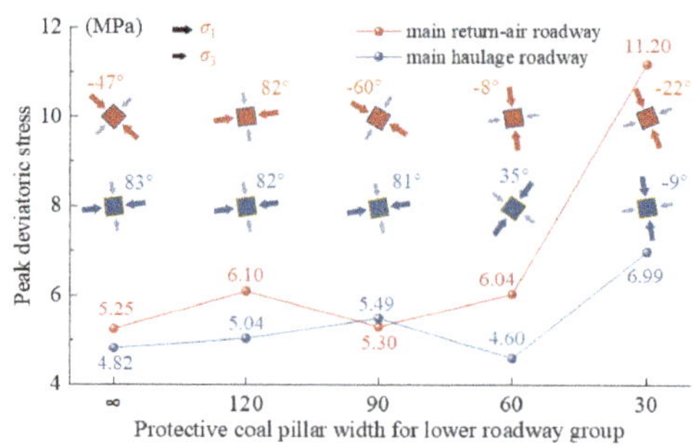

Figure 13. Stress distribution of the main roadway group in the lower coal seam.

10. Gao, F.; Stead, D.; Kang, H. Simulation of roof shear failure in coal mine roadways using an innovative UDEC Trigon approach. *Comput. Geotech.* **2014**, *61*, 33–41. [CrossRef]
11. Gao, F.Q.; Stead, D.; Kang, H.P. Numerical Simulation of Squeezing Failure in a Coal Mine Roadway due to Mining-Induced Stresses. *Rock Mech. Rock Eng.* **2015**, *48*, 1635–1645. [CrossRef]
12. Wang, X.F.; Wang, J.Y.; Chen, X.Y.; Chen, Z.C. A Roadway In Close Distance to coal seam in deep mine: Location selection and supporting practice based on creep characteristics of surrounding rocks. *Arch. Min. Sci.* **2021**, *66*, 407–419. [CrossRef]
13. Pan, S.Q.; Liu, S.T.; Cao, L.M.; Guo, J.Q.; Yuan, C. Deformation Failure and Support Test of Surrounding Rock in Deep Arched Roadway with Straight Wall. *Adv. Civ. Eng.* **2021**, *2021*, 11. [CrossRef]
14. Mohammadi, H.; Ebrahimi Farsangi, M.A.; Jalalifar, H.; Ahmadi, A.R. A geometric computational model for calculation of longwall face effect on gate roadways. *Rock Mech. Rock Eng.* **2016**, *49*, 303–314. [CrossRef]
15. Huang, B.; Zhang, N.; Jing, H.; Kan, J.; Meng, B.; Li, N.; Xie, W.; Jiao, J. Large deformation theory of rheology and structural instability of the surrounding rock in deep mining roadway. *J. China Coal Soc.* **2020**, *45*, 911–926.
16. Xiong, Y.; Kong, D.; Cheng, Z.; Wen, Z.; Ma, Z.; Wu, G.; Liu, Y. Instability Control of Roadway Surrounding Rock in Close-Distance Coal Seam Groups under Repeated Mining. *Energies* **2021**, *14*, 5193. [CrossRef]
17. Shan, R.; Peng, Y.; Kong, X.; Li, H. Research progress of coal roadway support technology at home and abroad. *Chin. J. Rock Mech. Eng.* **2019**, *38*, 2377–2403.
18. Chen, D.; Xie, S.; Wu, Y.; Guo, F.; He, F.; Liu, R. Reasonable location of stopping line in close-distance underlying coal seam and partition support of large cross-section roadway. *Int. J. Coal. Sci. Technol. 9*. [CrossRef]
19. Yuan, W.; Hong, K.; Liu, R.; Ji, L.; Meng, L. Numerical Simulation of Coupling Support for High-Stress Fractured Soft Rock Roadway in Deep Mine. *Adv. Civ. Eng.* **2022**, *16*, 1–10.
20. Yuan, Y.; Wang, W.J.; Yuan, C.; Yu, W.J.; Peng, W.Q. Large deformation failure mechanism of surrounding rock for gateroad under dynamic pressure in deep coal mine. *J. China Coal Soc.* **2016**, *41*, 2940–2950.
21. Xie, S.; Wang, E.; Chen, D.; Jiang, Z.; Li, H.; Liu, R. Collaborative control technology of external anchor-internal unloading of surrounding rock in deep large-section coal roadway under strong mining influence. *J. China Coal Soc.* **2022**, *47*, 1946–1957.
22. Liu, S.; Yang, K.; Tang, C. Asymmetric failure mechanism and control of downhill roadway group of soft rock in deep mine. *J. Min. Saf. Eng.* **2019**. [CrossRef]
23. Li, G.; Sun, Y.; Zhang, J.; Zhang, Q.; Sun, C.; Zhang, S.; Bi, R. Experiment and application of coalcrete on roadway stability: A comparative analysis. *Adv. Mater. Sci. Eng.* **2020**, *2020*, 1–14. [CrossRef]
24. Sun, Y.; Bi, R.; Sun, J.; Zhang, J.; Taherdangkoo, R.; Huang, J.; Li, G. Stability of roadway along hard roof goaf by stress relief technique in deep mines: A theoretical, numerical and field study. *Geomech. Geophys. Geo-Energy Geo-Resour.* **2022**, *8*, 1–16. [CrossRef]
25. Xie, S.; Guo, F.; Wu, Y. Control Techniques for Gob-Side Entry Driving in an Extra-Thick Coal Seam with the Influence of Upper Residual Coal Pillar: A Case Study. *Energies* **2022**, *15*, 3620. [CrossRef]
26. Wang, E.; Xie, S. Determination of coal pillar width for gob-side entry driving in isolated coal face and its control in deep soft-broken coal seam: A case study. *Energy Sci. Eng.* **2022**, *10*, 2305–2316. [CrossRef]
27. Zhang, Z.; Bai, J.; Chen, Y.; Yan, S. An innovative approach for gob-side entry retaining in highly gassy fully-mechanized longwall top-coal caving. *Int. J. Rock Mech. Min. Sci.* **2015**, *80*, 1–11. [CrossRef]
28. Chen, D.; Guo, F.; Xie, S.; Wang, E.; Wu, Y.; Jiang, Z.; Wang, L.; Cui, J.; Zhang, X.; Liu, R. Mining-induced failure characteristics and surrounding rock control of gob-side entry driving adjacent to filling working face in the deep coal mine. *Energy Sci. Eng.* **2022**, *10*, 2593–2611. [CrossRef]
29. Wu, R.; Zhang, P.; Kulatilake, P.H.S.W.; Luo, H.; He, Q. Stress and deformation analysis of gob-side pre-backfill driving procedure of longwall mining: A case study. *Int. J. Coal. Sci. Technol.* **2021**, *8*, 1351–1370.
30. He, F.L.; Lv, K.; Li, X.B.; Qin, B.B.; Li, L. Failure Mechanism and Control of Lower Retracement Channel in Close-Distance Double-Thick Coal Seams. *Shock Vib.* **2021**, *2021*, 1–19. [CrossRef]
31. Wu, W.-D.; Bai, J.-B.; Wang, X.-Y.; Yan, S.; Wu, S.-X. Numerical Study of Failure Mechanisms and Control Techniques for a Gob-Side Yield Pillar in the Sijiazhuang Coal Mine, China. *Rock Mech. Rock Eng.* **2018**, *52*, 1231–1245. [CrossRef]
32. Sun, Y.; Bi, R.; Chang, Q.; Taherdangkoo, R.; Zhang, J.; Sun, J.; Huang, J.; Li, G. Stability Analysis of Roadway Groups under Multi-Mining Disturbances. *Appl. Sci.* **2021**, *11*, 7953. [CrossRef]
33. Yang, S.; Li, G.; Bi, R.; Yao, B.; Feng, R.; Sun, Y. The Stability of Roadway Groups under Rheology Coupling Mining Disturbance. *Sustainability* **2021**, *13*, 12300. [CrossRef]
34. Wang, W.; Yuan, C.; Yu, W.; Wu, H.; Peng, W.; Peng, G.; Liu, X.; Dong, E. Stability control method of surrounding rock in deep roadway with large deformation. *J. China Coal Soc.* **2016**, *41*, 2921–2931.
35. Barber, J.R. *Elasticity*; Springer: Berlin/Heidelberg, Germany, 2002.
36. Zhilun, X. *Elasticity*; Beijing Higher Education Publishing House: Beijing, China, 2006.
37. Qian, M.; Xu, J.; Wang, J. *Ground Pressure and Strata Control*; China University of Mining and Technology Press: Beijing, China, 2021.
38. Xiong, L.-X.; YU, L.-J. Comparative study of three Mohr strength criteria. *J. Yangtze River Sci. Res. Inst.* **2016**, *33*, 81.
39. Cai, M.F.; He, M.C.; Liu, D. *Rock Mechanics and Engineering*; Science Press: Beijing, China, 2013.
40. Mohammad, N.; Reddish, D.; Stace, L. The relation between in-situ, and laboratory rock properties used in numerical modelling. *Int. J. Rock Mech. Min. Sci.* **1997**, *34*, 289–297. [CrossRef]

41. Medhurst, T.; Brown, E. A study of the mechanical behaviour of coal for pillar design. *Int. J. Rock Mech. Min. Sci.* **1998**, *35*, 1087–1105. [CrossRef]
42. Hoek, E.; Brown, E. The Hoek–Brown failure criterion and GSI–2018 edition. *J. Rock Mech. Geotech. Eng.* **2019**, *11*, 445–463. [CrossRef]
43. Hoek, E.; Brown, E.T. Empirical strength criterion for rock masses. *J. Geotech. Eng. Div.* **1980**, *106*, 1013–1035. [CrossRef]

energies

MDPI

Article

Study on Stability and Control of Surrounding Rock in the Stopping Space with Fully Mechanized Top Coal Caving under Goaf

Fulian He [1,2], Bingquan Liu [1], Deqiu Wang [1], Dongdong Chen [1,*], Yanhao Wu [1], Liming Song [1], Xiang Ma [1], Qiucheng Ye [1], Zaisheng Jiang [1], Fangfang Guo [1], Weiguang Wang [3] and Yiyi Wu [1]

1 School of Energy & Mining Engineering, China University of Mining & Technology (Beijing), Beijing 100083, China
2 Beijing Key Laboratory for Precise Mining of Intergrown Energy and Resources, China University of Mining & Technology (Beijing), Beijing 100083, China
3 Jinneng Holding Group, Datong 037000, China
* Correspondence: chendongbcg@163.com

Abstract: Under the condition of fully mechanized top coal caving in close-distance coal seams, the surrounding rock of the stopping space easily loses stability during the withdrawal of mining equipment in the working face because the lower coal seam working face is located under the goaf and the overburden rock has a large range of complex interaction. Field investigation, theoretical analysis, laboratory experiment, similar simulation experiment, numerical simulation, and field industrial tests are used to carry out the research on the stability and control of the surrounding rock in the large section stopping space under the goaf in this paper. The research conclusions are as follows. (1) It is determined that the lower coal seam working face can only stop mining under the goaf, and the reasonable stopping position under the goaf should ensure that the key block fracture line of the main roof is behind the support. (2) The interaction law between the main roof's key blocks of the upper and lower coal seams is analyzed, and the catastrophic conditions for sliding instability and rotary instability of the main roof's key blocks of the upper and lower coal seams are obtained. (3) "Anchorage with push and pull equipment-Embedded anchorages and trays" integral anchoring technology is developed. The dimensions of the push and pull equipment are determined. (4) Through numerical simulation of the distribution characteristics of the anchor cable pre-stress field, the asymmetric control scheme of "Partition long and short anchor cables + Integral polyurethane mesh + Embedded anchorages and trays for roof protection" is determined. The rock pressure observation shows that the withdrawal of the working face equipment is implemented safely.

Keywords: close-distance coal seams; stopping space; roof rocks; stability; asymmetric control

Citation: He, F.; Liu, B.; Wang, D.; Chen, D.; Wu, Y.; Song, L.; Ma, X.; Ye, Q.; Jiang, Z.; Guo, F.; et al. Study on Stability and Control of Surrounding Rock in the Stopping Space with Fully Mechanized Top Coal Caving under Goaf. *Energies* **2022**, *15*, 8498. https://doi.org/10.3390/en15228498

Academic Editor: Adam Smoliński

Received: 12 October 2022
Accepted: 10 November 2022
Published: 14 November 2022

1. Introduction

The coal seam group is the main occurrence condition of coal seams in China, and the mining of close-distance coal seams is one of the main problems faced by coal mining enterprises in China [1,2]. Thick coal seam reserves account for approximately 45% of all coal reserves in China. The mining of thick coal seams is the main part of coal mining in China, and fully mechanized top coal caving is the first choice for thick seam mining [3–5]. After the mining of the working face is completed, the retracement channel is reserved in order to facilitate the rapid withdrawal of hydraulic supports, shearers, and other equipment [6]. At present, equipment withdrawal is performed in mainly two ways: pre-driven retraction channel and non-pre-driven retraction channel [7]. It is not simple to control the deformation of surrounding rock under the influence of advance abutment pressure of the working face with pre-driven retracement channels [8]. In contrast, a non-pre-driven retraction channel can avoid the strong incoming pressure and can be set up flexibly, which has a better application value [9,10]. During the retraction of the equipment of the working face, a large section stopping space is constitutive of the retraction channel and the supporting area of supports. The stability of the surrounding rock in the stopping space guarantees the efficient and safe retraction of the equipment [11]. Many scholars have conducted research on the stability and control of the surrounding rock in the stopping space. Yang et al. [12] studied the influence of the main roof rupture position and overburden structure on the stability of surrounding rock in the stopping space in order to achieve the safe retraction of equipment at the working face of soft coal seams and proposed a differentiated support scheme for the roof and lane gang of retraction channels. In view of the complex fault structure and the difficulty of roof management during the late mining period of the working face, Song [13] selected a reasonable position of the stopping line through the study of roof transport law and analysis of mine pressure measurements. Through numerical simulation, Chen et al. [14] concluded that there are four different damage zones around the stopping space, namely the shear damage zone, tensile damage zone, partial elasticity zone, and plastic damage zone, and adopted various measures such as optimizing the position of the stopping line, reducing the width of the stopping section, and increasing the number of anchor cables to ensure the stability of the surrounding rock in the stopping space. Wu et al. [15] proposed a rapid withdrawal technology for non-pre-driven retraction channels in heavily mechanized top-coal caving faces by studying the law of mine pressure appearing on the working face and the reasonable position of stopping lines for the problem of crushed pre-driven retraction channels. Liu et al. [16] analyzed the most dangerous rupture patterns of the immediate roof and main roof in the stopping space in view of roof-fall accidents and withdrawal accidents in the deep stop mining space. The study showed that the immediate roof presented showed a zonal progressive rupture pattern under strong loads, and the main roof formed a lateral rotation structure with the retraction of the support and triggered the instability of the support. Wang et al. [17] analyzed the position of the stopping line and the key block of the main roof to investigate the stability of the surrounding rock of the retraction channel using UDEC numerical simulation software and clarified the reasonable position relationship between the stopping line and the key block of the main roof.

The above research is of great significance to ensure the stability of surrounding rock in the stopping space and the efficient withdrawal of equipment, but they are all concentrated on the condition of a single coal seam with fully mechanized mining. There are few studies on the stability of surrounding rock in the stopping space under the condition of fully mechanized top coal caving in close-distance seams. Under the condition of close-distance coal seam mining, if the upper coal seam is not mined, the stopping space of the lower coal seam is located under the solid coal, and its ground pressure law and surrounding rock control are not different from that of single coal seam mining, as shown in Figure 1. If the upper coal seam has been mined, the stopping space of the lower coal seam is located under the goaf. As shown in Figure 2, the ground pressure is relatively severe due to the influence of the large-scale joint movement of overburden, and the stopping space is difficult to control. [18–22] The top coal above the stopping space of the fully mechanized top coal caving face is easily broken. It is difficult to push the anchoring agent when using the anchor cable support, and the anchor cable anchorage in the stopping space is generally exposed. During the last mining and hanging of the net, the anchorage will be damaged when moving the support. This causes the failure of the support components and brings difficulties to the surrounding rock control [23–25].

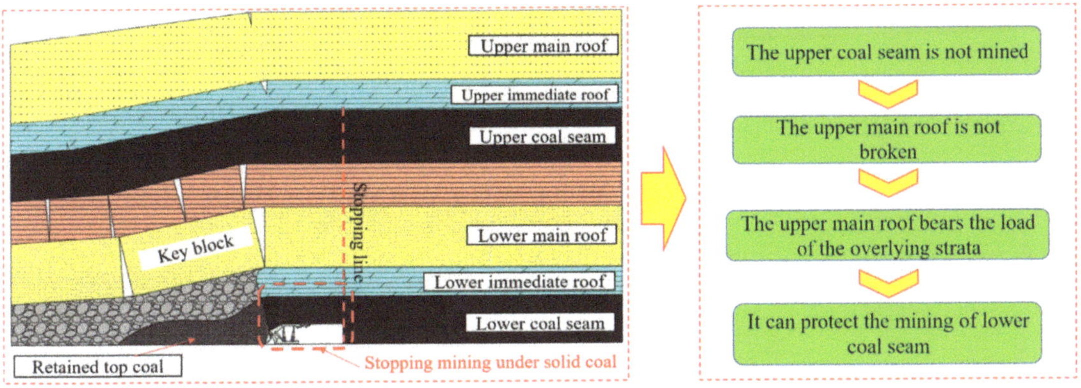

Figure 1. Schematic diagram of stopping mining under solid coal.

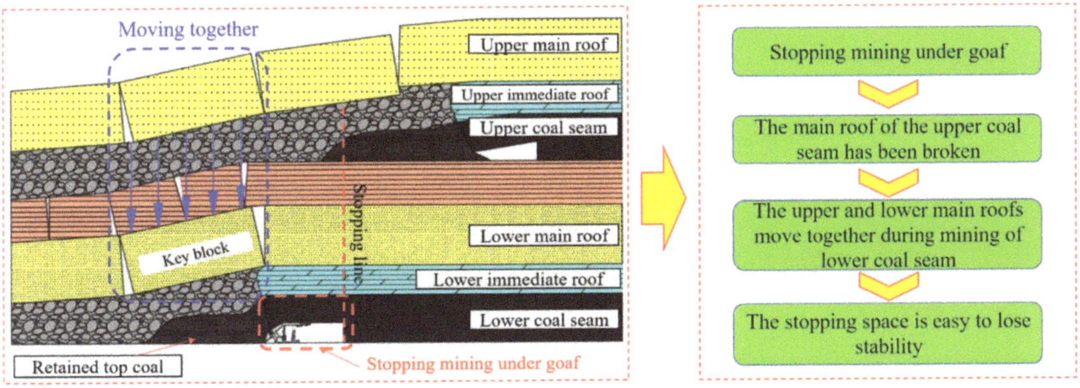

Figure 2. Schematic diagram of stopping mining under goaf.

Based on the above discussion, it is necessary to study the stability and control of the surrounding rock in the large section stopping space under the goaf with fully mechanized top coal caving. By means of field investigation, theoretical analysis, laboratory experiment, similar simulation experiment, numerical simulation, and industrial field tests, this paper clarifies that the stopping space of the lower coal seam can only be located under the goaf. The reasonable stopping position under the goaf should ensure that the fracture line of the main roof's key block is behind the support. The catastrophic conditions for sliding instability and rotary instability of the main roof's key blocks of upper and lower coal seams are obtained. The integral anchoring technology of "Anchorage with push and pull equipment + embedded anchorages and trays" is developed. The asymmetric control scheme of surrounding rock in the stopping space is proposed, and the industrial field test is finally carried out.

2. Engineering Geological Conditions

A mine is currently mining the overlying No. 4 coal seam and the underlying No. 3 coal seam, both of which are flat seams. The mining method is fully mechanized top coal caving mining. The average thickness of No. 4 coal seams and No. 3 coal seams are about 7.5 m and 7 m, respectively. The average distance between the two coal seams is 25 m. The strike length of the No. 8201 working face is 2517 m, the dip length is 180 m, and the average dip angle is 2°, which corresponds to the No. 8210 goaf of the overlying No. 4 coal seam. The coal seam structure of the No. 8201 working face is complex, with gangue distributed. The lithology of the immediate roof of No. 8201 working face is kaolinite, carbonaceous mudstone, fine sandstone, etc., mainly composed of quartz, and joint fissures are not developed. The lithology of the main roof is medium-grained sandstone, consisting mainly of quartz, with low joint fissure rate, dense sand body cementation, and good stability. The geological histogram of the No. 8201 working face and the position relationship diagram of the upper and lower working faces are shown in Figure 3.

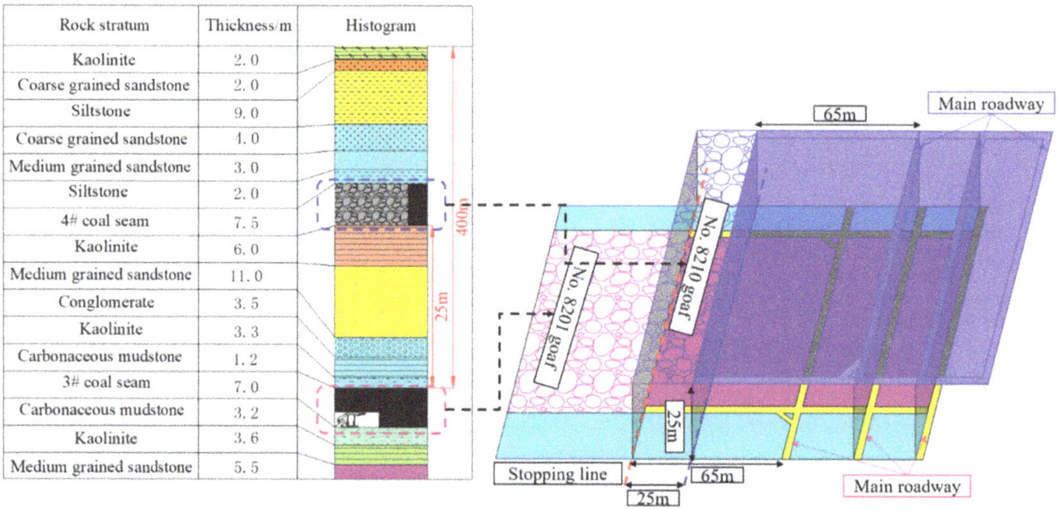

Figure 3. Geological histogram of panel and position relationship diagram of the upper and lower working faces.

According to the field survey, the width of the coal pillar between the stopping line and the main roadway of the No. 8210 working face is 65 m. To avoid the serious deformation and instability of the main roadway caused by the mining disturbance of the working face, the distance between the stopping line and the main roadway in the No. 8201 working face of the underlying No. 3 coal seam should not be less than 65 m. Therefore, the stopping line of the No. 8201 working face can only be located below the goaf of the upper coal seam. The stopping caving coal distance of the No. 8201 working face is 35 m. If the top coal caving is stopped before the stopping of the fully mechanized top coal caving face, the top coal that has not been caving can be used as the bearing body of the roof rock to reduce the amount of roof rotary subsidence, which is conducive to the stability of the large section stopping space, as shown in Figure 4a [26,27]. If the top coal caving is not stopped, the height of the immediate broken roof cannot reach the position of the main roof, and the turning angle of the key blocks of the main roof becomes larger, which very easily loses stability. It is a threat to the safety of the lower support and the stopping space, as shown in Figure 4b. Therefore, in order to ensure the safety of the stopping space, coal caving is generally stopped within a certain length from the stopping line [28–30].

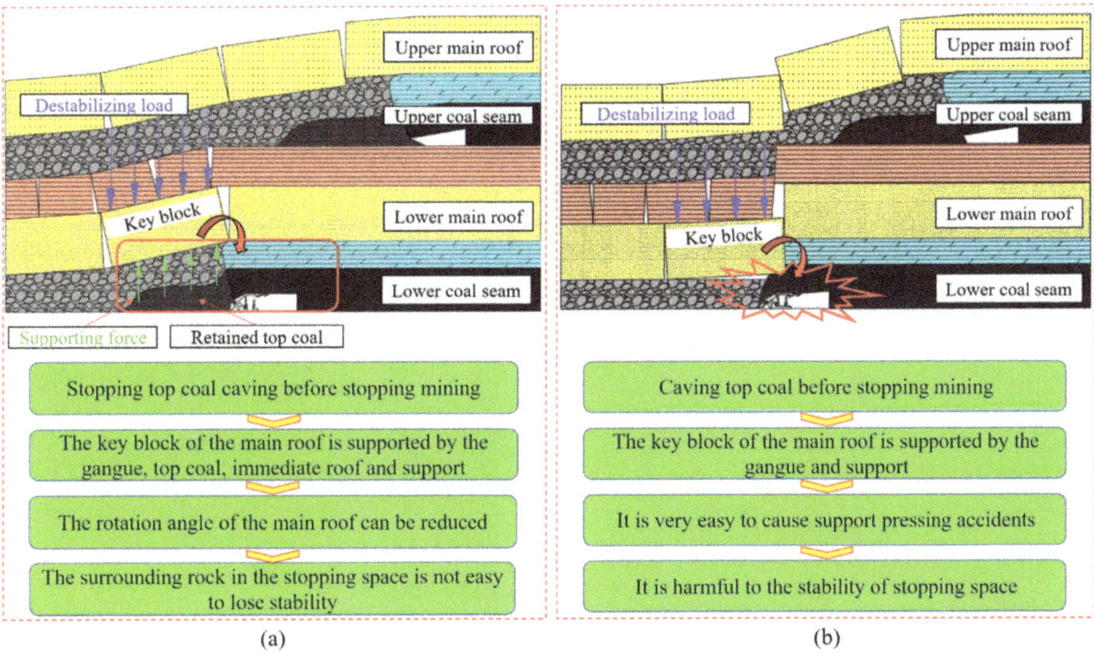

Figure 4. Schematic diagram of no top coal caving and top coal caving before stopping mining. (**a**) No top coal caving before stopping mining. (**b**) Top coal caving before stopping mining.

3. Analysis of Stopping Position under Goaf

Different stopping positions under the goaf will also affect the stability of the stop mining space [31]. According to the location relationship between the main roof fracture line and the support, the stopping positions can be divided into three categories. (1) When the main roof fracture line is in front of the support, the key block is supported by the immediate roof, the top coal, and the support. With the withdrawal of the lower support, the support ability of the support to the main roof is weakened, and the key block is prone to instability, which brings great hidden dangers to the safety of the stopping mining space, as shown in Figure 5a. (2) When the main roof fracture line is above the support, the key block is supported by the gangue, the top coal, the immediate roof, and the support. To ensure the stability of the key block, enough top coal should be reserved behind the support. If the parking distance is short, the key block easily loses stability, and the support below the key block is difficult to withdraw, as shown in Figure 5b. (3) When the main roof fracture line is behind the support, the key block is supported by the gangue and the top coal, and its rotary instability has little impact on the stopping mining space. At this time, the stopping mining space is under the protection of the cantilever beam structure of the main roof, which is the best stopping position, as shown in Figure 5c. In comparison, the stopping positions (1) and (2) should be avoided in actual production. The reasonable stopping position under the goaf should ensure that the fracture line of the key block of the main roof is behind the support.

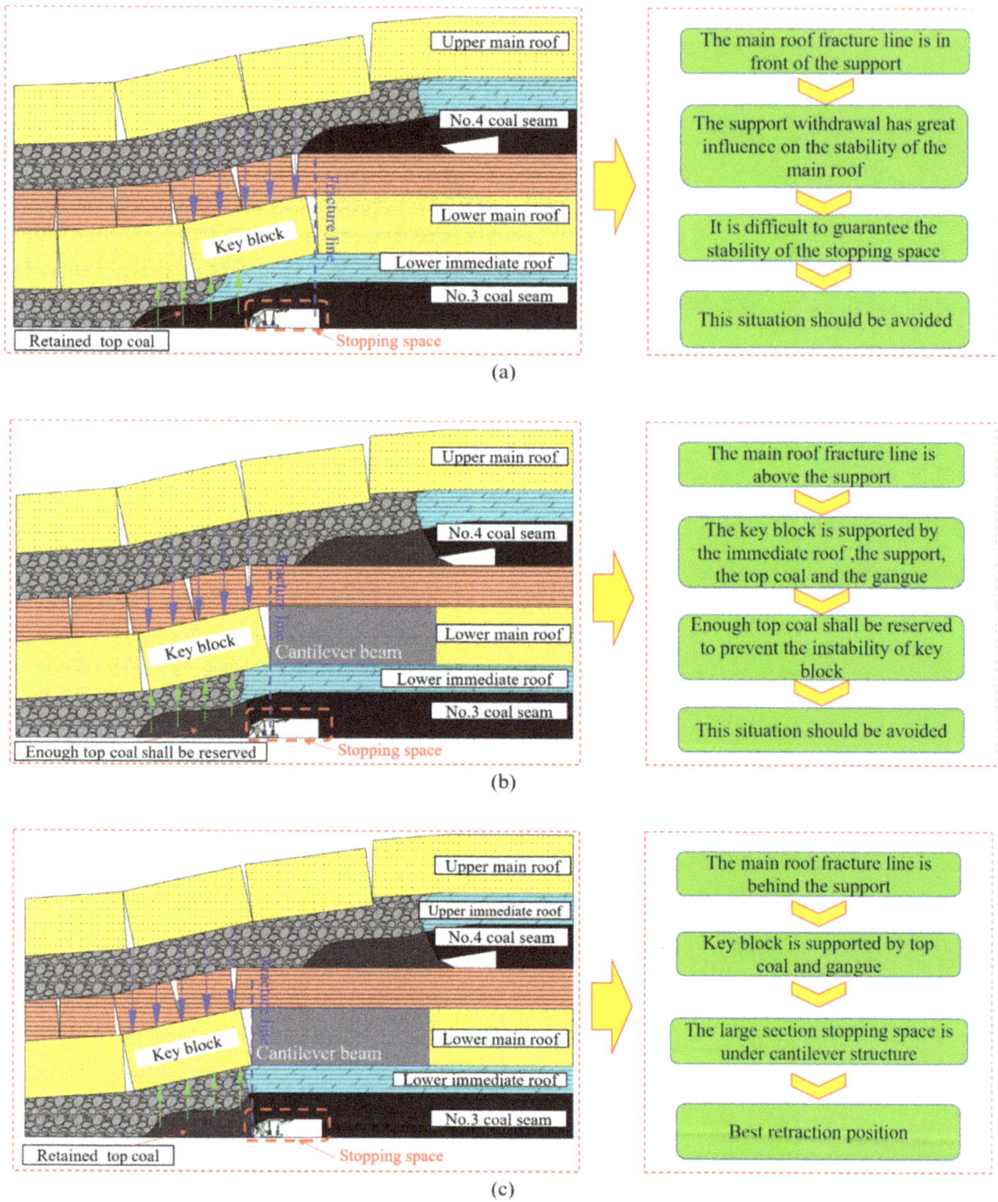

Figure 5. Schematic diagram of different stopping positions under the goaf. (**a**) The main roof fracture line is in front of the support. (**b**) The main roof fracture line is above the support. (**c**) The main roof fracture line is behind the support.

4. Stability Analysis of Key Blocks in Close Distance Coal Seam Mining

The No.4 coal seam was mined first, its main roof collapsed, and the goaf has been compacted. The No.3 coal seam is being mined. The distance between the two coal seams is 25 m, which belongs to a close-distance coal seam. Therefore, it is necessary to consider the damage of the overlying coal seam mining to its floor strata. According to the measured data of the upper coal seam panel mining, the floor damage depth is 8.5 m. When the lower coal seam is mined, its key block is locally damaged and interacts with the key block of the upper coal seam, as shown in Figure 6.

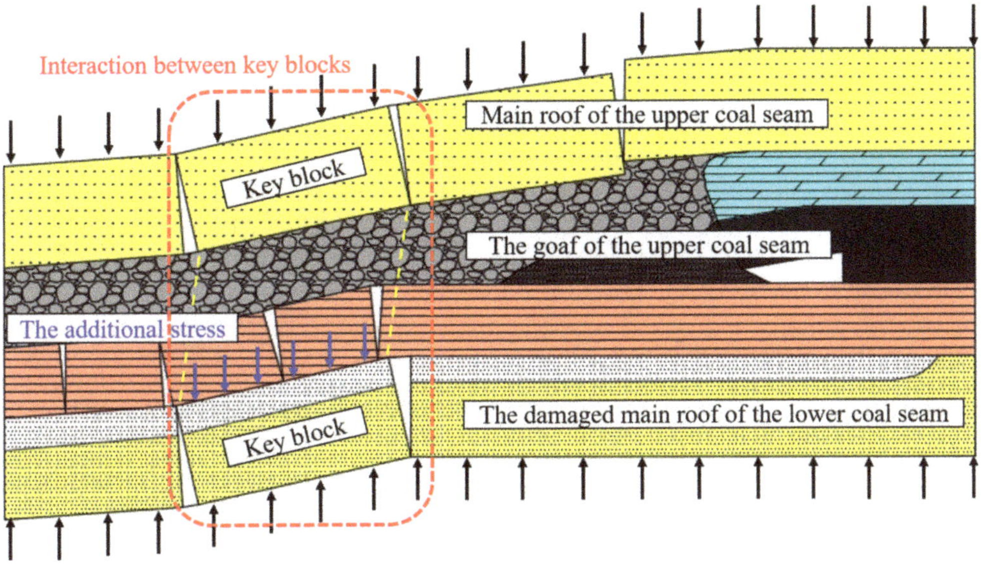

Figure 6. Interaction between key blocks of the upper main roof and lower main roof.

The main roof of the upper coal seam shall bear the rock strata load with a thickness of 20 m, about 0.52 MPa. The stability of the key block in the upper coal seam is analyzed. The average length of the key block is 19.7 m. In the periodic weighting stage, the critical loads for sliding instability and rotary instability are q_{uh} and q_{uz}, and the expression is as follows [32]:

$$\begin{cases} i_{uh} = \frac{\cos\theta_1 \tan\varphi + \sin\theta_1(1+\tan\theta_1\tan\varphi)}{1-\tan\theta_1\tan\varphi} \\ q_{uh} = \frac{5i_{uh}^2[\sigma_t]}{15-i_{uh}^2} \end{cases} \tag{1}$$

$$q_{uz} = \frac{\eta[\sigma_c](i_u - \sin\theta_1)^2}{2[\cos\theta_2 + (i_u + \sin\theta_1)\tan\theta_1]} \tag{2}$$

In the formula:

i_u is the lumpiness of the key block in the upper coal seam;
i_{uh} is the critical lumpiness for sliding instability of the key block;
θ_1 is the rotating angle of the key block, °;
$[\sigma_c]$ is the ultimate compressive of rock mass, MPa;
$[\sigma_t]$ is the ultimate tensile strength of rock mass, MPa;
tanφ and η are the friction coefficient between rock blocks, taken as 0.3.

Now, carry out a mechanical analysis on the damaged main roof of the lower coal seam. The top 2.5 m of the main roof rock is damaged by the mining of the upper coal seam, and the sequence degree is 0.77. Meanwhile, when the working face is mining, the average length of its key block is 19.1 m. In the periodic pressure stage, the key block's critical loads q_{lh} and q_{lz} for sliding instability and slewing instability are as follows: (3), (4) [33,34]:

$$\begin{cases} i_{lh} = \frac{\cos\theta_2 \tan\varphi + \sin\theta_2(1+\tan\theta_2\tan\varphi)}{1-\tan\theta_2\tan\varphi} \\ q_{lh} = \frac{5i_{lh}^2[\sigma_t]\psi}{15-i_{lh}^2} \end{cases} \tag{3}$$

$$q_{lz} = \frac{\eta[\sigma_c]\psi(i_l - \sin\theta_2)^?}{2[\cos\theta_2 + (i_l + \sin\theta_2)\tan\theta_2]} \tag{4}$$

In the formula:

i_l is the lumpiness of the key block in the lower coal seam;
i_{lh} is the critical lumpiness of the key block for sliding instability;
θ_2 is the rotation angle of the key block, °;
ψ is sequence degree, and its value is 0.77.

From this, we can obtain the curves of critical load, lumpiness, and rotation angle of the damaged key block in the lower coal seam, as shown in Figure 7.

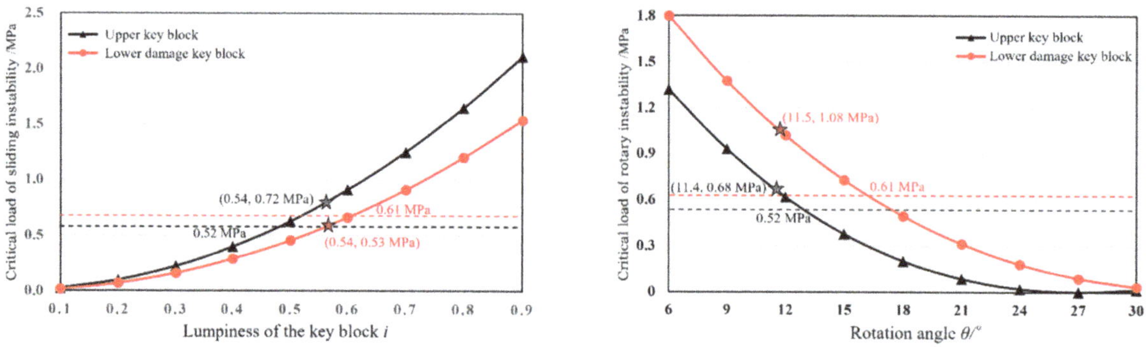

Figure 7. Linkage analysis of upper and lower main roof key blocks.

Obtained by calculation: the critical loads of the key block in the upper coal seam are 0.72 MPa and 0.68 MPa, which are both greater than the load value of the overlying load layer of 0.52 MPa. Therefore, when mining the lower coal seam, it is only necessary to consider the stability of the damaged main roof key blocks and its load layer. The load of the damaged key block in the lower coal seam is 0.53 MPa. The balance force of the damaged key block cannot bear the load of 0.61 MPa. However, because the hydraulic support provides a bearing capacity close to 1 MPa, the key block of the damaged basic roof will not slide and lose stability when mining in the lower coal seam. At the same time, the load of the damaged key block in the lower coal seam is 1.08 MPa, which can balance itself. Therefore, in view of the damage to the main roof strata of the lower coal seam caused by the mining of the upper coal seam, when mining in the lower coal seam working face, the key blocks will not slide and rotate instability, which can ensure safe mining.

5. "Anchorage with Push and Pull Equipment-Embedded Anchorages and Trays" Integral Anchoring Technology

5.1. Development of Anchorage with Push and Pull Equipment

The stopping space of a fully mechanized top coal caving working face is characterized by a large section span, large top coal thickness, and unsynchronized deformation of roof coal and rock strata, which easily causes layer separation and dislocation. Moreover, the strength of the top coal is low, and cracks are easily produced under the influence of mining pressure. Especially, the top coal above the hydraulic support is seriously crushed by the extrusion of the upper roof and support. To accurately understand the broken situation of the top coal in the stopping mining space, the top coal above the stopping mining space in the final mining stage is peeked by the borehole peep method, as shown in Figure 8. The coal body in the borehole is severely broken, the hole-forming effect is poor, the cracks are seriously developed, and there are obvious phenomena of delamination dislocation and hole collapse. The broken top coal has severely challenged the control of surrounding rock in the stopping mining space.

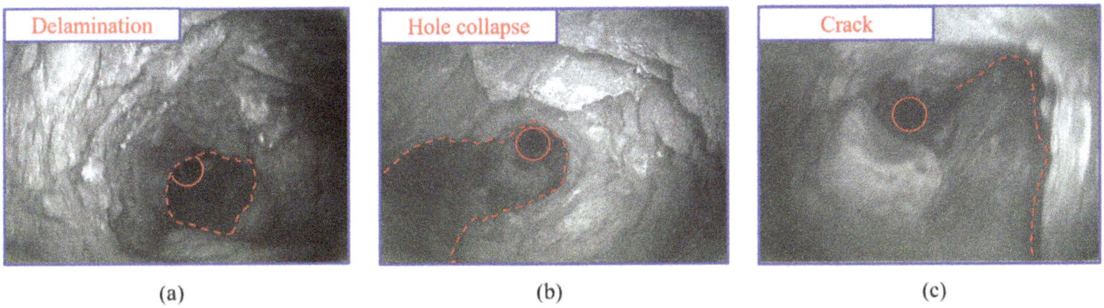

Figure 8. Borehole peep of top coal. (**a**) Delamination occurred in the borehole. (**b**) Collapse occurred in the borehole. (**c**) Cracks occurred in the borehole.

Bolts-anchor cable support is a common means to realize effective control of surrounding rock, and the installation of an anchoring agent is an important link in the supporting operation. At present, the anchoring agent pushing generally adopts the method of unconstrained whole pushing, which can improve the installation efficiency of the anchoring agent compared with staged pushing. However, when there is delamination or hole collapse in the roof drilling hole, the front-end anchoring agent can easily get stuck in the delamination or hole collapse. If it continues to push in, the anchoring agent will bend and then be cut and leaked by sharp coal rock. Because the anchoring agent leaked in advance and failed to reach the anchoring point at the bottom of the hole, the anchoring effect could not be achieved. The unconstrained pushing experiment of the anchoring agent was carried out using a transparent acrylic pipe to simulate the drilling hole, as shown in Figure 9.

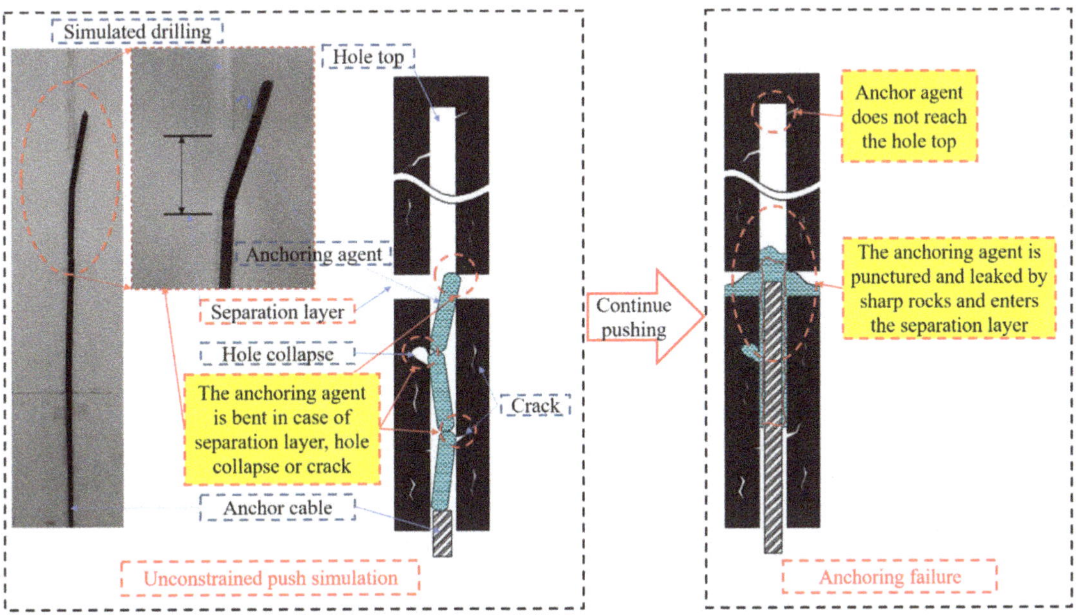

Figure 9. Disadvantages of unrestrained integral pushing anchoring agent.

If push and pull technology is adopted, the problem of anchoring agent pushing can be effectively solved. The push and pull anchoring equipment is shown in Figure 10. The anchorage with push and pull equipment comprises a push–pull tray and a U-shaped clip.

The pushing force of the anchor cable acts on the bottom tray and is transmitted upwards through the U-shaped clip, thus forming a double force of pushing and guiding the middle anchoring agent in the same direction. The U-shaped clip has a certain rigidity, which can dredge the drilling hole and guide the anchoring agent to reach the top of the drilling hole smoothly through the delamination or hole collapse area. Then, the anchor machine is used to push the anchor cable. When the anchor cable pierces the tray, stir the anchoring agent. The effect of anchoring with push and pull is shown in Figure 11. The equipment has a simple structure and good stability and fundamentally solves the problems of easy hole plugging, difficult pushing, and poor anchoring effect of the anchoring agent.

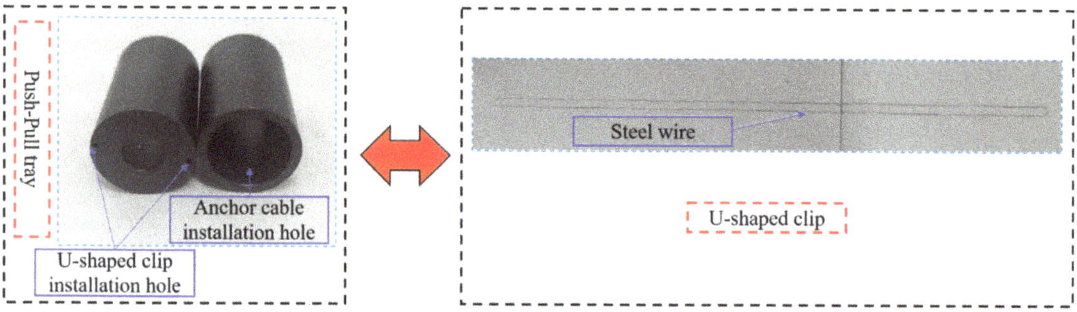

Figure 10. Schematic diagram of anchorage with push and pull equipment.

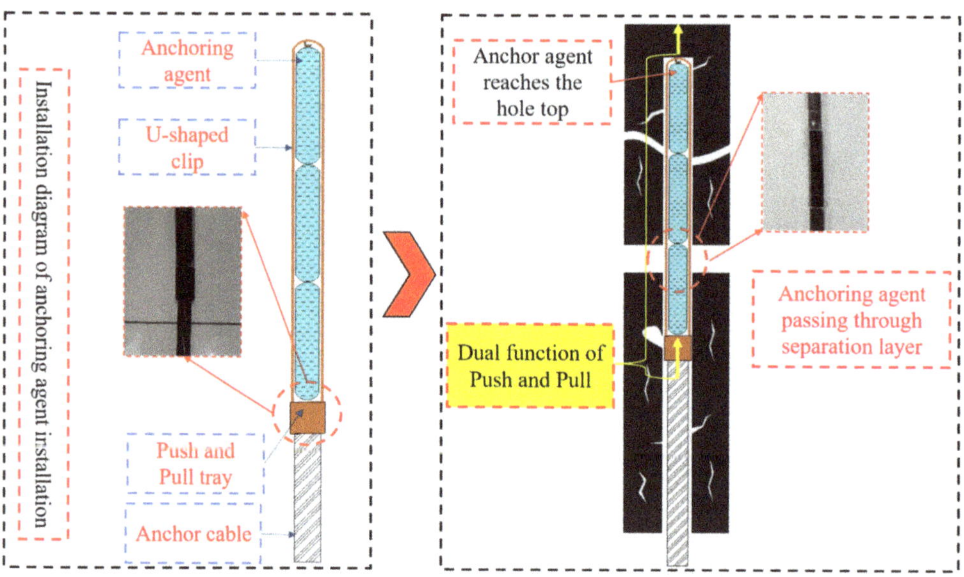

Figure 11. Effect diagram of anchorage with push and pull equipment.

The push–pull tray is the key equipment to bear the push force of the anchor cable, fix the U-shaped clip, and realize the double force of "push" and "pull". The push–pull tray in kind is shown in the figure. The push–pull tray should not be damaged during normal pushing to ensure that the anchoring agent can reach the hole top smoothly. After that, the top of the tray is damaged by a manual pushing force, and the anchor cable passes through the tray to stir the anchoring agent. To meet the above requirements, it is necessary to test the mechanical properties of the top plate of the push–pull tray. The push–pull tray is made of brittle and easily cracked plastic. To ensure that it is pierced by the anchor cable after lifting the anchoring agent, the top plate of the pushing tray needs a reasonable thickness. Therefore, the thickness of the top plate is 0.3 mm, 0.6 mm, and 1 mm, respectively, for the mechanical loading test, and the test process is shown in Figure 12.

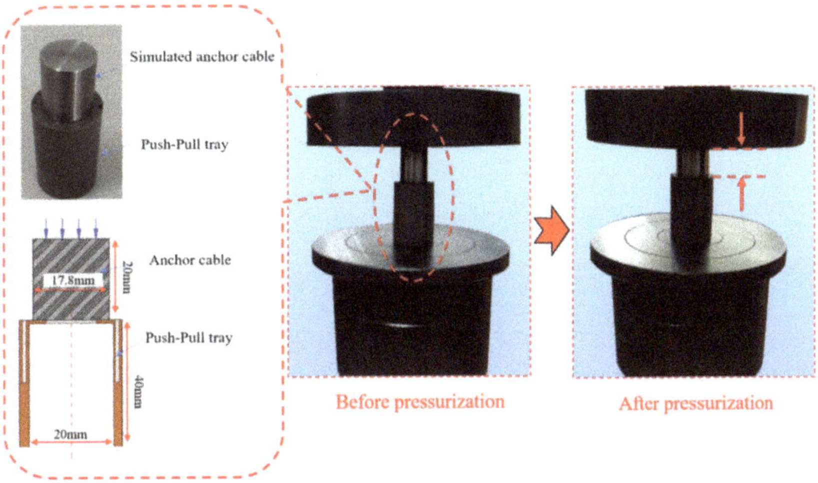

Figure 12. Mechanical loading test of push–pull tray.

The failure pattern of the top plate after the loading test is shown in Figure 13. the top plate of 0.3 mm and 0.6 mm suffered shear failure, whereas that of 1 mm suffered tensile shear failure. The time–load curves for different top plate thicknesses are shown in Figure 14. The maximum load that the 0.3 mm thick top plate can carry is 68.9 N; for the 0.6 mm thick top plate, the maximum load is 340.6 N; the 1.0 mm thick top plate is loaded with a maximum load of 1222.1 N.

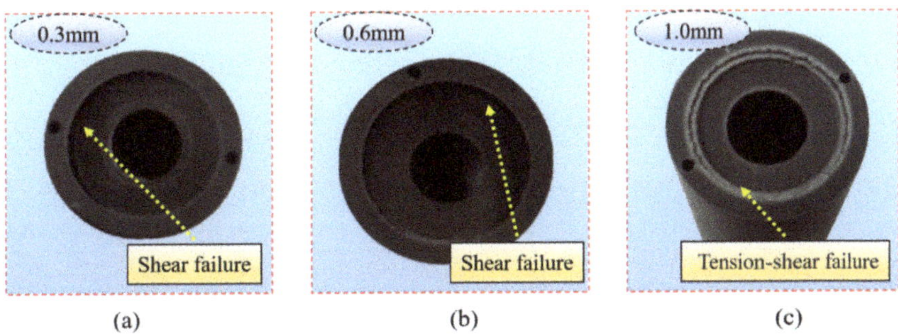

Figure 13. Failure patterns of top plates of different thicknesses. (**a**) shear failure; (**b**) shear failure; (**c**) tensile shear failure.

The measured mass of the push–pull tray and U-shaped clip M_1 = 70 g, the mass of the anchor agent M_2 = 1500 g, taking the acceleration of gravity g = 10 N/kg, resulting in the total gravity of the push–pull device and anchor agent G = 15.7 N. Considering the resistance during the pushing process and the unevenness of the anchor cable end, the bearing capacity of the top plate should not be less than 150 N. To ensure that the final tray can be pierced by the anchor cable, the load-bearing capacity of the top plate should not be too large. Combining the peak pressure and the fitting curve of the top plate thickness, as shown in Figure 15, the reasonable thickness of the top plate was finally determined to be 0.5 mm, and the whole size of the push–pull anchorage device is shown in Figure 16.

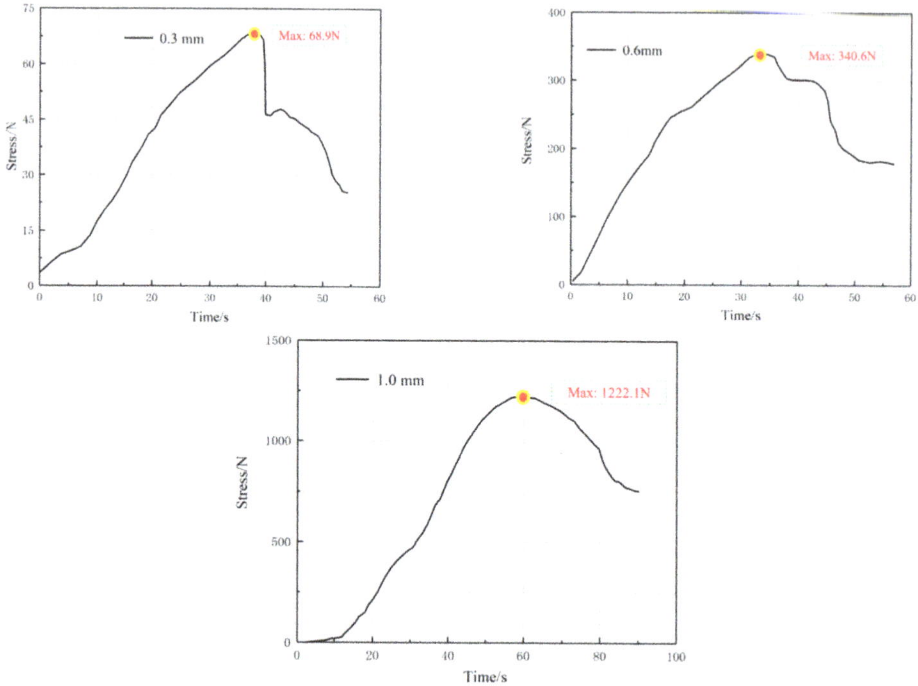

Figure 14. Loading curves for different thicknesses of the top plate. (**a**) Plate thickness 0.3 mm; (**b**) Plate thickness 0.6 mm; (**c**) Plate thickness 1.0 mm.

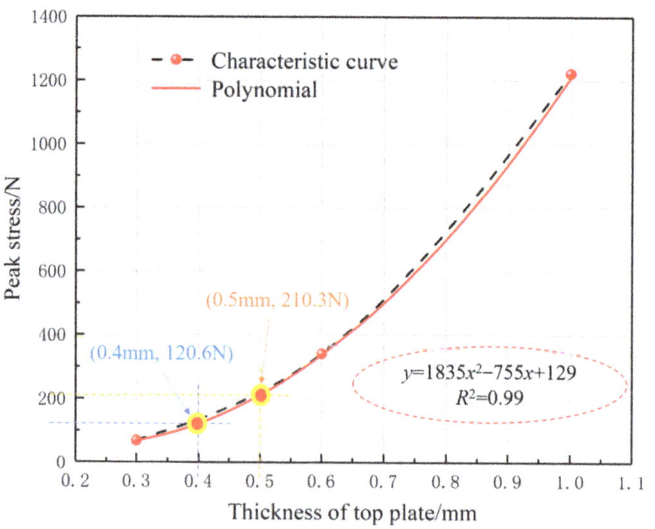

Figure 15. Peak stress with fitting curve for top tray thickness.

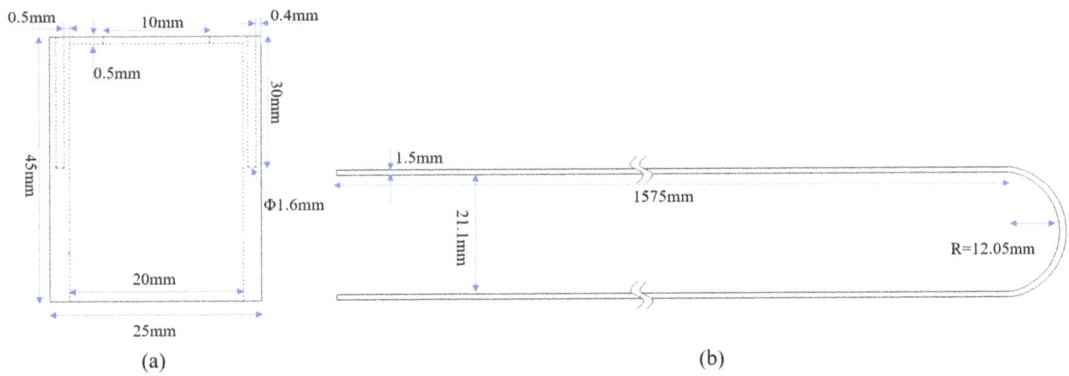

Figure 16. Dimensional drawing of the push–pull anchorage device (**a**) Home view of push–pull tray. (**b**) Home view of U-shaped clip.

In order to test the effect of anchorage with push and pull equipment, pull-out experiments were carried out on the anchor cable in the field. The anchor cable can be divided into a free section and an anchorage section in the borehole, as shown in Figure 17. According to the theory of elasticity, the total elongation of the anchor cable should be the elongation of the free section under the pull-out force. If the total elongation of the anchor cable is greater than the theoretical elongation of the free section of the anchor cable, it proves that the stress in the anchorage section is shifted to the deeper part, and the anchorage section is partially failed.

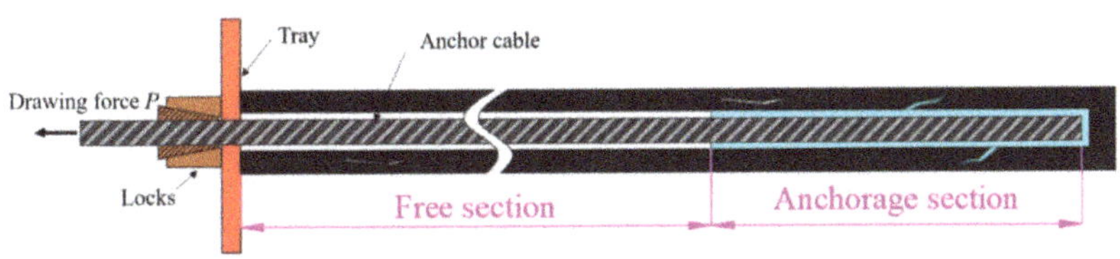

Figure 17. Schematic diagram of the anchor cable in the borehole.

The field test data are shown in Figure 18. The anchor cable anchorage reliability is higher with the use of the integral push–pull anchorage installation process. No. 1 and No. 2 anchor ropes of unconstrained anchorage showed a significant increase in expansion when the pull-out force was greater than 160 kN, and the total elongation of the anchor cable was significantly greater than the theoretical elongation of the anchor cable. There are two reasons for this phenomenon. On the one hand, it is due to the partial failure of the anchor section, resulting in the growth of the free section length of the anchor cable; on the other hand, it is caused by the relative sliding of the anchor cable under the pulling load due to the reduced frictional resistance between the grouted body (bonded body) of the anchor section and the borehole wall. No. 3 and No. 4 anchor cables were installed with a pull-out force of 200 kN, and the elongation of the anchor cable was approximately equal to the theoretical elongation of the free end of the anchor cable, which indicates that no partial anchor failure or relative slippage occurred in the anchor cable anchorage section.

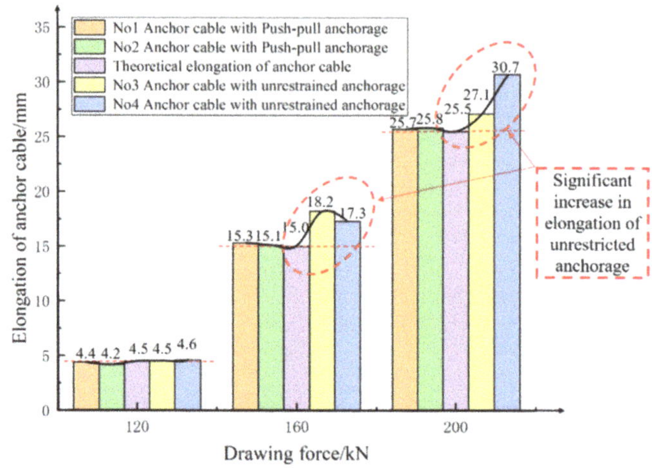

Figure 18. Pull-out test results for anchor cables.

5.2. Anchoring Technology with Embedded Anchorages and Trays

The anchorages of the anchor cables in the stopping space are generally exposed. When the anchor cables exist in the top coal in front of the support during the terminal mining hanging net shifting, the exposed anchorages will prevent the support from shifting the frame with stress, and if the frame continues to be shifted, the anchorages will be damaged, causing the anchor cables to separate from the tray and the support force of the anchor cables to be sharply reduced, resulting in the failure of the support function of the anchor cables, as shown in Figure 19.

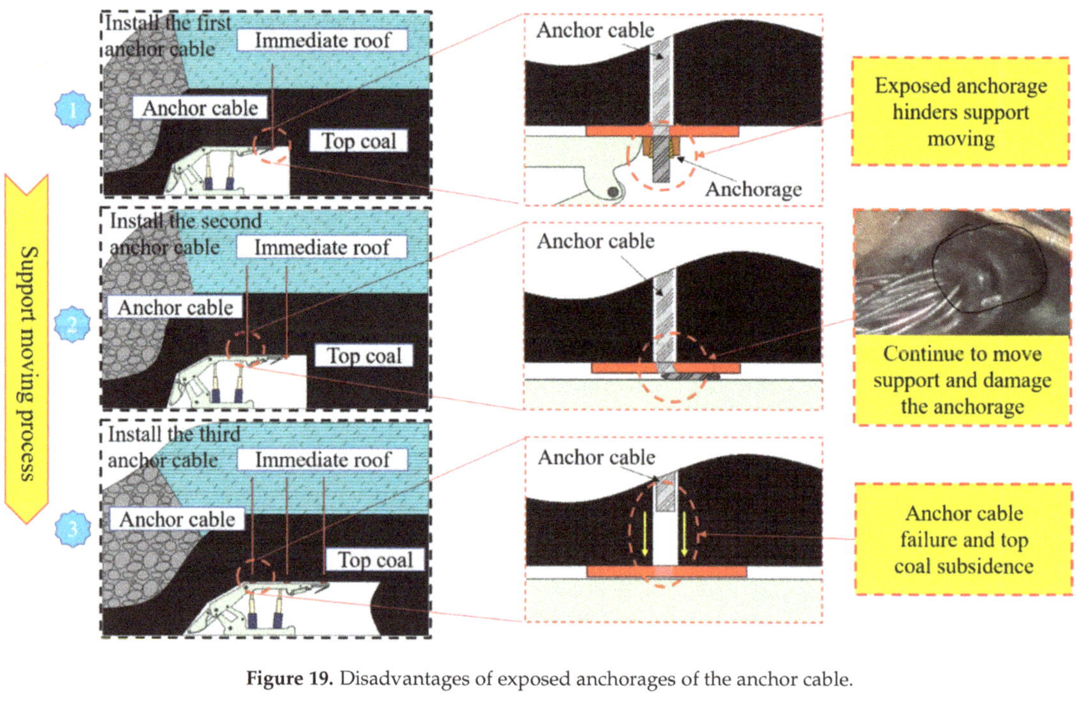

Figure 19. Disadvantages of exposed anchorages of the anchor cable.

In order to overcome the disadvantages of exposed anchorages, a new anchoring method with embedded anchorages and trays has been developed, as shown in Figure 20. The top coal is drilled with a reaming bit to form an embedded slot, the depth of which is such that the anchorages and trays are fully embedded and do not interfere with the support movement. Compared with the exposed anchorages, the technology with embedded anchorages and trays can effectively solve the problems of support movement with pressure, damage to the anchorages, and weakening or invalidating the support system. The effect of embedded anchorages and tray support is shown in Figure 21.

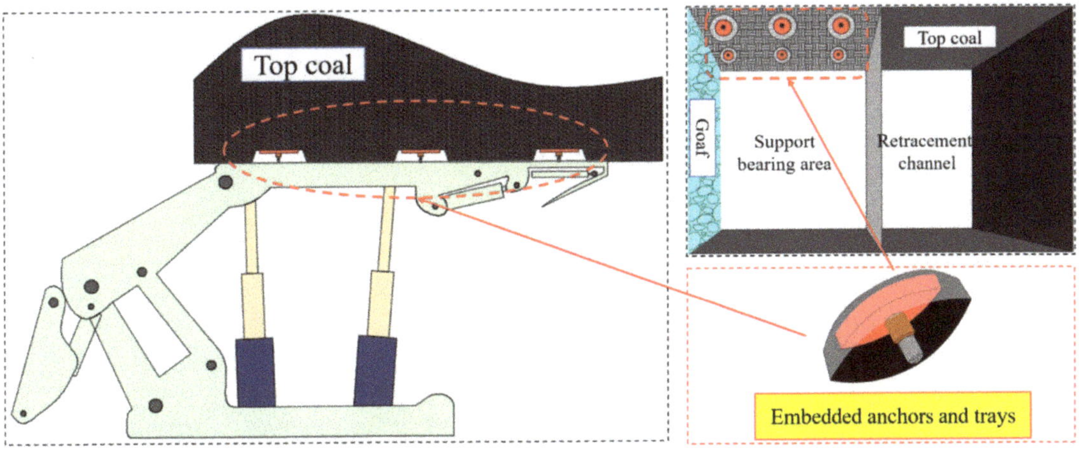

Figure 20. Schematic diagram of embedded anchorages and trays.

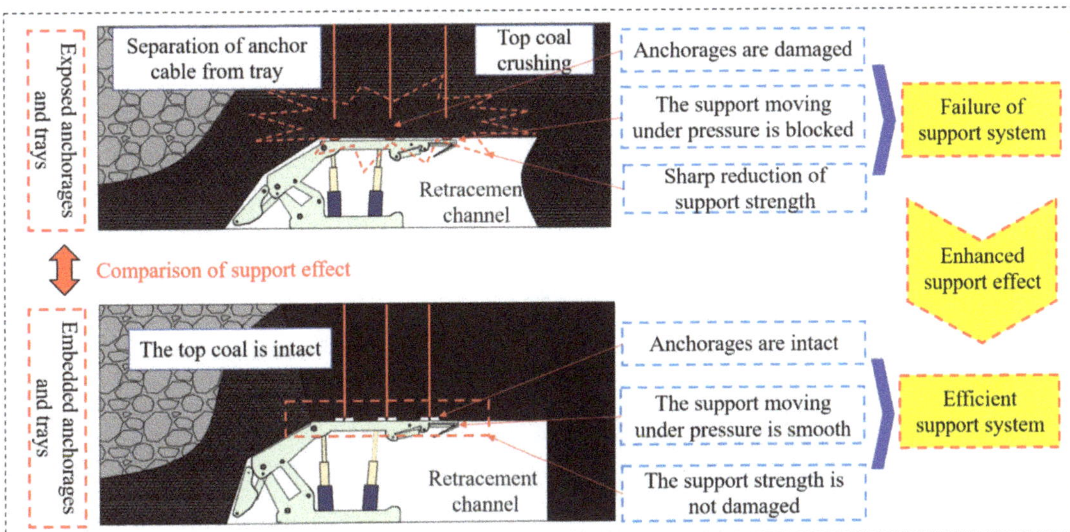

Figure 21. Comparison of the effect between exposed and embedded anchorages and trays.

"Anchorage with push and pull equipment-Embedded anchorages and trays" integral anchoring technology in the stopping space can improve the efficiency of anchor cable installation, enhance the effect of anchor cable, strengthen the support capacity of anchor cable, and realize the effect of anchor cable on the surrounding rock.

6. Support Schemes for Stopping Space

Unlike a traditional roadway, the large section stopping space has a special rock structure, so it is necessary to analyze the structural characteristics of the stopping space when designing the support scheme. As shown in Figure 22, the stopping space consists of the retracement channel, the support bearing area, and the front coal wall area. In terms of the overburden structure, it is safest for the stopping space to be under the protection of the main roof cantilever beam structure. Namely, the main roof fracture line is located behind the support. To satisfy this condition, the working face needs to be stopped just after passing the fracture line (in effect, the stopping point is the end of the working face pressure cycle), followed by the construction of the retracement channel.

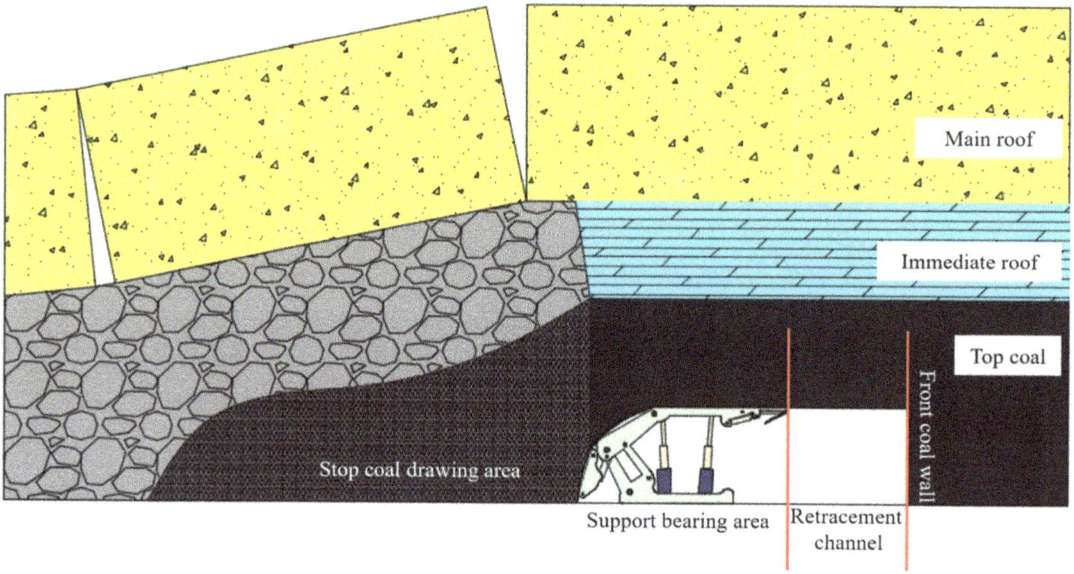

Figure 22. Illustration of the stopping space zoning in fully mechanized top coal caving mining method.

According to the characteristics of the surrounding rock structure of the stopping space and the on-site support withdrawal process, different zones of the stopping space have the following different requirements for the support effect. (1) The roof of the retracement channel needs strong support because this area is the key passage area for support withdrawal, so it is necessary to ensure that it is stable before and during support withdrawal. (2) The roof of the support area needs relatively weak support because after the support is withdrawn, the roof of the support area cannot be suspended in a large area and needs to be withdrawn and collapsed at the same time, but it still needs a certain bearing capacity; otherwise, the collapsed roof is close to the support, which easily causes the support to be difficult to pull out. (3) The front coal mass needs surface protection and support to avoid a large area of wall spalling during withdrawal.

Four support schemes are designed according to the actual situation on the site. In order to evaluate the support effects of different support designs, the distribution of the pre-stress field of large-section spatial support for stopping mining is simulated by FLAC3D numerical simulation software. The model is 100 m long, 50 m wide, and 100 m high. The displacement condition is used around the model to fix the boundary. Mohr-Coulomb criterion is selected as the constitutive relation of the model. The anchor cable shall be applied with a pre-tightening force of 120 kN, and the anchor bolt shall be applied with a pre-tightening force of 80 kN. The distribution of the pre-stress field is shown in Figure 23.

From the pre-stress field, it can be seen that the bolt (cable) support has changed the stress distribution of the surrounding coal body. The boundary stress of the pre-stress field is 0.02 MPa. The high-stress area is mainly concentrated in the tray and anchorage section. Scheme (a) is that the support area and the retracement channel are supported by anchor cables, and the front coal wall area is supported by bolts. The length of anchor cables is 4500 mm, the spacing is 2500 mm, and the length of bolts is 1700 mm, the spacing is 1600 mm. From the distribution of pre-stress, it can be seen that the high pre-stress is concentrated at the end of the anchor cable and distributed sporadically. The upper roof and the front coal wall do not realize the connection of the pre-stress field. In scheme (b), the spacing of roof anchor cables above the support area is 1800 mm, the spacing of roof anchor cables above the retracement channel is 1200 mm, the length of front coal wall bolts is 1700 mm, and the spacing is 1000 mm. It can be seen from the distribution of pre-stress that the high pre-stress is connected between the upper roof anchor cables, and the low pre-stress field of the upper roof and the front coal wall is connected. In scheme (c), the length of the roof anchor cable above the support area is 4500 mm, and the spacing is 1800 mm. The length of the roof anchor cable above the retracement channel is 6300 mm, and the spacing is 1200 mm. The length of the front coal wall bolt is 1700 mm, and the spacing is 1000 mm. From the distribution of the pre-stress field, it can be seen that the range of the pre-stress field of the roof anchor cable above the retracement channel is larger than that above the support area, and the lower pre-stress field of the upper roof and the front coal wall are interconnected. In scheme (d), the included angle between the right anchor cable of the roof above the retracement channel and the vertical direction is 15 °, and the included angle between the uppermost bolt of the front coal wall and the horizontal direction is 15 °. The other parameters are the same as scheme (c).

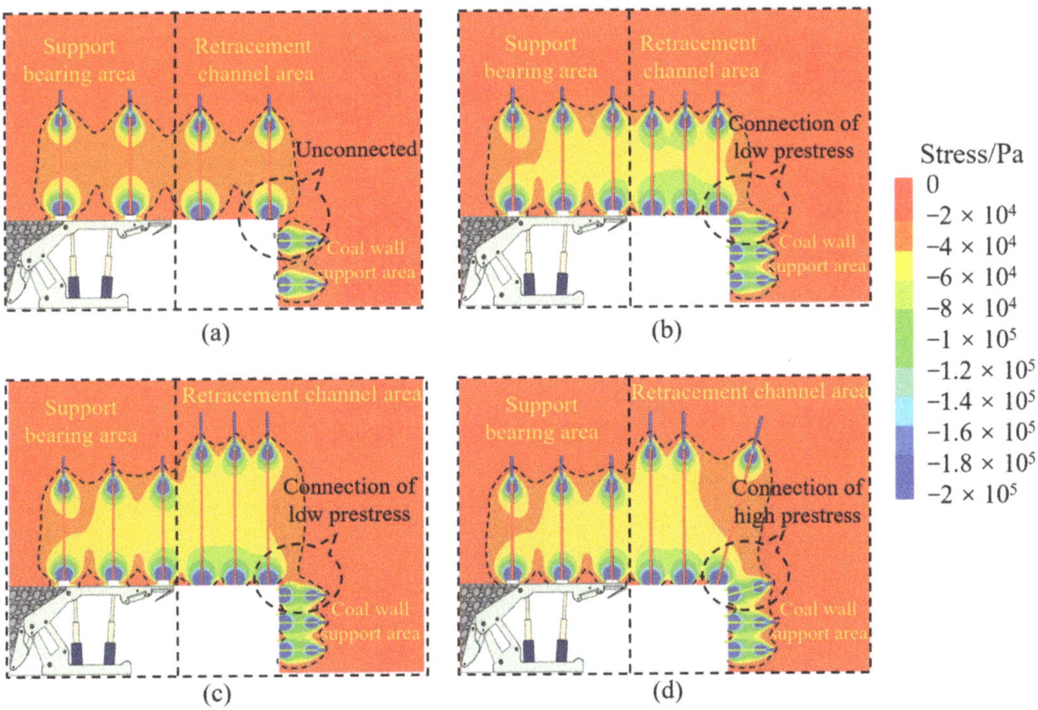

Figure 23. Schematic Diagram of Prestressing Field of Different Support Schemes (**a**–**d**).

From the distribution of the pre-stress field, it can be seen that the control range of the pre-stress field of the roof anchor cable of the retracement channel is large, and the support effect is better than that of the roof of the support area, which reflects the asymmetric support characteristics of relatively strong support in the retracement channel and relatively weak support in the roof above the support area. In addition, the high pre-stress field of the upper roof and the front coal wall are interconnected, which realizes the effective control of the anchor (cable) support over the entire space for mining suspension.

Combined with the distribution of the support pre-stress field and the actual situation on site, and referring to the support design of the stopping space in other working faces of this mine, the asymmetric support scheme of "Partition long and short anchor cables + Integral polyurethane mesh + Embedded anchorages and trays for roof protection" was finally determined, as shown in Figures 24 and 25.

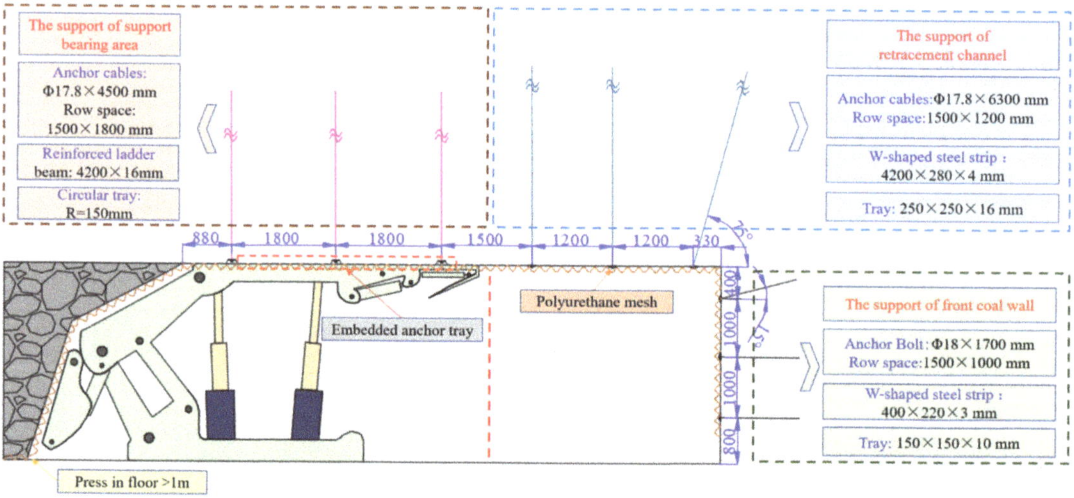

Figure 24. Asymmetric support parameters of stopping space.

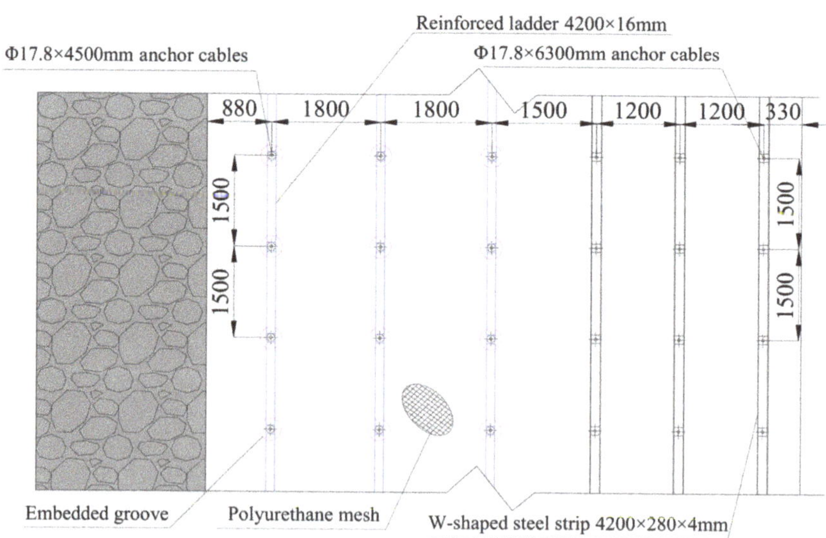

Figure 25. Top view of stopping space support.

7. Industrial Test

When the 8201 working face enters the final mining stage, the above support scheme will be applied to the field practice. Stop top coal caving before stopping mining of the working face. After stopping top coal caving for 5 m, start to lay the overall polyurethane mesh of the full section, and then the working face will continue to advance without caving until the polyurethane mesh is inserted into the broken top coal that is not placed behind the support for about 1 m, and then drive the retracement channel to form a large section space of about 9 m for stopping mining. "Anchorage with push and pull equipment-Embedded anchorages and trays" integral anchoring technology is used to install the anchor (cable), forming an efficient asymmetric support system. After field comparison, the new anchoring agent installation method can save half the time compared with the traditional installation method. The installation efficiency of the anchoring agent is doubled. The on-site support effect is shown in Figure 26.

Figure 26. Effect picture of on-site support.

In order to evaluate the surrounding rock control effect of the stopping space, measuring points are set to monitor the displacement of the roof and coal wall in the retracement channel. The layout plan of measuring points is shown in Figure 27. The monitoring data are shown in Figure 28. Within 10 days of the arrangement of the measuring points, the displacement of the coal wall in front of the retracement channel maintained a growth trend, but the growth rate gradually slowed down. After 10 days, the growth rate gradually stabilized, and the maximum displacement of the coal wall was 150 mm. At the initial monitoring stage, the roof displacement increased significantly. Due to the support of the hydraulic support, the roof displacement tends to be stable quickly. After 8 days of monitoring, the roof displacement increases and tends to be flat. The maximum roof displacement is 127 mm.

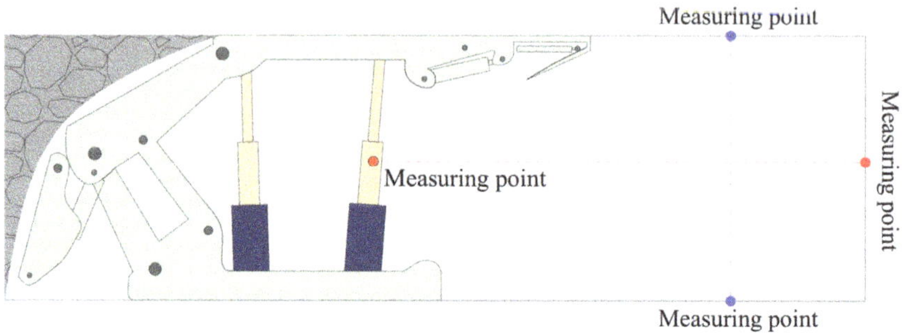

Figure 27. Layout of measuring points.

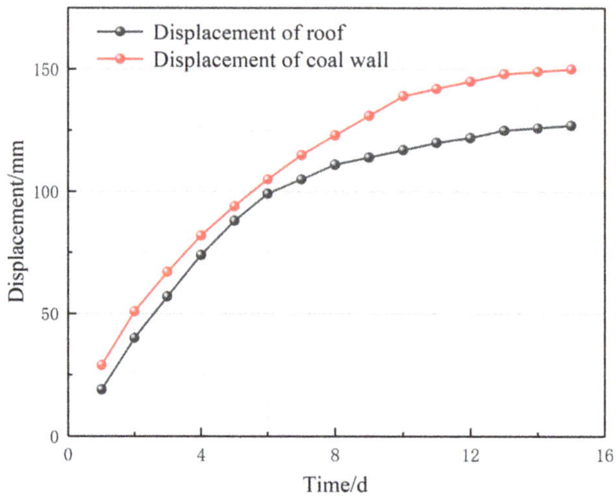

Figure 28. Deformation of surrounding rock.

8. Conclusions

(1) It is determined that the stopping line of the lower coal seam can only be located below the goaf of the upper coal seam, and the reasonable stopping position is that the fracture line of the main roof key block is located behind the support.

(2) It is concluded that the key block of the upper coal seam main roof will not lose stability during the mining of the lower coal seam, and the damaged key block of the lower coal seam cannot bear the load of overlying rocks, but it will not lose stability under the supporting action of the support.

(3) Anchorage with push and pull equipment is developed. It is determined that the height of the push and pull tray is 45 mm, the thickness of the top plate is 0.5 mm, and the length of the U-shaped clip is 1575 mm. A new anchoring method of embedded anchorages and trays is proposed to realize the synergetic control of the surrounding rock.

(4) The asymmetric control scheme of "Partition long and short anchor cables + Integral polyurethane mesh + Embedded anchorages and trays for roof protection" is determined. The rock pressure observation shows that it realizes the safe withdrawal of the working face equipment.

Author Contributions: Conceptualization, F.H. and D.C.; Data curation, B.L., Y.W. (Yanhao Wu), and L.S.; Formal analysis, D.W., Z.J., F.G., W.W. and Y.W. (Yiyi Wu); Funding acquisition, F.H. and D.C.; Methodology, Y.W. (Yanhao Wu) and L.S.; Project administration, D.C.; Software, X.M., B.L. and Q.Y.; Supervision, F.H.; Writing—original draft, D.C. and D.W.; Writing—review and editing, D.C., B.L., D.W. and X.M. All authors have read and agreed to the published version of the manuscript.

Funding: This work was supported by the National Natural Science Foundation of China (Grant No.52004286), the National Natural Science Foundation of China (Grant No.51974317), the Fundamental Research Funds for the Central Universities (2021YJSNY01, 2022YJSNY09).

Conflicts of Interest: The authors declare no conflict of interest.

References

1. Shang, Y.Q.; Kong, D.Z.; Pu, S.J.; Xiong, Y.; Li, Q.; Cheng, Z.B. Study on Failure Characteristics and Control Technology of Roadway Surrounding Rock under Repeated Mining in Close-Distance Coal Seam. *Mathematics* **2022**, *10*, 2166. [CrossRef]
2. Chen, D.; Guo, F.; Li, Z.; Ma, X.; Xie, S.; Wu, Y.; Wang, Z. Study on the Influence and Control of Stress Direction Deflection and Partial-Stress Boosting of Main Roadways Surrounding Rock and under the Influence of Multi-Seam Mining. *Energies* **2022**, *15*, 8257. [CrossRef]
3. Bhattacharya, M.; Rafiq, S.; Bhattacharya, S. The role of technology on the dynamics of coal consumption–economic growth: New evidence from China. *Appl. Energy* **2015**, *154*, 686–695. [CrossRef]
4. Xie, S.; Wang, E.; Chen, D.; Jiang, Z.; Li, H.; Liu, R. Collaborative Control Technology of External Anchor-Internal Unloading of Surrounding Rock in Deep Large-Section Coal Roadway under Strong Mining Influence. *J. China Coal Soc.* **2022**, *47*, 1946–1957.
5. Zhang, B.S.; Yang, Z.P.; Ji, C.X.; Guo, Z.F.; Li, H.Y. Research on the Influence of the Key Stratum Position on the Support Working Resistance during Large Mining Height Top-Coal Caving Mining. *Adv. Civ. Eng.* **2021**, *2021*, 6690280. [CrossRef]
6. Wang, B.N.; Mu, L.; He, M.M.; Gu, S.C. Mechanism Analysis of Roof Deformation in Pre-Driven Longwall Recovery Rooms Considering Main Roof Failure Form. *Sustainability* **2022**, *14*, 9093. [CrossRef]
7. Zhu, C.; Yuan, Y.; Chen, Z.S.; Liu, Z.H.; Yuan, C.F. Study of the Stability Control of the Rock Surrounding Double-Key Strata Recovery Roadways in Shallow Seams. *Adv. Civ. Eng.* **2019**, *2019*, 9801637. [CrossRef]
8. Wang, B.N.; Dang, F.N.; Gu, S.C.; Huang, R.B.; Miao, Y.P.; Chao, W. Method for determining the width of protective coal pillar in the pre-driven longwall recovery room considering main roof failure form. *Int. J. Rock. Mech. Min.* **2020**, *130*, 104340. [CrossRef]
9. Kang, H.P.; Lv, H.W.; Zhang, X.; Gao, F.Q.; Wu, Z.G.; Wang, Z.C. Evaluation of the Ground Response of a Pre-driven Longwall Recovery Room Supported by Concrete Cribs. *Rock. Mech. Rock. Eng.* **2016**, *49*, 1025–1040. [CrossRef]
10. Fei, L.; Jiang, Z.X. Research on Deformation Mechanism of Retracement Channel during Fully Mechanized Caving Mining in Superhigh Seam. *Adv. Civ. Eng.* **2018**, *2018*, 1368965. [CrossRef]
11. Xu, J.H.; Miao, X.X.; Pu, H.; Cao, S.G. Determination of reasonable position and stability analysis of finishing cut in fully-mechanized coal face with topped caving. *Chin. J. Rock Mech. Eng.* **2004**, *23*, 1981–1985.
12. Yang, R.S.; Li, Y.L.; Zhu, Y.; Xiao, C.L.; Zhao, Y.; Shi, G.L. Study on stability control of equipment removal channel for high cutting mining face under special condition. *Coal Sci. Technol.* **2017**, *45*, 10–15.
13. Song, B. Instant Removal Gateway for Large Height Working Face in Soft Coal Seam. *Saf. Coal Mines* **2014**, *45*, 136–139.
14. Chen, Z.S.; Yuan, Y.; Zhu, C.; Wang, W.M. Stability Mechanism and Control Factors on Equipment Removal Area under "Goaf-Roof-Coal" Structure. *Adv. Civ. Eng.* **2021**, *2021*, 6628272. [CrossRef]
15. Wu, J.N.; Feng, X.W.; Liu, B.K. Rapid removing technology of heavy fully mechanized top coal caving mining face without preset removing gateway. *Coal Sci. Technol.* **2008**, *36*, 14–17.
16. Liu, Y.M.; Liu, Y.H.; Gao, M.S. Roof Fracture Characteristics of Deep Mining-pause Space Without Pre-excavation Roadway. *Saf. Coal Mines* **2016**, *47*, 197–199.
17. Wang, X.; Ji, W.P.; Li, X.M.; Liu, F.; Zhang, N.B. Study on Position of Withdrawal Entry in Fully Mechanized Caving Face of Extremely Thick Coal Seam. *Coal Technol.* **2010**, *29*, 63–65.
18. Ning, J.G.; Wang, J.; Tan, Y.L.; Xu, Q. Mechanical mechanism of overlying strata breaking and development of fractured zone during close-distance coal seam group mining. *Int. J. Min. Sci. Technol.* **2020**, *30*, 207–215. [CrossRef]
19. Guo, G.C.; Yang, Y.K. The study of key stratum location and characteristics on the mining of extremely thick coal seam under goaf. *Adv. Civ. Eng.* **2021**, *2021*, 8833822. [CrossRef]
20. Xie, S.; Wu, Y.; Guo, F.; Zou, H.; Chen, D.; Zhang, X.; Ma, X.; Liu, R.; Wu, C. Application of Pre-Splitting and Roof-Cutting Control Technology in Coal Mining: A Review of Technology. *Energies* **2022**, *15*, 6489. [CrossRef]
21. Pan, W.D.; Zhang, S.P.; Liu, Y. Safe and efficient coal mining below the goaf: A case study. *Energies* **2020**, *13*, 864. [CrossRef]
22. Zhang, W.; Zhang, D.S.; Qi, D.H.; Hu, W.M.; He, Z.M.; Zhang, W.S. Floor failure depth of upper coal seam during close coal seams mining and its novel detection method. *Energy Explor. Exploit.* **2018**, *36*, 1265–1278. [CrossRef]
23. Vu, T.T. Solutions to prevent face spall and roof falling in fully mechanized longwall at underground mines, Vietnam. *Min. Min. Dep.* **2022**, *16*, 127–134. [CrossRef]
24. Małkowski, P.; Niedbalski, Z.; Majcherczyk, T.; Bednarek, Ł. Underground monitoring as the best way of roadways support design validation in a long time period. *Min. Min. Dep.* **2020**, *14*, 1–14. [CrossRef]

25. Zhu, C.; Xu, Y.Z.; Wu, Y.X.; He, M.C.; Zhu, C.Q.; Meng, Q.X.; Lin, Y. A hybrid artifi cial bee colony algorithm and support vector machine for predicting blast-induced ground vibration. *Earthq. Eng. Eng. Vib.* **2022**, *21*, 861–876.
26. Xie, S.R.; Wu, Y.Y.; Ma, X.; Chen, D.D.; Guo, F.F.; Jiang, Z.S.; Li, H.; Zou, H.; Liu, R.P.; Zhang, X. Reasonable stopping method and retracement channel support at fully mechanized top coal caving working face of 15 m extra-thick coal seam: A case study. *Energy Sci. Eng.* **2022**. [CrossRef]
27. Chen, D.; Wu, Y.; Xie, S.; Guo, F.; He, F.; Liu, R. Reasonable Location of Stopping Line in Close-Distance Underlying Coal Seam and Partition Support of Large Cross-Section Roadway. *Int. J. Coal Sci. Technol.* **2022**, *9*, 55. [CrossRef]
28. Singh, G.S.P.; Singh, U.K. Assessment of goaf characteristics and compaction in longwall caving. *Min. Technol.* **2011**, *120*, 222–232. [CrossRef]
29. Alehossein, H.; Poulsen, B.A. Stress analysis of longwall top coal caving. *Int. J. Rock Meck. Min.* **2010**, *47*, 30–41. [CrossRef]
30. Klishin, V.I.; Fryanov, V.N.; Pavlova, L.D.; Opruk, G.Y. Modeling top coal disintegration in thick seams in longwall top coal caving. *J. Min. Sci.* **2019**, *55*, 247–256. [CrossRef]
31. Wang, X.Z.; Ju, J.F.; Xu, J.L. Theory and applicable of yield mining at ending stage of fully-mechanized face in shallow seam at Shendong mine area. *J. Min. Saf. Eng.* **2012**, *3*, 151–156.
32. Qian, M.G.; Miao, X.X.; Xu, J.L. Theoretical Study of Key Sratum in Ground Control. *J. China Coal Soc.* **1996**, *21*, 2–7.
33. Xie, S.R.; Wu, Y.Y.; Guo, F.F.; Chen, D.D.; Wang, E.; Zhang, X.; Zou, H.; Liu, R.P.; Ma, X.; Li, S.J. Interaction Mechanism of the Upper and Lower Main Roofs with Different Properties in Close Coal Seams: A Case Study. *Energies* **2022**, *15*, 5533. [CrossRef]
34. He, S.S.; Xie, S.R.; Song, B.H.; Zhou, D.Q.; Sun, Y.D.; Han, S. Breaking laws and stability analysis of damage main roof in close distance hypogynous seams. *J. China Coal Soc.* **2016**, *41*, 2596–2605.

energies

Article

New Technology of Pressure Relief Control in Soft Coal Roadways with Deep, Violent Mining and Large Deformation: A Key Study

Shengrong Xie [1,2], Hui Li [1], Dongdong Chen [1,*], Shaohua Feng [1], Xiang Ma [1], Zaisheng Jiang [1] and Junqi Cui [1]

1 School of Energy and Mining Engineering, China University of Mining & Technology-Beijing, Beijing 100083, China
2 Beijing Key Laboratory for Precise Mining of Intergrown Energy and Resources, China University of Mining & Technology-Beijing, Beijing 100083, China
* Correspondence: chendongbcg@163.com

Abstract: Previous studies have shown that the influence of deep dynamic pressure on the surrounding rock control of a coal roadway is one of the difficulties in mine roadway support. Based on the investigation of the headgate 11231 in a coal mine, this study analyzes the damage characteristics of coal roadway surrounding rock affected by deep dynamic pressure, expounds on the difficulties of controlling the roadway surrounding rock, and creatively proposes a cooperative control technology of external anchor–internal unloading for regulating large deformation of roadways. The vertical stress distribution and transfer law of surrounding rock with different hole-making depths, spacing, and lengths after roadway excavation were simulated and studied, and an appropriate parameter range of hole-making space in the stage without dynamic pressure influence was obtained. Considering the influence of mining dynamic pressure, the surrounding rock pressure relief effect of each optimized hole-making parameter was analyzed. In addition, the optimal hole-making parameters (the hole-making depth, spacing, and length were 8 m, 3.2 m, and 3 m, respectively) that can effectively reduce the high stress of roadway shallow surrounding rock in two stages (without and with dynamic pressure) and ensure integrity of the shallow surrounding rock were obtained. The actual field application shows that the new technology can reduce the higher rib deformation by approximately 850 mm and achieve a good surrounding rock control effect. The research and practice show that the pressure relief control for soft coal roadways with deep, violent mining and large deformation has achieved success, providing technical support for the maintenance of the same type of roadway.

Keywords: deep mine; coal roadway; numerical simulation; pressure relief; external anchor–internal unloading; surrounding rock control technology

Citation: Xie, S.; Li, H.; Chen, D.; Feng, S.; Ma, X.; Jiang, Z.; Cui, J. New Technology of Pressure Relief Control in Soft Coal Roadways with Deep, Violent Mining and Large Deformation: A Key Study. *Energies* **2022**, *15*, 9208. https://doi.org/10.3390/en15239208

Academic Editor: Adam Smoliński

Received: 29 October 2022
Accepted: 21 November 2022
Published: 5 December 2022

1. Introduction

With the deep mining of coal resources gradually becoming the norm [1–3], the engineering response problems of discontinuous, uncoordinated large deformation and large-scale instability of roadway surrounding rock caused by the typical deep "three high" occurrence environment and coal mining dynamic pressure have become urgent engineering problems that need to be solved [4,5]. Research shows that the complex stress field [6], deformation brittle–ductile transition [7], continuous deterioration [4,8], and strong rheology of deep coal and rock mass are the major reasons for the continuous large deformation of deep roadway surrounding rock. In China, coal roadways excavated by underground coal mines account for approximately 80% of the total roadway excavation volume [9]. Coal science and technology workers have conducted a lot of research on the surrounding rock properties [10,11], stress environment [12,13], and roadway deformation characteristics [14,15] of deep coal roadways through field investigation, theoretical analysis, experimental research, and numerical simulation.

In recent years, a lot of work has been carried out to study the impact of mining on the rock mass and to recommend measures for reducing strain changes of rock mass and minimizing the stress on the surface. Since mining production has a significant effect on the stress-strain behavior of rock mass, the issues of reducing this influence are very relevant and scientists around the world are trying to minimize it. Adigamov, A.E., et al. [16] established the stress-strain behavior model of disturbed rock mass with regard to anisotropy and discontinuities, which can be used to calculate the strength of underground rock mass; Khayrutdinov, A.M., etc., [17] studied the change of stress-strain characteristics of rock mass after using different-strength backfill and carried out research on the stress-strain relationship of disturbed rock mass under different conditions. On the basis of studying the influence of mining activities on the stress-strain relationship of rock mass [18,19], scholars at home and abroad have developed many targeted coal roadway surrounding rock support technologies. High prestressed strong bolt support [20], high strength, high stiffness and high prestressed bolt support [21], and butted long bolt support [22] technologies can effectively control the generation and development of separation, sliding, and cracks in shallow surrounding rock as well as improve the post-peak strength of deformed surrounding rock and its mechanical parameters. Support technologies, such as the constant resistance and large deformation bolt support technology [23] and high-strength pressure relief anchor box beam support system [24] can release the uncontrollable deformation energy in the surrounding rock after the roadway support is completed, and then, give full play to the support capacity of the bolt, so that the bolt and the pressure relief surrounding rock can form a stable support bearing body. The anchor cable truss support technology [25] can simultaneously provide extrusion stresses in the horizontal and vertical directions of the roadway roof and rib, thus effectively reducing the maximum tensile stress and maximum shear stress of the surrounding rock in the anchorage zone. The anchor-grouting combined support technology [26] can re-condense the broken surrounding rock of the coal roadway, effectively improve the mechanical properties of the surrounding rock, and ensure the integrity and safety of the roadway surrounding rock. In addition, the active–passive coupling support technology [27] involves combining bolts (cables) with the passive support components to maintain stability of the coal roadway surrounding rock under special geological conditions and fully utilize the initial support resistance from the active support and the high deformation resistance of the passive support. To control the deformation of deep coal roadway surrounding rock, while seeking innovation of the support technology, it has become an important research direction for the surrounding rock control of deep coal roadways to realize the release or transfer of high concentrated stress in the coal roadway surrounding rock through stress control (surrounding rock pressure relief) to achieve stability of the surrounding rock. Currently, many roadway pressure relief control methods are available, such as arranging the roadway in the stress reduction zone [28], physical pressure relief of surrounding rock (e.g., by drilling [29] and blasting [30,31]), excavating the pressure relief roadway [32], and roof cutting pressure relief [33]. The pressure relief control technology can solve the problem of controlling the surrounding rock under various working conditions to a certain extent.

Analyzing the above, it can be noted that the surrounding rock control in soft coal roadways with deep, violent mining and large deformation is a very topical issue, but the existing research results have not made a big breakthrough in the surrounding rock control of this type of roadway. Given the above problems, this study creatively proposes a cooperative control technology of external anchor–internal unloading of surrounding rock in deep coal roadways. First, the roadway shallow surrounding rock is strengthened through an anchor-grouting combined support. Then, a hydraulic hole-making machine is used to create pressure relief space with appropriate spacing in a certain range of the deep coal roadway to transfer the high concentrated stress from the shallow to the deep rock and effectively improve the stress environment of the roadway surrounding rock. The drawings in Figure 1 show how by combining the characteristics of new and old pressure relief technologies, both conventional pressure relief drilling [29] and hole-making pressure relief can effectively realize the transfer of high concentrated stress in the shallow surrounding rock to the deep rock. However, the conventional pressure relief dense drilling makes the shallow surrounding rock more broken, reducing its strength. In contrast, the hole-making pressure relief technology uses steel pipes to effectively solidify the borehole surrounding rock and ensures the integrity and effectiveness of the shallow surrounding rock and support structure. The new cooperative control technology was first applied to the deep dynamic pressure mining roadway and achieved success, ensuring the roadway rib did not expand during the service period. This has important research value for the surrounding rock pressure relief support of a high stress and large deformation coal roadway.

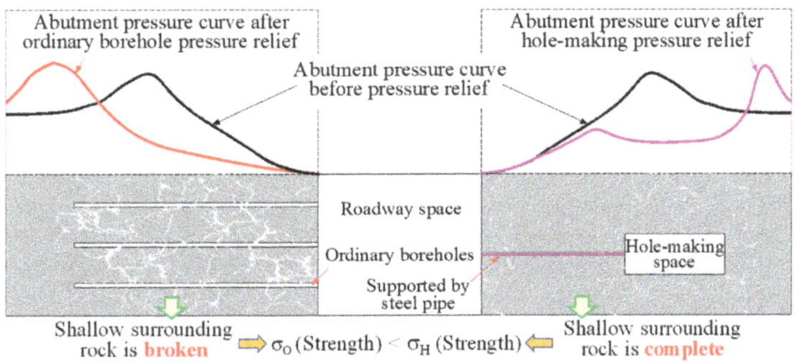

Figure 1. Comparison between the internal hole-making pressure relief technology and conventional pressure relief technology of surrounding rock.

2. Project Overview

2.1. Engineering Geology and Problems

The headgate 11231 of a coal mine is located in the 1100 southern mining area. In the south of the headgate is solid coal panel 11231, in the northern part is fault SF_{86}, in the eastern part is the −760 m horizontal main roadway, and in the western part is the concentrated roadway of the 1100 southern mining area. The average buried depth of the headgate is 740 m, and the roadway size is 5.0 × 3.5 m (width × height), with an anchor mesh cable combined support. The average thickness of the No. 2 coal seam is 4.5 m, and the average dip angle is 14°. Panel 11231 is a gangue backfilling panel, and the compressed ratio of the gangue after compaction is approximately 80%. A drawing of the location and columnar of panel 11231 is shown in Figure 2.

In the process of advancing each panel in the 1100 southern mining area, although the goaf is filled with gangue, the surrounding rock of the section roadway within 90–120 m in front of the panel still has large deformation owing to the mining dynamic pressure. As the mining progresses, the surrounding rock is continuously renovated by a team to ensure normal mining, as illustrated in Figure 3. According to statistics, during the service period of the 1100 southern mining area, the section roadways need to be renovated at least once, and some sections even require two to three renovations. The high frequency of renovation work has greatly increased the cost of roadway support and the labor intensity of workers. Moreover, it has seriously restricted the safe and efficient production of the mine. To solve the engineering problem caused by the repeated roadway renovation and improve the production efficiency of the mine, this study selects a reasonable position of headgate 11231 in a coal mine as a test roadway for applying the new support technology. The position of the test roadway is displayed in Figure 2.

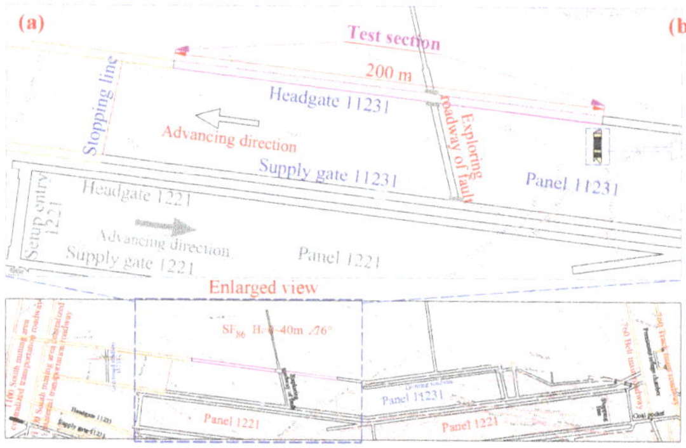

(b)

Lithology	Columnar	Thickness/m	Depth/m
No.1 coal seam		0.74	722.14
Siltstone		3.78	725.92
Medium grained sandstone		6.05	731.97
Siltstone		3.78	735.75
Mudstone		0.10	735.85
No.2 coal seam		4.50	740.35
Siltstone		1.00	741.35
No.2 lower coal seam		1.60	742.95
Siltstone		3.20	746.15
Medium fine sandstone		3.91	750.06

Figure 2. Location of panel 11231 and the columnar section of the coal and rock strata. (**a**) Location of panel 11231; (**b**) Columnar section of the coal and rock strata.

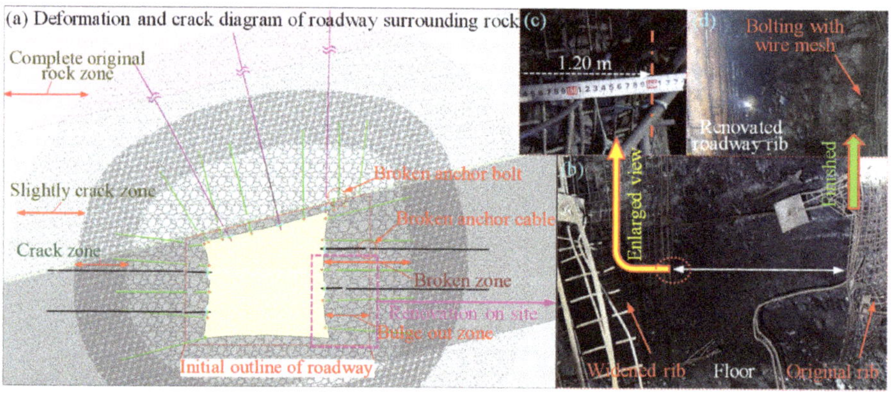

Figure 3. Renovation site of headgate 11231. (**a**) Deformation and crack diagram of roadway surrounding rock; (**b**) Roadway rib renovation site; (**c**) Repair depth of roadway rib; (**d**) The roadway rib has been renovated.

As indicated in the curve chart in Figure 4, the accumulated approach of the two ribs is approximately 610 mm (no mining influence stage) between the completion of headgate 11231 and 9 months before the mining dynamic pressure. The influence range of the increased dynamic pressure of panel 11231 is approximately 90 m. Because of the influence of mining dynamic pressure on the panel, the deformation of the surrounding rock increases sharply. The data show that within the period (approximately 21 days) from the influence of dynamic pressure on the roadway surrounding rock to 30 m from the panel, the accumulated approach of the two ribs of the roadway surged to 1240 mm (mining influence stage). Then, the roadway was renovated, and the displacement monitoring was performed again. The accumulated approach of the two ribs of the renovated roadway to the completion of mining of the panel was approximately 520 mm (mining influence stage). Therefore, the total approach of the two ribs of the roadway is approximately 1760 mm. The continuous large-scale extrusion of the two ribs of headgate 11231 (Figure 4(II)-a–c) led to damage of the support structure (Figure 4(I)-a–c). The reduction in support strength of the roadway rib coal aggravated the continuous large deformation of the coal body, forming a vicious cycle of large deformation of the surrounding rock and damage of the support body, which posed a great threat to the safety of underground personnel and equipment.

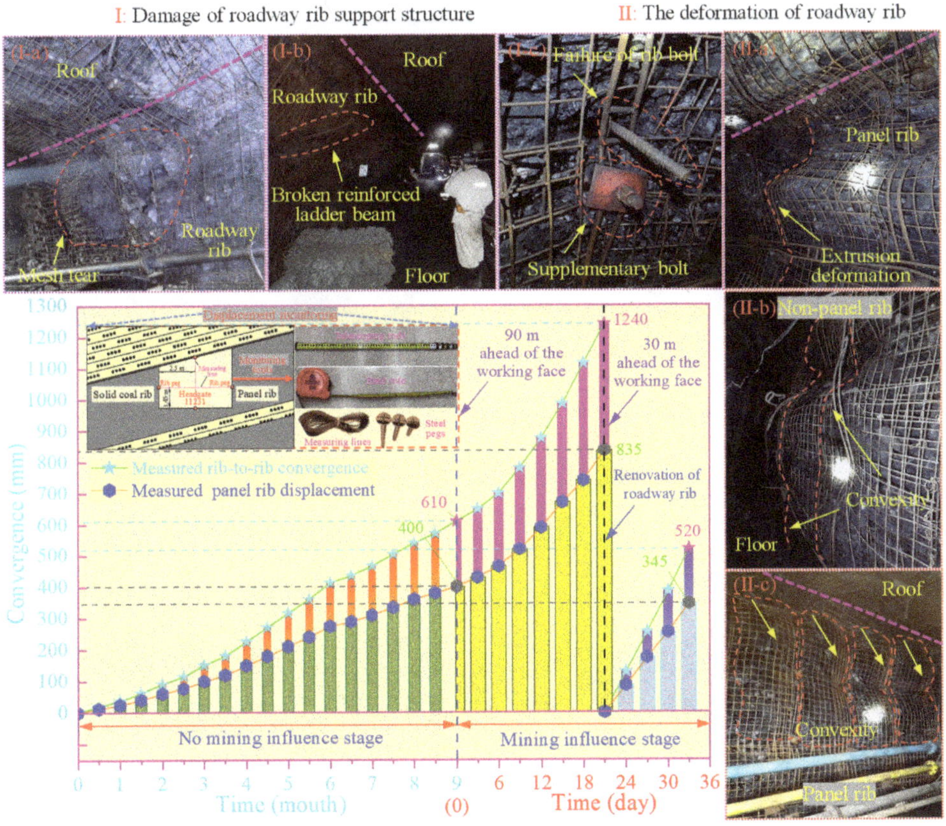

Figure 4. Deformation overview of the two ribs of headgate 11231 surrounding rock. (**I-a**) Tearing of metal mesh; (**I-b**) Large deformation of roadway rib; (**I-c**) Anchor bolt failure; (**II-a**) Large deformation of the panel rib; (**II-b**) Large deformation of non-panel rib; (**II-c**) Roadway rib bulges out.

2.2. Analysis of Control Difficulties of Roadway Surrounding Rock

Through investigation and analysis of the mining and production geological conditions of panel 11231, the reasons for the large deformation of headgate 11231 are as follows:

(1) High in situ stress. The average buried depth of headgate 11231 is 740 m, and the primary rock stress reaches 20 MPa. After roadway excavation, the shallow surrounding rock changed from a three-dimensional to a two-dimensional stress state. Moreover, the stress was redistributed, which increased the surrounding rock stress by two to three times at a certain depth. The surrounding rock of the roadway deteriorates rapidly owing to the high in situ stress and stress state change (as depicted in Figure 5a, the roadway in this section (near panel 11231) experiences deformation of surrounding rock dominated by vertical stress, which is mainly manifested in large roof subsidence).

(2) Complex tectonic stress. The geological structure of the 1100 southern mining area is complex, and the tectonic stress caused by tectonic movement has a great influence on the stability of the roadway surrounding rock. As illustrated in Figure 5b, this section of the roadway (near panel 11231) undergoes deformation of surrounding rock dominated by tectonic stress, which is mainly manifested by the large amount of movement of the two ribs getting closer.

(3) The coal body is soft and broken. The strength of the coal body in a coal mine is low, owing to the loose and soft properties of the coal body. The borehole peep (Figure 5c-I) shows that the coal body is broken when the roadway rib depth is 3 m. It can be seen from Figure 5c-II that when the roadway rib has no large deformation, the soft coal body in the roadway rib peels off under an effective support. The low strength and weak self-supporting capacity of the coal body is the main reason for the large deformation of headgate 11231.

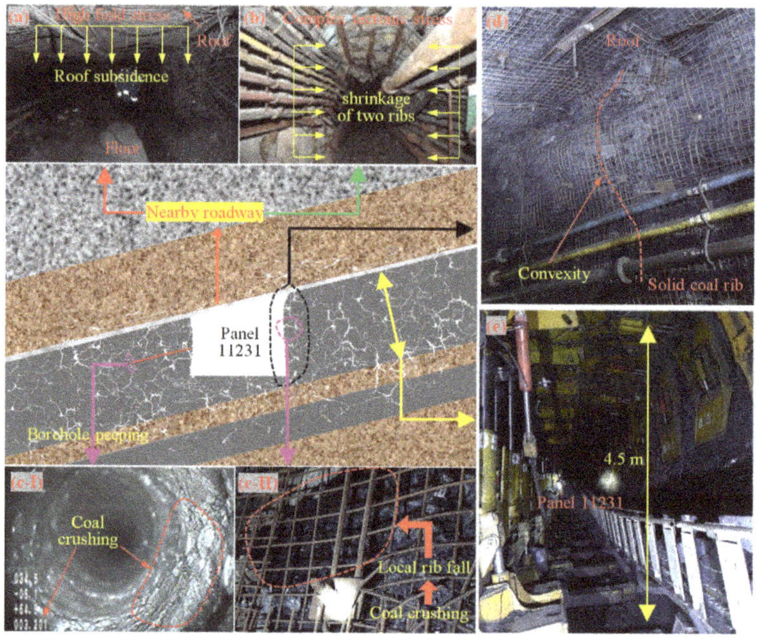

Figure 5. Schematic for the analysis of surrounding rock failure difficulties of headgate 11231. (**a**) Roadway deformation caused by high field stress; (**b**) Roadway deformation caused by complex tectonic stress; (**c-I**) Borehole peeping; (**c-II**) Crushed coal; (**d**) Large deformation of roadway rib; (**e**) Large mining height.

(4) The roadway section is large. Headgate 11231 has a center height of 3.5 m and a roadway width of 5 m. Research shows that the larger the roadway section, the easier it is to cause greater stress concentration, and the more serious the deformation and damage of the roadway surrounding rock. In addition, headgate 11231 is a trapezoidal roadway driven along the coal seam roof, and the height of the roadway panel rib (higher rib) reaches 4.2 m. The irregular cross-sectional shape causes the surrounding rock of the roadway to be subjected to unbalanced pressure, and the surrounding rock of the roadway undergoes asymmetric deformation. The displacement monitoring data show that the deformation of the higher rib of the headgate is larger than that of the non-panel rib (lower rib). As displayed in the curve data in Figure 4, the deformation of the higher rib accounts for 65.6 and 67.0% of the accumulated approach of the two ribs in the stage without and with dynamic pressure, respectively.

(5) Strong influence of mining. The average mining height of panel 11231 is 4.5 m. As indicated in Figure 5e, although the goaf is filled with gangue, owing to the low compressed ratio and large mining height, the overburden movement is strong after mining, and the strong disturbance range of the panel is up to 90 m. The broken coal body of the roadway surrounding rock within the disturbance range can easily undergo large deformation.

3. Principle and Key Technical Parameters of the Cooperative Control Technology of External Anchor–Internal Unloading of Coal Roadway Surrounding Rock

According to the above analysis, to solve the problem of repeated renovation of mining roadways caused by their large deformation, technical research should be conducted from the following two perspectives: (1) complex stress environment in coal roadways (difficulties 1, 2, and 5) and (2) soft and broken characteristics of surrounding rock (difficulty 3). Based on this, the cooperative control technology of external anchor–internal unloading of coal roadway surrounding rock is proposed in this study.

3.1. Principle of Cooperative Control Technology of External Anchor–Internal Unloading of Coal Roadway Surrounding Rock

The cooperative control technology of external anchor–internal unloading refers to strengthening the shallow surrounding rock of the roadway through the combined strategy of anchor-grouting and then using physical means to make pressure relief holes with reasonable spacing in a certain range of the deep part of the roadway to improve the stress environment of the roadway surrounding rock. The pressure relief holes can also provide a large compensation space for the deep coal body to transfer to the roadway space and effectively block the intermediate source of large deformation of roadway surrounding rock. In addition, the strengthening of the surrounding rock in the shallow part of the roadway can restrict the shallow coal body from moving to the pressure relief space. Generally, the internal pressure relief method involves making large-diameter holes with reasonable spacing in a certain range in the deep part of the roadway by using hydraulic hole-making equipment. The method mainly includes determining reasonable technical parameters using geological steel pipes with an appropriate diameter to effectively support the ordinary boreholes in the shallow part of the roadway and creating large-diameter holes in the deep part of the roadway.

Figure 6 illustrates the principle of cooperative control technology of external anchor–internal unloading. This technology mainly includes two aspects, namely "external anchor" and "internal unloading". As mentioned above, the external anchor is a high-efficiency and strong pre-tightening support structure of anchor cable truss beams formed by strong anchor cables and a channel steel or steel belt beam. The shallow coal roadway is grouted to form an anchor-grouting reinforced bearing body of shallow surrounding rock of the roadway (Figure 6A), which creates a good surrounding rock environment for internal hole-making and pressure relief. Internal unloading mainly includes two objectives. (1) The pressure relief space in the deep part of the roadway makes the peak area of the abutment stress caused by the roadway rib substantially transfer to the deep part (Figure 6B,C), which reduces continuous damage (due to high concentrated stress) to the shallow coal and rock mass and improves its stress environment. (2) Continuous large-diameter holes provide a large compensation space for the transfer of deep coal body to the roadway space and effectively block the intermediate source of large deformation of roadway surrounding rock (Figure 6D). It should be pointed out that the structural integrity of the surrounding rock in the shallow anchorage zone should not be damaged during internal hole-making and pressure relief, and the pipe-fixing method can be used to effectively support the shallow ordinary drilling area to ensure that the strength of the coal and rock mass in this area is not reduced by the hole-making.

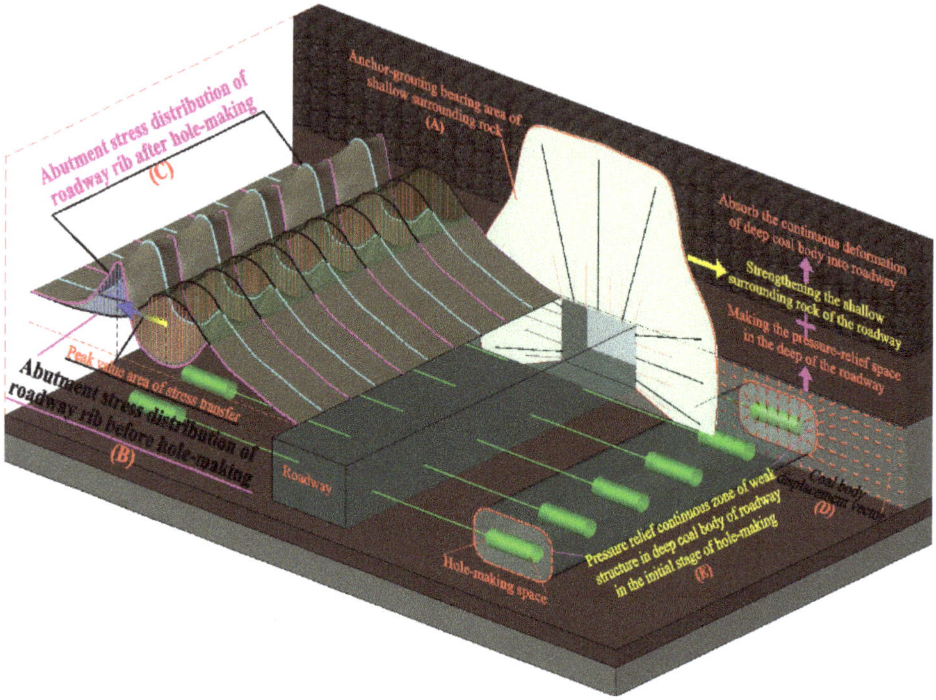

Figure 6. Principle of cooperative control technology of external anchor–internal unloading of roadway surrounding rock.

3.2. Key Technical Parameters of Internal Hole-Making and Pressure Relief

Effective implementation of the internal hole-making and pressure relief technology mainly involves the reasonable selection of three technical parameters, as shown in Figure 7.

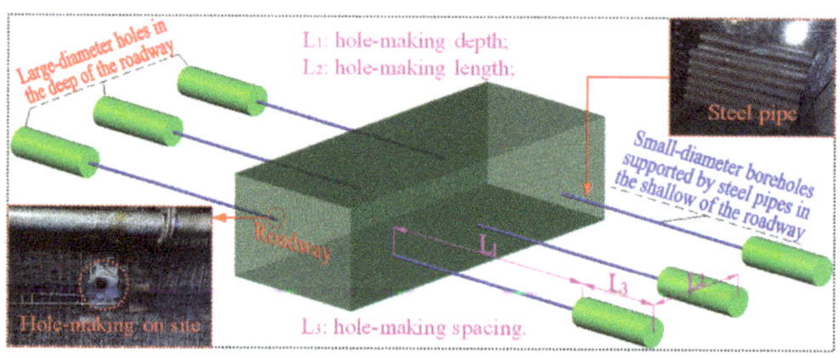

Figure 7. Schematic for the layout of hole-making in the two ribs of the roadway.

(1) Hole-making depth L_1. The hole-making depth affects the abutment stress distribution of the roadway rib and the anchorage range of the shallow surrounding rock after the completion of roadway excavation. When the hole-making depth is extremely large, it cannot effectively relieve the pressure, and even more seriously, it may increase the stress peak in the shallow anchorage zone and affect the stability of the shallow surrounding rock. On the other hand, when the hole-making depth is extremely small, the stress transfer effect is not significant, and the hole-making space can easily cause damage to the shallow anchorage zone.

(2) Hole-making spacing L_2. After the hole-making space is created, the coal body outside the hole-making space migrates to this space, forming a certain range of surrounding

rock loose areas. Appropriate spacing of hole-makings can connect the loose areas of two adjacent holes-making along the axial direction of the roadway, and the loose areas of continuous hole-making connect with each other in the axial direction of the roadway to form a pressure relief continuous zone (Figure 6E). When the spacing is extremely large, the pressure relief continuous zone cannot be formed. In addition, the coal body between the hole-making spaces is still in a state of high concentrated stress, and the surrounding rock pressure relief effect is poor. In contrast, when the spacing is extremely small, dense ordinary boreholes cause great damage to the shallow surrounding rock, which affects its integrity, and at the same time increases the labor of workers and reduces construction efficiency.

(3) Hole-making length L_3. The hole-making length has a significant influence in terms of two aspects: (1) the greater the hole quantity and length, the greater the coal output and the more obvious the compensation effect of the hole-making space on deep coal; (2) pressure relief amplitude, namely, the transfer distance of highly concentrated stress. In general, when the hole-making depth is appropriate, the greater the hole-making length, the greater the transfer distance of the peak value of the roadway rib abutment stress to the deep rock. When the hole-making length is extremely small, the pressure relief range is small, the pressure relief effect is not sufficient, and the pressure relief hole-making is closed in a short time, preventing its blocking effect on the transfer of deep coal body to the roadway space. On the other hand, when the hole-making length is extremely large, the deeper hole-making space has little effect on the pressure relief of the shallow surrounding rock of the roadway.

Owing to the large dip angle of the coal seam and limited by the technical capability of the hydraulic hole-making equipment, the coal output effect of the internal hole-making space in the lower rib of the headgate 11231 test section is poor and cannot meet the technical requirements. Meanwhile, comparing the displacement monitoring data of headgate 11231, the higher rib deformation of the headgate accounts for more than 65% of the total displacement of the two ribs of the roadway. Therefore, effectively limiting the deformation of the higher rib surrounding rock of the headgate can greatly relieve the renovation pressure of headgate 11231 and meet the engineering requirements of the panel.

4. Study on Key Technical Parameters of Pressure Relief by Internal Hole-Making

4.1. Establishment of Numerical Model and Research Ideas

To determine the key technical parameters of pressure relief by internal hole-making in deep coal roadways affected by dynamic pressure, a numerical simulation study was carried out according to the actual situation on-site and the technical principle of pressure relief by internal hole-making combined with the FLAC3D finite element software. The numerical model is depicted in Figure 8. The numerical model size is 130 × 70 × 90 m. The length of panel 11231 is 59 m. The mining height is 4.5 m, and the dip angle of the coal and rock strata is 14°. Considering that panel 11231 is a gangue backfilling panel, the final compressed ratio is approximately 80%. It is considered that the equivalent mining height of the panel is 0.9 m. The mining roadways are arranged along the coal seam roof. The roadway width is 5 m, the center height of the roadway is 3.5 m, the distance from the borehole to the roadway floor is 1.5 m, and the upward angle of the hole-making in different schemes ranges from 3.0 to 8.5°. Each coal and rock stratum in the numerical model adopts the Mohr-Coulomb constitutive model. The left and right boundaries of the model are fixed with horizontal displacement along the x-direction, the front and rear boundaries of the model are fixed with horizontal displacement along the y-direction, and the bottom boundary of the model is fixed with z-displacement along the vertical direction. Moreover, a load of 17.115 MPa is applied to the top boundary of the model to simulate the overburden weight. The coefficient of lateral pressure of the model is 1.2. Based on the basic mechanical parameters of coal rock mass measured in the laboratory and data from the literature, the mechanical parameters were calculated [34–36]. Table 1 presents the physical and mechanical parameters of each coal and rock stratum.

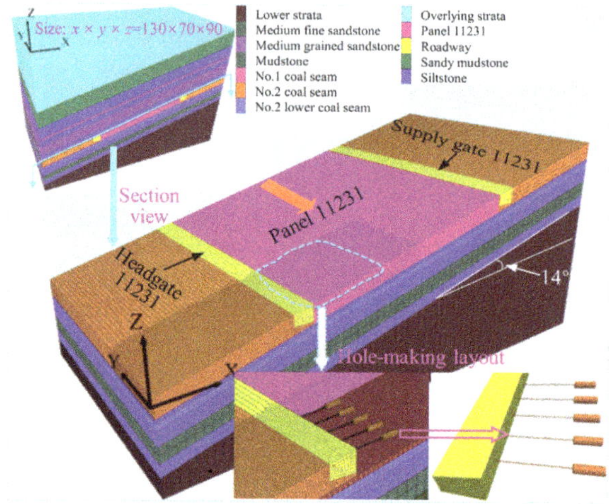

Figure 8. Numerical simulation model.

Table 1. Mechanical parameters of coal and rock strata.

Rock Strata	K/GPa	G/GPa	C_m/MPa	σ_{tm}/MPa	φ_m/(°)	D/(kg·m^{-3})
No. 1 coal seam	2.60	1.60	1.00	1.32	25.0	1.40
Siltstone	4.77	4.59	3.60	3.15	32.5	2.65
Sandy mudstone	4.87	4.18	3.20	3.10	35.0	2.40
Medium grained sandstone	4.50	4.10	3.15	3.05	33.0	2.65
No. 2 coal seam	2.60	1.60	1.00	1.32	25.0	1.40
Mudstone	4.20	4.10	2.00	2.00	28.0	2.40
No. 2 lower coal seam	2.60	1.60	1.00	1.32	25.0	1.40
Medium fine sandstone	4.77	4.88	4.10	3.80	33.0	2.65
Overlying strata	4.87	4.18	3.20	3.10	35.0	2.40
Lower strata	4.77	4.59	3.60	3.15	32.5	2.65

Given the key technical parameters (L_1, L_2, and L_3) of pressure relief by internal hole-making of headgate 11231 in a coal mine, a variety of numerical simulation schemes were set up by using a control variable method. The goals were to study the vertical stress distribution law of the surrounding rock under different hole-making parameters and determine the final hole-making parameters of the roadway. As presented in Table 2, there are 15 numerical simulation schemes in total (the technical parameters of Schemes 3, 8, and 13 are the same).

Table 2. Numerical simulation schemes.

<table>
<tr><th>Scheme Number</th><th colspan="7">Technical Parameter of Hole-Making</th></tr>
<tr><td rowspan="2">Schemes 1–5</td><td>L_2/m</td><td>L_3/m</td><td colspan="5">L_1/m</td></tr>
<tr><td>4.0</td><td>3.0</td><td>4.0</td><td>6.0</td><td>8.0</td><td>10.0</td><td>12.0</td></tr>
<tr><td rowspan="2">Schemes 6–10</td><td>L_1/m</td><td>L_3/m</td><td colspan="5">L_2/m</td></tr>
<tr><td>8.0</td><td>3.0</td><td>2.4</td><td>3.2</td><td>4.0</td><td>4.8</td><td>5.6</td></tr>
<tr><td rowspan="2">Schemes 11–15</td><td>L_1/m</td><td>L_2/m</td><td colspan="5">L_3/m</td></tr>
<tr><td>8.0</td><td>4.0</td><td>1.0</td><td>2.0</td><td>3.0</td><td>4.0</td><td>5.0</td></tr>
</table>

According to the field investigation, the surrounding rock deformation of headgate 11231 can be divided into two stages: the no mining influence stage and the mining influence stage. Based on the displacement monitoring data in Figure 4, the displacement of the two ribs and deformation of the higher rib, respectively, account for 34.1 and 33.9% of the total deformation in the no mining influence stage. That is, the deformation of the two ribs of the coal roadway in the no mining influence stage accounts for approximately 1/3 of the total deformation, and the roadway first goes through the no mining influence stage after the support is completed. In this stage, the surrounding rock of the deep coal roadway is mainly affected by the high abutment stress of the roadway rib formed after roadway excavation. Therefore, it is necessary to study the pressure relief law of the coal roadway surrounding rock under different hole-making parameters in the no mining influence stage after roadway excavation to determine the appropriate hole-making parameters and reduce the deformation of the roadway rib at this stage. The abutment pressure caused by coal mining has changed the original stress distribution state of the roadway surrounding rock. The superposition of the advanced abutment pressure and the original high abutment stress of the roadway rib makes the stress of the roadway rib (panel rib) complex and changeable. Based on the key technical parameters of pressure relief by hole-making determined in the stage without mining influence, the parameters of pressure relief by hole-making are continuously optimized so that it can effectively relieve the surrounding rock pressure and reduce the influence of the superimposed high stress on the roadway rib on the surrounding rock in the mining influence stage.

4.2. Key Technical Parameters of Pressure Relief by Internal Hole-Making in the No Mining Influence Stage

(1) Hole-making depth

The vertical stress distributions in the higher rib of the roadway with different hole-making depths are illustrated in Figure 9b–f. When the hole-making depths are 4, 6, and 8 m, the peak stress of the higher rib of the roadway is located in the deeper part of the hole-making space. When the hole-making depth increases from 8 m to 10 m and 12 m, a new high stress peak zone is formed between the roadway space and the hole-making space, which is not conducive to the stability of the roadway surrounding rock. The stress monitoring lines are arranged at the higher rib of the roadway along the direction of hole-making, as displayed in Figure 9(A_1-A_1) to (F_1-F_1). According to the stress distribution value of the higher rib of the roadway under the conditions of no hole-making and different hole-making depths, a comparison diagram of the stress curve of the roadway rib is drawn, as indicated in Figure 10.

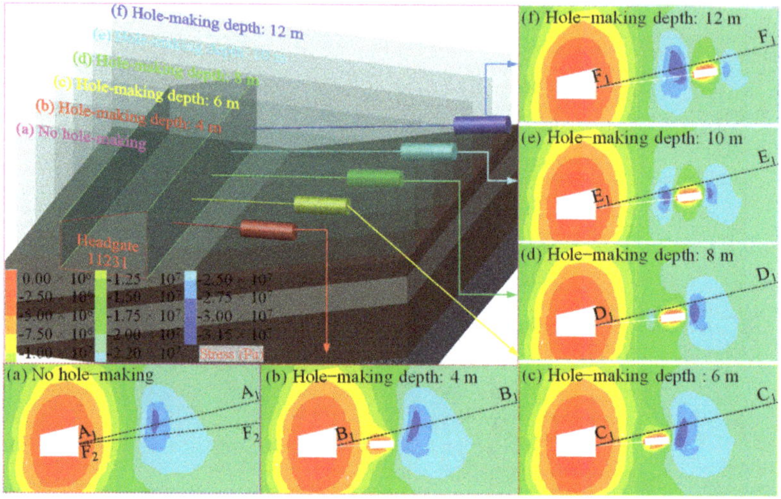

Figure 9. Layout scheme under different hole-making depths and distribution nephogram of vertical stress.

To effectively evaluate the pressure relief effect of different hole-making depths on the surrounding rock, two evaluation indices were selected according to the stress transfer law. First is the transfer amplitude of the high concentrated stress, which is the transfer distance from the original stress peak position to the deep part of the roadway. Second is the reduction in the high concentrated stress in the shallow surrounding rock of the roadway, that is, the pressure relief effect on the surrounding rock in the high abutment stress zone of the roadway rib after roadway excavation. The pressure relief effect from high to low is categorized as excellent, good, fair, no, and poor. As shown in Figure 10a, when the hole-making depth is 4 m, the transfer distance from the original stress peak position to the deep part is 0.5 m. The pressure relief space only reduces the stress in the low stress zone of the shallow roadway, which cannot achieve the purpose of inward movement of the original high concentrated stress of the roadway rib, and there is almost no pressure relief effect. As depicted in Figure 10b, when the hole-making depth is 6 m, the transfer distance is 1.5 m. The pressure relief space makes the stress in the original stress peak zone of the roadway rib decrease significantly: the stress at the peak position decreases by 7.6 MPa, and the high stress zone transfers to the deep part. Thus, the pressure relief effect is evident. As illustrated in Figure 10c, when the hole-making depth is 8 m, the transfer distance is 3.0 m. The pressure relief space makes the stress in the original stress peak zone of the roadway rib decrease overall: the stress at the peak position decreases by 17.3 MPa, and the high stress zone transfers to the deep part considerably. Therefore, the pressure relief effect is remarkable. As displayed in Figure 10d,e, when the hole-making depths are 10 and 12 m, the stress distributions on the original stress peak zone and shallower surrounding rock are not changed significantly after pressure relief. Furthermore, the original stress peak position is not moved inward effectively, and even the shallow abutment stress increases locally, aggravating the damage on the surrounding rock.

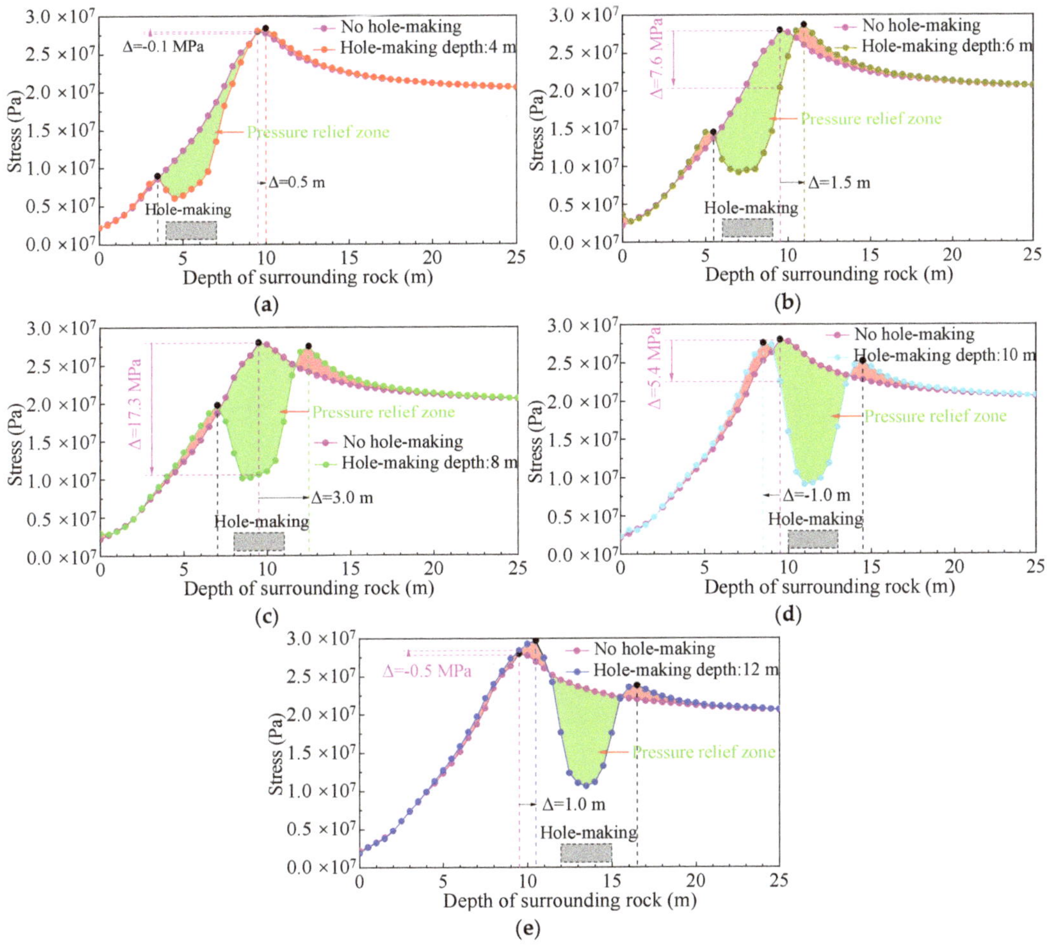

Figure 10. Comparison diagram of the roadway rib stress curve at different hole-making depths. (**a**) Hole−making depth = 4 m; (**b**) Hole−making depth = 6 m; (**c**) Hole−making depth = 8 m; (**d**) Hole−making depth = 10 m; (**e**) Hole−making depth = 12 m.

(2) Hole-making spacing

The vertical stress distributions on the higher rib of the roadway with different values of hole-making spacing are indicated in Figure 11a–e. When the hole-making spacing is 2.4 and 3.2 m, a good pressure relief zone can be formed between the hole-making spaces, so that multiple holes form a continuous pressure relief zone along the axis of the roadway. Moreover, the high concentrated stress in the shallow part of the surrounding rock can be uniformly transferred to the deep surrounding rock. Thus, the pressure relief effect is good. When the hole-making spacing is increased to 4.0 m, the transfer effect of the high concentrated stress of the coal mass between the two holes to the deep surrounding rock becomes worse, and a stress concentration zone (Zones A and B) near the inner and outer ends of the hole-making area appears. When the hole-making spacing is 4.8 m, Zones A and B are connected, and the stress of the coal mass between the two holes is restored to the original stress state, resulting in no pressure relief effect. When the hole-making spacing continues to increase to 5.6 m, the high stress moving inward to the deep part accumulates in the coal mass between the two holes, forming a stress-increasing area, which not only has no pressure relief effect, but also causes great damage to the coal mass.

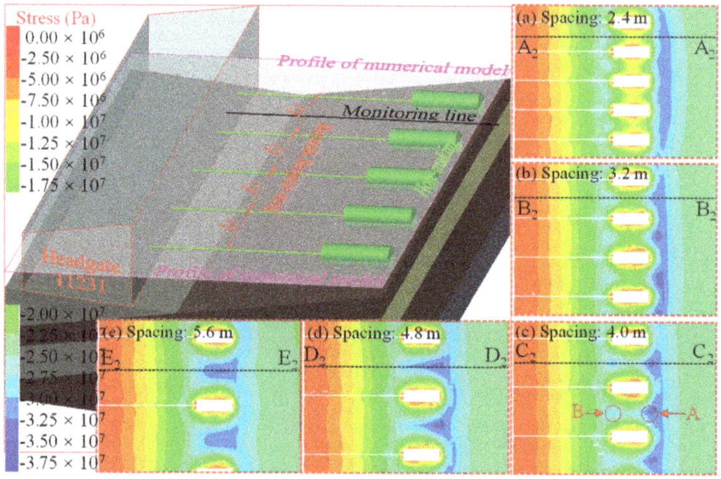

Figure 11. Layout scheme under different hole-making spacing and distribution nephogram of vertical stress.

To accurately evaluate the pressure relief effect of different values of hole-making spacing on the surrounding rock, the two evaluation indices proposed above were used. It can be seen from Figure 11 that the coal mass between the two holes (the position with the worst pressure relief effect) has the greatest impact on the stress change of the surrounding rock with different values of hole-making spacing. Therefore, the vertical stress monitoring lines ((F2-F2) in Figure 9 and (A2-A2) to (E2-E2) in Figure 11) were arranged at the center of the two holes and parallel to the holes in the different hole-making spacing schemes. Figure 12 presents the stress curve comparison diagram of the roadway rib under different values of hole-making spacing.

As shown in Figure 12a, when the hole-making spacing is 2.4 m, the transfer distance from the original stress peak position to the deep part is 3.0 m. After pressure relief, the stress in the original stress peak zone decreases significantly, and the stress at the peak position decreases by 13.5 MPa. Thus, the pressure relief effect is excellent. As depicted in Figure 12b, when the hole-making spacing is 3.2 m, the transfer distance is 2.5 m. After pressure relief, the stress at the original stress peak position decreases by 6.0 MPa, and the pressure relief effect is good. When the hole-making spacing is increased to 4.0 m, the stress in the peak zone of the original stress does not decrease significantly after pressure relief but increases slightly in local areas, and the pressure relief effect is generally fair. As illustrated in Figure 12d,e, when the hole-making spacing values are 4.8 m and 5.6 m, the stress in the original stress peak zone increases significantly after pressure relief. The stress at the peak position increases by 3.8 MPa and 4.5 MPa, respectively, and the transfer distances are 2.0 m and 1.0 m, respectively. Hence, the pressure relief effect is fair.

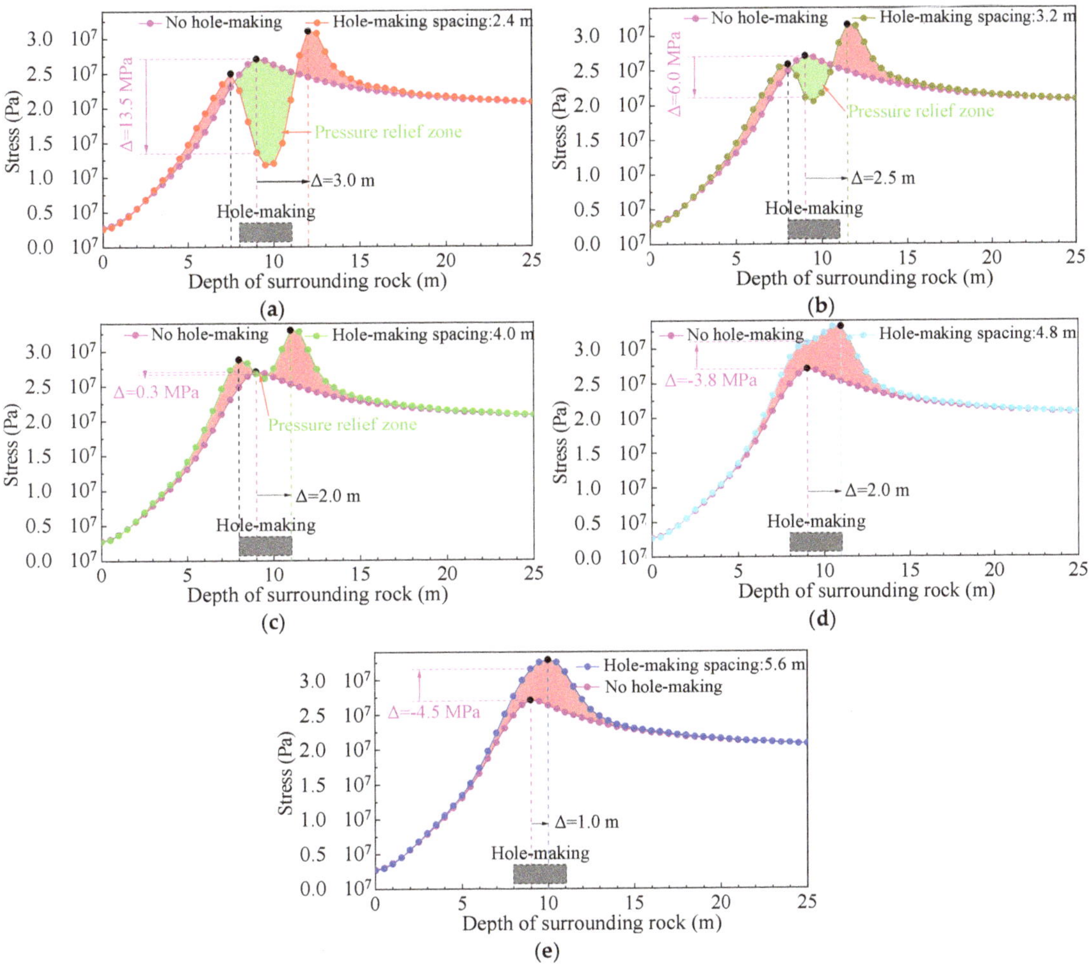

Figure 12. Comparison diagram of the roadway rib stress curve at different vales of hole-making spacing. (**a**) Hole−making spacing = 2.4 m; (**b**) Hole−making spacing = 3.2 m; (**c**) Hole−making spacing = 4.0 m; (**d**) Hole−making spacing = 4.8 m; (**e**) Hole−making spacing = 5.6 m.

(3) Hole-making length

The vertical stress distributions of the higher rib of the roadway with different hole-making lengths are indicated in Figure 13b–f. With an increase in the hole-making length, the transfer distance from the peak stress zone of the roadway rib to the deep part gradually increases, and the range of the stress peak zone of the roadway rib gradually decreases. When the hole-making length increases to 3 m, the range of the stress peak zone of the increased hole-making length does not change significantly.

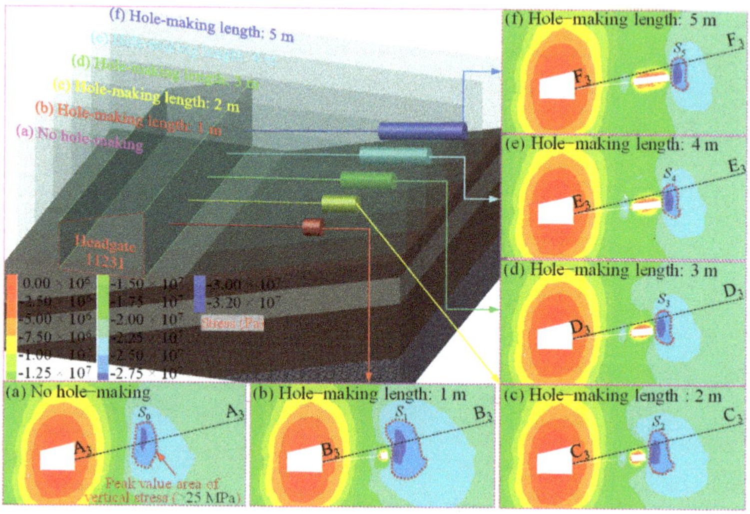

Figure 13. Layout scheme under different hole-making lengths and distribution nephogram of vertical stress.

The layout of stress monitoring lines is shown in Figure 13(A_3-A_3) to (F_3-F_3), and Figure 14 presents a comparison of the roadway rib stress curves with different hole-making lengths. The two indices proposed above were used to evaluate the pressure relief effect of the surrounding rock with different hole-making lengths. In addition, to analyze the influence of the different hole-making lengths on the range of the peak stress zone, the stress of 25 MPa was defined as the stress boundary value to measure the range of the stress peak zone. Thus, the reduction coefficient K of the surrounding rock stress peak zone under different hole-making lengths is

$$k = \frac{S_i}{S_0} \times 100\% \tag{1}$$

where S_i is the area of the stress peak zone (greater than 25 MPa) when the hole-making length is 1–5 m, i = 1–5; and S_0 is the area of the stress peak zone (greater than 25 MPa) without hole-making. By defining the area of the original stress peak zone (greater than 25 MPa) as 1, measuring Figure 13, and combining with Equation (1), we obtain k_i = 1.90, 0.90, 0.56, 0.51, and 0.54 (i = 1–5). Analysis shows that the smaller the value of k, the greater the pressure relief effect.

As depicted in Figure 14a–e, when the hole-making lengths are 1, 2, 3, 4, and 5 m, the transfer distances from the original stress peak position to the deep part are 1.5, 2.0, 3.0, 3.5, and 5.0 m, respectively, and the stress reduction values at the peak position are 4.8, 14.8, 17.3, 19.31, and 20.3 MPa, respectively. From the stress curve, it can be observed that the different hole-making lengths have good pressure relief effects. To effectively evaluate the cost performance K of the different hole-making lengths in transferring the original stress peak point to the deep part, the ratio of the transfer distance from the original stress peak position to the deep part to the hole-making length (which can represent the construction cost, that is, the greater the hole-making length, the greater the construction cost) is calculated as follows:

$$K = \frac{l_i}{L_{2-i}} \times 100\% \tag{2}$$

where l_i is the transfer distance (in m) from the original stress peak position to the deep part, i = 1–5; and L_{2-i} is the hole-making length (in m), i = 1–5. Substituting the known data into Equation (2), we obtain K_i = 1.5, 1.0, 1.0, 0.875, and 1.0, respectively. It can be observed from the analysis that the greater the K value, the higher the cost performance of hole-making and pressure relief.

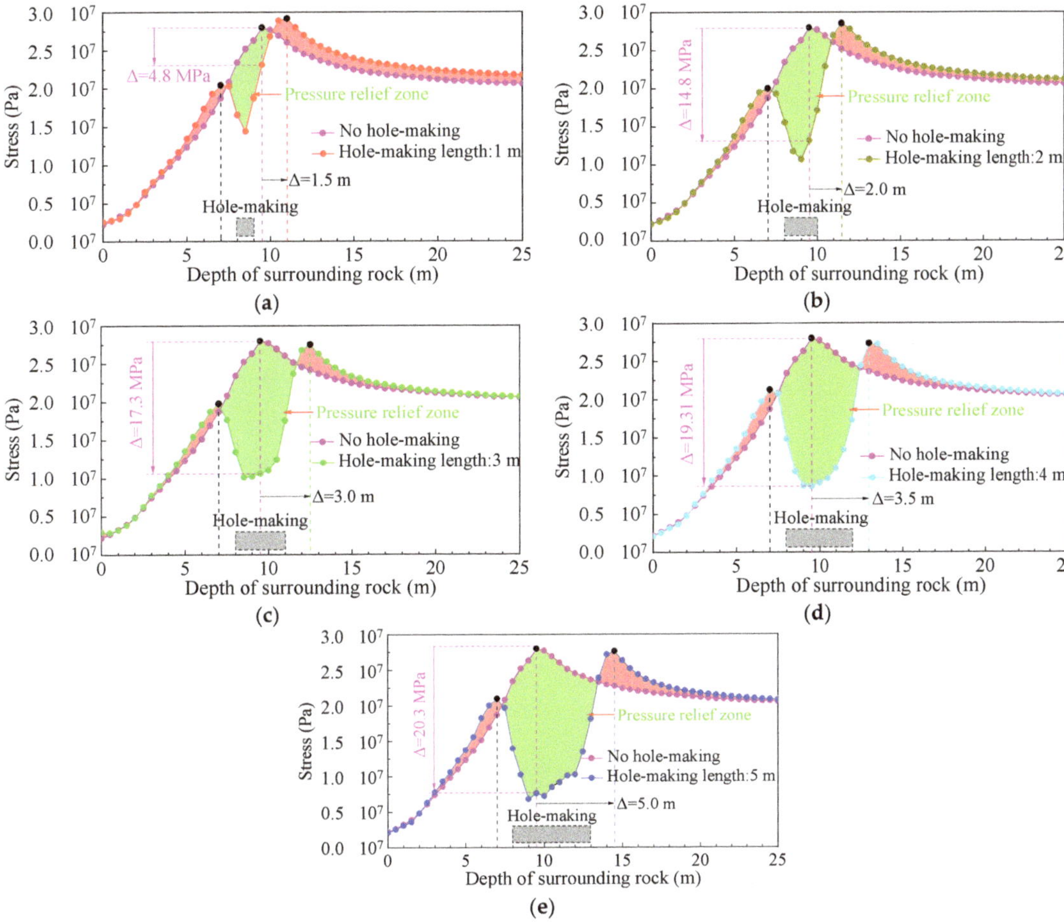

Figure 14. Comparison of the roadway rib stress curves with different hole-making lengths. (**a**) Hole-making length = 1 m; (**b**) Hole-making length = 2 m; (**c**) Hole-making length = 3 m; (**d**) Hole-making length = 4 m; (**e**) Hole-making length = 5 m.

(4) Research and analysis of key technical parameters

According to the above analysis, the pressure relief effect of the different hole-making schemes was evaluated with the vertical stress as the main index. The comprehensive evaluation results of each key technical parameter evaluation index on the pressure relief effect of the surrounding rock of different hole-making schemes are listed in Table 3. According to the analysis, Schemes 2 and 3 can be selected with respect to the hole-making depth, that is, the hole-making depths are 6 and 8 m, respectively. Schemes 6, 7, and 8 can be selected in terms of the hole-making spacing; thus, the hole-making spacing values are 2.4, 3.2, and 4.0 m, respectively. When the hole-making length is 1 m, the reduction coefficient of the range of the stress peak zone $k_1 > 1$. In addition, the transfer distance from the original stress peak position to the deep part is small, so this scheme should be excluded. For the other hole-making length schemes (Schemes 12–15), it is not appropriate to analyze the pressure relief effect simply from the perspective of stress, as it should be analyzed in combination with other factors.

Table 3. Evaluation results of surrounding rock pressure relief for different hole-making schemes.

Evaluation Index		Scheme Number (Hole-making depth)				
		Scheme 1	Scheme 2	Scheme 3	Scheme 4	Scheme 5
1	Pressure relief effect	No	Good	Excellent	No	Poor
2	Transfer distance to deep/m	0.5	1.5	3.0	−1.0	1.0
Evaluation index		Scheme number (Hole-making spacing)				
		Scheme 6	Scheme 7	Scheme 8	Scheme 9	Scheme 10
1	Pressure relief effect	Excellent	Good	Fair+	Fair	Fair
2	Transfer distance to deep/m	3.0	2.5	2.0	2.0	1.0
Evaluation index		Scheme number (Hole-making length)				
		Scheme 11	Scheme 12	Scheme 13	Scheme 14	Scheme 15
1	Pressure relief effect	Fair	Good	Good	Good	Good
2	Transfer distance to deep/m	1.5	2.0	3.0	3.5	5.0
3	k	1.90	0.90	0.56	0.51	0.54
4	K	1.5	1.0	1.0	0.875	1.0

4.3. Optimization Analysis of Key Technical Parameters of Pressure Relief by Internal Hole-Making in the Mining Influence Stage

Panel 11231 adopts step-by-step mining with an interval of 5 m. Figure 15 shows the nephogram of abutment stress distribution and the stress monitoring curve in front and behind the panel near the rib of the headgate. By analyzing Figure 15, the following can be inferred: (1) The stress reduction zone is from the coal rib of the panel to 4 m in front of the panel. Moreover, the stress rise zone is 4 m in front of the panel and further away, and the distance between the stress peak position and the coal rib of the panel is 12 m. (2) Affected by the dynamic pressure of coal mining, the stress value in the higher rib of the headgate increases sharply. The vertical stress values of the coal body at depths $\geq$ 6 m are greater than the primary rock stress, forming a high concentrated stress zone. The stress peak zone of the roadway rib shifts deeper under the influence of mining, with a transfer distance of 4–5 m. (3) The advanced abutment pressure caused by coal mining and the high abutment stress of the roadway rib are superimposed on each other, which jeopardizes the stability of the roadway rib coal in the area affected by dynamic pressure. To summarize, transferring the high concentrated stress in the shallow coal body of the roadway using hole-making and reducing the continuous damage caused by the high stress to the shallow coal and rock are essential for maintaining the stability of the surrounding rock of the coal roadways affected by mining.

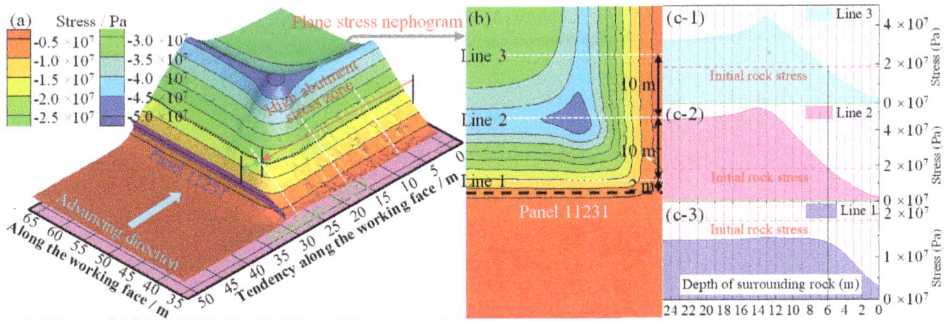

Figure 15. Distribution law of abutment stress in front of panel 11231. (**a**) 3D stress distribution of roadway rib; (**b**) Line layout; (**c-1**) 22 m in front of the panel (**c-2**) 12 m in front of the panel; (**c-3**) 2 m in front of the panel.

Based on the key technical parameters of hole-making determined at the stage without dynamic pressure influence and in combination with the dynamic pressure disturbance law of panel 11231, we continued to optimize the hole-making parameters and arranged stress monitoring lines at reasonable positions in the stress rise zone in front of the panel to analyze the stress distribution law of surrounding rock with different hole-making schemes under the influence of superimposed stress.

(1) Hole-making depth

As illustrated in Figure 16, when the hole-making depth is 6 m or 8 m, the stress decreases in the area where the stress on the higher rib of the panel increases during the dynamic pressure influence stage. Comparing Figures 15a and 16b, the following can be observed: (1) When the hole-making depth is 8 m, the range of the stress reduction zone and the pressure relief degree are larger than that when the hole-making depth is 6 m. (2) When the hole-making depth is 8 m, the pressure relief space transfers the peak zone of the superimposed stress of the roadway rib to the deeper part, and the pressure relief effect is good. (3) When the hole-making depth is 6 m, the hole-making space causes great damage to the coal body at the roadway depth of 5 m, which affects the stability of the rock mass of bolt-grouting coal in the shallow part of the roadway. Based on the above analysis, Scheme 3 is preferred with respect to the hole-making depth of headgate 11231; that is, the hole-making depth is 8 m.

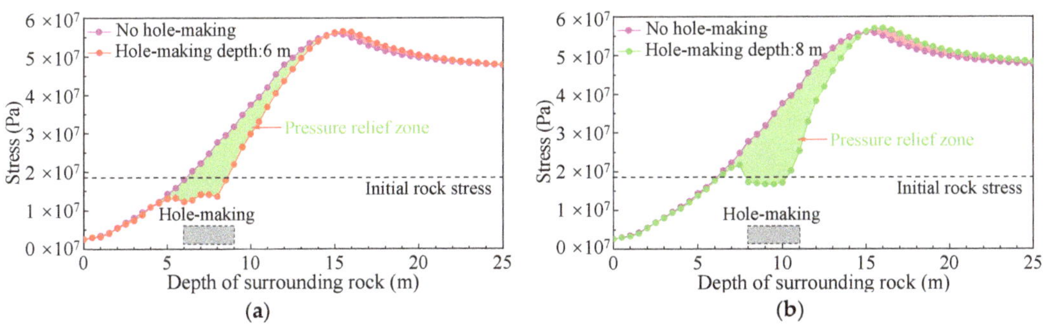

Figure 16. Comparison of stress curves of the roadway rib at different hole-making depths in the mining influence stage. (**a**) Hole-making depth = 6 m; (**b**) Hole-making depth = 8 m.

(2) Hole-making spacing

Figure 17(a-1), (b-1), and (c-1) show the vertical stress nephogram of the higher rib of the roadway with different values of hole-making spacing in the advanced dynamic pressure zone. The center positions of the two holes are respectively selected to monitor their stress values. The corresponding stress curves are displayed in Figure 17(a-2), (b-2), and (c-2). Figure 17(a-2), (b-2) indicate that the pressure relief zone is still formed at the center of the two holes. When the hole-making spacing increases to 4.0 m, stress concentration between the two holes occurs, and the hole-making space threatens the stability of the roadway rib coal body, which is not conducive to the stability of the roadway rib surrounding rock in the advanced dynamic pressure zone. The hole-making spacing is extremely small, and the boreholes are dense. The dense boreholes cause great damage to the coal body in the shallow anchorage zone of the roadway, which is not conducive to the stability of the roadway surrounding rock. Through comprehensive comparison and analysis, it is determined that the optimal hole-making spacing is 3.2 m.

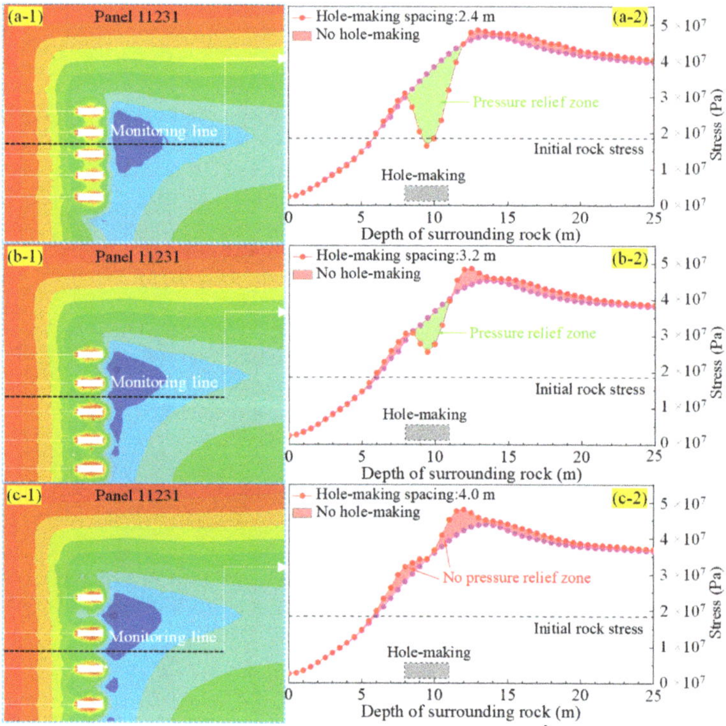

Figure 17. Stress distribution on roadway rib at different values of hole-making spacing in the mining influence stage. (**a-1**) Stress nephogram with spacing of 2.4 m; (**b-1**) Stress nephogram with spacing of 3.2 m; (**c-1**) Stress nephogram with spacing of 4.0 m; (**a-2**) Stress curve with spacing of 2.4 m; (**b-2**) Stress curve with spacing of 3.2 m; (**c-2**) Stress curve with spacing of 4.0 m.

(3) Hole-making length

As shown in Figure 18, with an increase in the hole-making length, the pressure relief range increases, the peak position of the superimposed stress shifts deeper into the surrounding rock, and the pressure relief effect becomes more evident. However, considering the construction capacity of hole-making machines and the use efficiency of the hole-making space, the hole-making length cannot be increased without limit. Comparing Figure 18a–d, the hole-making lengths of 3 and 4 m can meet the pressure relief requirements of the surrounding rock in the mining influence stage. As depicted in Figure 18b,c, when the hole-making lengths are 3 and 4 m, the effective pressure relief ranges of the surrounding rock are 4.5 m and 5.0 m, respectively. As the hole-making length increases by

1 m, the effective pressure relief range does not increase significantly. In combination with equation (2), $K_3 = 1.0 > K_4 = 0.875$. Comparing the pressure relief effect and construction quantities of the two schemes, the hole-making length of 3 m can better meet the pressure relief requirements, reduce the construction quantities, and obtain the optimal solution of hole-making pressure relief and efficient construction.

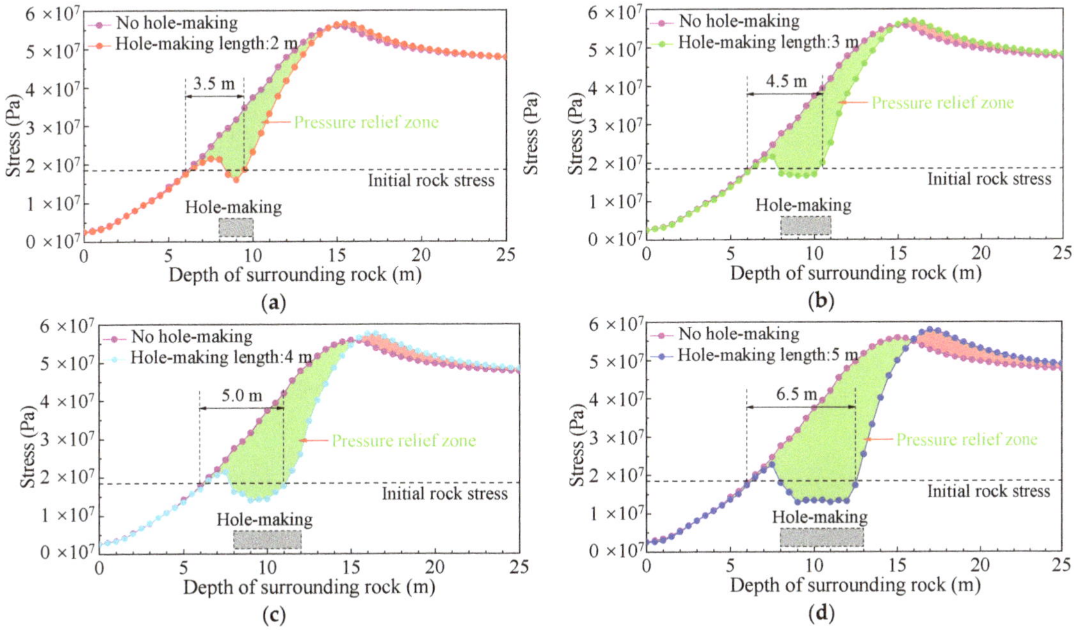

Figure 18. Comparison of stress curves of roadway rib at different hole-making lengths in the mining influence stage. (**a**) Hole-making length = 2 m; (**b**) Hole-making length = 3 m; (**c**) Hole-making length = 4 m; (**d**) Hole-making length = 5 m.

(4) Determination of key technical parameters

According to the stress distribution law of the surrounding rock with different hole-making schemes in the advanced dynamic pressure zone of panel 11231 and the comprehensive analysis of factors such as construction benefit, cost, and construction period, the key parameters for pressure relief of higher rib by hole-making in headgate 11231 were determined as follows: the hole-making depth, spacing, and length are 8 m, 3.2 m, and 3 m, respectively.

5. Cooperative Control Technology of External Anchor–Internal Unloading of Surrounding Rock in Deep Coal Roadways

5.1. Technical Parameters of External Anchor–Internal Unloading of Surrounding Rock in the Test Section of Headgate 11231

The cooperative control technology of external anchor–internal unloading of surrounding rock in the test section of headgate 11231 was implemented in two stages, namely, strengthening of the shallow surrounding rock and pressure relief by hole-making in the deep part. As illustrated in Figure 19I-a–c, reinforcement of the shallow surrounding rock is first carried out for the test section of the headgate, and three steel strand anchor cables with dimensions of φ21.8 × 10,500 mm and a row spacing of 2.0 × 1.6 m are added to the roof. Moreover, three steel strand anchor cables with dimensions of φ21.8 × 4500 mm are added to the two ribs. The row spacing between the higher ribs is 1.3 × 1.6 m, whereas that between the lower ribs is 1.1 × 1.6 m. Each anchor cable uses three resin anchor agents of model Z2360 for each hole. The roof and rib anchor cables are connected by H-type double steel belt beams supporting large square pallets (400 × 400 mm) and small pallets (200 × 200 mm) to form the roof and rib anchor cable truss beam structure, which can resist the overall outward heave of the two ribs of soft coal in the roadway. The pre-tightening force of the anchor cables is not less than 250 kN. After completing the anchor cables, shotcrete was applied on the higher rib surface of the roadway in the test section, and C20 concrete was sprayed with a thickness of 30 mm. After completion of the surrounding rock shotcrete, coal grouting was conducted on the roadway rib. Three grouting holes were arranged on the higher rib of the roadway with a spacing of 1.2 m, and two grouting holes were arranged on the lower rib with a spacing of 1.4 m. The grouting material was cement slurry and soluble silicate, in which the water–cement ratio was 1.3:1–1.4:1 (mass ratio), and the ratio of the cement slurry to soluble silicate was 1:0.5–1:1.0 (volume ratio). The cement used was P.O42.5 normal Portland cement, and the Baume density of soluble silicate (liquid sodium silicate) was 35° Bé. The depth of the grouting hole was 3 m, the grouting pressure was not less than 2.5 MPa, and the row distance of the grouting hole was 3.0 m.

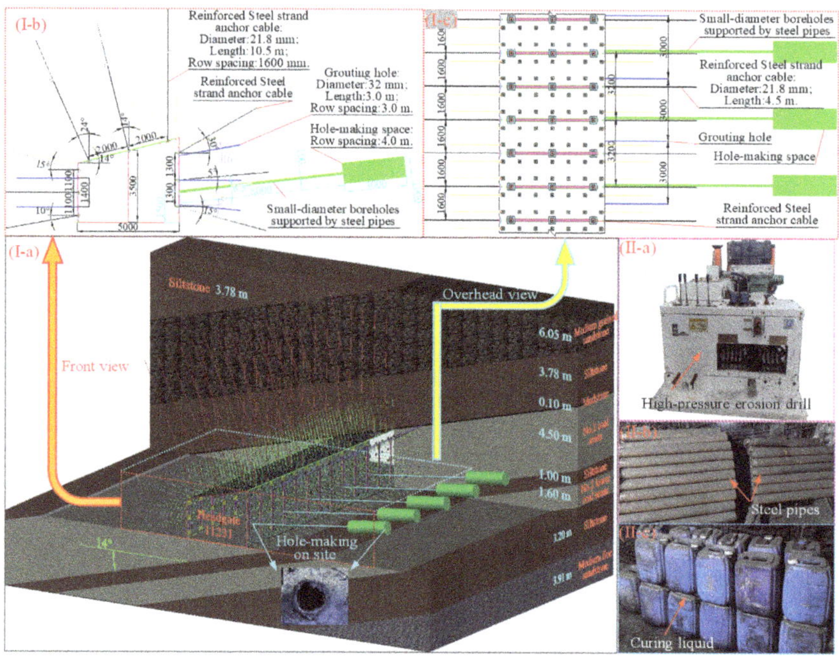

Figure 19. Schematic of cooperative control technology of external anchor–internal unloading of coal roadway surrounding rock. (**I-a**) 3D schematic diagram of roadway support; (**I-b**) Front view; (**I-c**) Top view; (**II-a**) Drilling rig; (**II-b**) Steel pipe; (**II-c**) Grouting materials.

After completing the bolt-grouting support of the shallow surrounding rock of the roadway, deep coal hole-making was performed to relieve the pressure, as indicated in Figure 19. The distance from the borehole to the roadway floor is 1.5 m, and the upward angle is 6.5°. The diameter of the shallow drilling hole is 100 mm. To protect the integrity of the shallow surrounding rock and reduce the damage caused by hole-making to the shallow coal, steel pipes were imbedded in shallow boreholes (Figure 19(II)-a), and grouting was conducted for pipe fixing. The deep part is the hole-making space, and its key parameters are as described above: the hole-making depth, spacing, and length are 8 m, 3.2 m, and 3 m, respectively.

5.2. Analysis on the Effect of Cooperative Control Technology of External Anchor–Internal Unloading in Roadway Surrounding Rock

(1) Analysis of the prestressed field of the surrounding rock reinforced anchor cables in the coal roadway test section

According to the support parameters of the original bolts and reinforced anchor cables in the test section of headgate 11231, the numerical simulation software was used to simulate the prestressed field formed by the roadway shallow surrounding rock and hole-making space, and the effective control range and control principle of the cooperative control technology of external anchor–internal unloading were comprehensively analyzed. As shown in Figure 20, two supporting bodies (anchor coal mass and hole-making space, as depicted in Figure 20a) and three zones (external anchor zone, buffer protection zone for anchorage zone, and internal unloading zone, as illustrated in Figure 20b) are formed in the roadway surrounding rock after adopting the cooperative control technology of external anchor–internal unloading. After implementing the anchor-grouting support in the roadway shallow surrounding rock, anchored coal mass is formed in the range of 0–5 m depth of the surrounding rock. The strength of the surrounding rock is improved effectively, which provides a basic protection environment for resisting large deformation of the surrounding rock caused by deep high ground pressure and strong mining. After the hole-making space is formed, a pressure relief deterioration zone is created within 7–12 m depth of the surrounding rock, which transfers the high stress in the shallow surrounding rock to the deep part, alleviating the continuous damage due to the high stress in the shallow surrounding rock. At the same time, the large-diameter hole-making space can effectively absorb the continuous deformation caused by the transfer of the deeper coal body to the roadway space due to the high horizontal stress. The depth of the surrounding rock (5–7 m) is the buffer protection space, whose function is to avoid damage to the shallow anchored coal mass caused by the hole space.

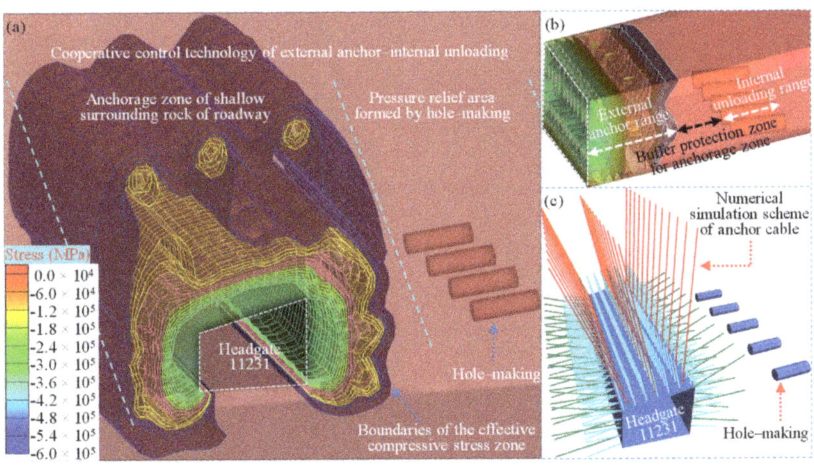

Figure 20. Schematic of the external anchor prestressed field and the hole−making space of coal roadway surrounding rock. (**a**) Three-dimensional prestress field; (**b**) Three zones of roadway rib; (**c**) Anchor bolt (cable) simulation scheme.

(2) Observation results and analysis of roadway surrounding rock displacement

The roadway surrounding rock displacements in the test section (using the new technology) and in the non-test section were monitored on-site. The results are displayed in Figure 21. The values of the deformation of the higher rib of the roadway in the non-test section are 400 mm and 780 mm in the no mining and mining influence stages, respectively. The deformation of the higher rib of the roadway in the test section is 190 mm after hole-making, and the accumulated reduction in the higher rib deformation of the roadway is approximately 850 mm. As indicated in Figure 22, after the new support technology is adopted, the available width of the roadway is not less than 4.2 m, which can always meet the ventilation and transportation requirements of the roadway. Furthermore, there is no need to renovate the roadway rib. This technology effectively limits the continuous deformation of the rib surrounding rock in the test section of headgate 11231 and ensures the stability of the surrounding rock of the mining roadway affected by the deep dynamic pressure.

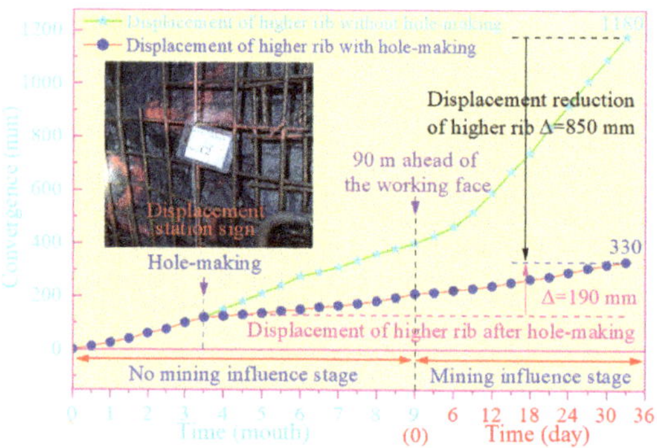

Figure 21. Deformation curve of coal roadway surrounding rock.

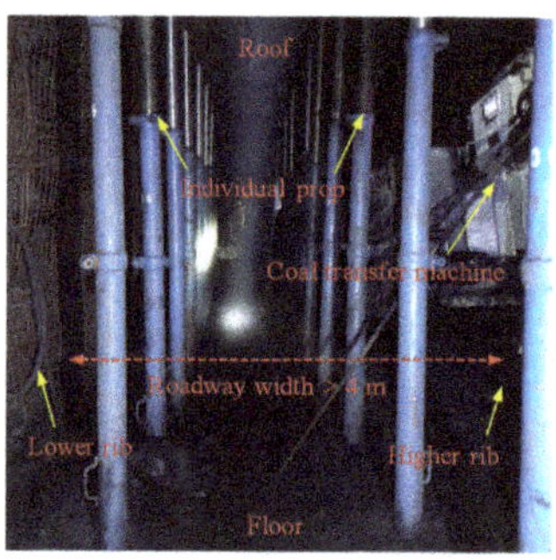

Figure 22. Advanced support section of the panel after adopting the external anchor–internal unloading technology.

6. Conclusions

(1) Given the surrounding rock control problem of a large deformation mining roadway disturbed by deep dynamic pressure in a coal mine, the cooperative control technology of external anchor–internal unloading of surrounding rock of a large deformation coal roadway was proposed.

(2) A numerical model conforming to the production of the geological conditions of panel 11231 of the coal mine was constructed, and the vertical stress distribution and transfer rule of surrounding rock under the conditions of five schemes for three key hole-making parameters were studied. The pressure relief degree and effect of each scheme's hole-making space in the no dynamic pressure influence stage were analyzed using a variety of evaluation indices, and the reasonable parameter range of the surrounding rock hole-making space in this stage was obtained.

(3) The pressure relief effect of each optimized scheme under the mining dynamic pressure condition of panel 11231 was studied, and the effective pressure relief range of each scheme was compared and a comprehensive evaluation of factors, such as construction benefit, cost, and construction period, was performed. Based on the

results, the major technical parameters for pressure relief during hole-making were finally determined as follows: the hole-making depth, spacing, and length were 8 m, 3.2 m, and 3 m, respectively.

(4) The field practice was carried out in the test section of headgate 11231. The monitoring results showed that the displacement of the hole-making rib was reduced by 850 mm, making the available width of the roadway no less than 4.2 m, which meets the ventilation and transportation requirements of the roadway, and eliminates roadway rib renovation, which ensures the stability of the surrounding rock. Thus, the new technology is of great significance for the further development of strategies for controlling the surrounding rock of deep coal roadways affected by dynamic pressure.

Author Contributions: Conceptualization, S.X.; Methodology, H.L. and D.C.; Software, S.F. and X.M.; Validation, S.F., X.M. and Z.J.; Writing—original draft, H.L.; Writing—review & editing, J.C.; Visualization, Z.J.; Supervision, D.C.; Project administration, H.L.; Funding acquisition, S.X. and D.C. All authors have read and agreed to the published version of the manuscript.

Funding: This work was supported by the National Natural Science Foundation of China (No. 52074296).

Conflicts of Interest: The authors declare no conflict of interest.

References

1. Jia, Z.; Xie, H.; Zhang, R.; Li, C.; Wang, M.; Gao, M.; Zhang, Z.; Zhang, Z. Acoustic Emission Characteristics and Damage Evolution of Coal at Different Depths Under Triaxial Compression. *Rock Mech. Rock Eng.* **2020**, *53*, 2063–2076. [CrossRef]
2. Xie, H.; Li, C.; He, Z.; Li, C.; Lu, Y.; Zhang, R.; Gao, M.; Gao, F. Experimental study on rock mechanical behavior retaining the in situ geological conditions at different depths. *Int. J. Rock Mech. Min. Sci.* **2021**, *138*, 104548. [CrossRef]
3. Xie, S.; Wu, Y.; Guo, F.; Zou, H.; Chen, D.; Zhang, X.; Ma, X.; Liu, R.; Wu, C. Application of Pre-Splitting and Roof-Cutting Control Technology in Coal Mining: A Review of Technology. *Energies* **2022**, *15*, 6489. [CrossRef]
4. Huang, B.; Zhang, N.; Hong, J.; Kan, J.; Meng, B.; Li, N.; Xie, W.; Jiao, J. Large deformation theory of rheology and structural instability of the surrounding rock in deep mining roadway. *J. China Coal Soc.* **2020**, *45*, 911–926.
5. Zhao, Y.; Zhou, J.; Liu, W. Characteristics of ground pressure and mechanism of coal burst in the gob side roadway at Xinjie deep mining area. *J. China Coal Soc.* **2020**, *45*, 1595–1606.
6. He, M. Research progress of deep shaft construction mechanics. *J. China Coal Soc.* **2021**, *46*, 726–746.
7. Xie, H.; Liu, T.; Gao, M.; Chen, L.; Zhou, H.; Ju, Y.; Gao, F.; Peng, X.; Li, X.; Peng, R.; et al. Research on in-situ condition preserved coring and testing systems. *J. China Coal Soc.* **2021**, *18*, 1840–1859. [CrossRef]
8. Hou, C. Key technologies for surrounding rock control in deep roadway. *J. China Univ. Min. Technol.* **2017**, *46*, 970–978.
9. Kang, H. Sixty years development and prospects of rock bolting technology for underground coal mine roadways in China. *J. China Univ. Min. Technol.* **2016**, *45*, 1071–1081.
10. Xie, S.; Wu, Y.; Chen, D.; Liu, R.; Han, X.; Ye, Q. Failure analysis and control technology of intersections of large-scale variable cross-section roadways in deep soft rock. *Int. J. Coal Sci. Technol.* **2022**, *9*, 19. [CrossRef]
11. Wu, C.; Qin, T.; Wang, L.; Liu, Z. Research on Surrounding Rock Control Technology of Dongbaowei Deep Mining Roadway. *Adv. Civ. Eng.* **2021**, *2021*, 6660989. [CrossRef]
12. Wang, S.; Ju, W.; Pan, J.; Lu, C. Mechanism of energy partition evolution of excavation roadway rockburst in coal seam under tectonic stress field. *J. China Coal Soc.* **2019**, *44*, 2000–2010.
13. Jing, H.; Wu, J.; Yin, Q.; Shi, X.; Zhao, Z. Particle flow simulation of rock burst and roof fall of deep coal roadway under dynamic disturbance. *Chin. J. Rock Mech. Eng.* **2020**, *39*, 3475–3487.
14. Li, Q.; Hou, J.; Han, T.; Liu, H.; Wang, S. Failure characteristics and support techniques of surrounding rock for deep rectangular roadway in Yangzhuang mine. *J. China Univ. Min. Technol.* **2016**, *45*, 1124–1131.
15. Jiang, Y.; Xu, Z.; Liu, Q.; Ma, H. An improved numerical manifold method for investigation of fracturing expan-sion and squeezing deformation of surrounding rock mass in deep coal roadway. *J. China Coal Soc.* **2020**, *45*, 579–589.
16. Adigamov, A.E.; Yudenkov, A.V. Stress-strain behavior model of disturbed rock mass with regard to anisotropy and discontinuities. *Min. Inf. Anal. Bull.* **2021**, *8*, 93–103. [CrossRef]
17. Khayrutdinov, A.M.; Kongar-Syuryun, C.B.; Kowalik, T.; Tyulyaeva, Y.S. Stress-strain behavior control in rock mass using different-stregth backfill. *Min. Inf. Anal. Bull.* **2020**, *2020*, 42–55. [CrossRef]
18. Rybak, J.; Khayrutdinov, M.M.; Kuziev, D.A.; Kongar-Syuryun, C.B.; Babyr, N.V. Prediction of the geomechanical state of the rock mass when mining salt deposits with stowing. *J. Min. Inst.* **2022**, *253*, 61–70. [CrossRef]
19. Kongar-Syuryun, C.B.; Ubysz, A.; Faradzhov, V. Models and algorithms of choice of development technology of deposits when selecting the composition of the backfilling mixture. *IOP Conf. Ser. Earth Environ. Sci.* **2021**, *684*, 012008. [CrossRef]

20. Wu, Y.; Xie, S.; Zhang, Y. Research on stability control of roadway intersections with nested variable cross-section in deep mine. *J. Min. Sci. Technol.* **2022**, *7*, 720–729.
21. Zhang, N.; Wang, C.; Gao, M.; Zhao, Y. Roadway support difficulty classification and controlling techniques for Huainan deep coal mining. *Chin. J. Rock Mech. Eng.* **2009**, *28*, 2421–2428.
22. Ma, N.; Zhao, Z.; Feng, J. Technology of butt long bolt on roadway supporting in difficult conditions. *Coal Sci. Technol.* **2013**, *41*, 117–121.
23. Xie, S.; Wu, Y.; Ma, X.; Chen, D.; Guo, F.; Jiang, Z.; Li, H.; Zou, H.; Liu, R.; Zhang, X. Reasonable stopping method and retracement channel support at fully mechanized top coal caving working face of 15 m extra-thick coal seam: A case study. *Energy Sci. Eng.* **2022**, 1–22. [CrossRef]
24. Li, S.; Wang, Q.; Li, W.; Wang, D.; Li, Z.; Jiang, B.; Wang, H.; Wang, H. Comparative field test study of pressure relief anchor box beam support system in deep thick top coal roadway. *Chin. J. Rock Mech. Eng.* **2012**, *31*, 656–666.
25. He, F.; Gao, F.; Sun, Y.; Li, S.; Song, B.; Yang, Y. Multiple-cable-girder-truss asymmetric support mechanism and its application in the roadway of fully mechanized top coal caving face with narrow coal pillar. *J. China Coal Soc.* **2015**, *40*, 2296–2302.
26. Yu, H.; Jia, H.; Liu, S.; Liu, Z.; Li, B. Macro and micro grouting process and the influence mechanism of cracks in soft coal seam. *Int. J. Coal Sci. Technol.* **2021**, *8*, 969–982. [CrossRef]
27. Huang, W.; Gao, Y.; Wen, Z.; Gao, L. Technology of gob-side entry retaining using concrete-filled steel tubular column as roadside supporting. *J. China Univ. Min. Technol.* **2015**, *44*, 604–611.
28. Xie, S.; Pan, H.; Zeng, J.; Wang, E.; Qiao, S. A case study on control technology of surrounding rock of a large section chamber under a 1200-m deep goaf in Xingdong coal mine, China. *Eng. Fail. Anal.* **2019**, *104*, 112–125. [CrossRef]
29. Zhang, S.; Chen, L.; Jia, H. The surrounding rock of deep borehole pressure relief and let the pressure bolt coupling analysis. *Appl. Mech. Mater.* **2014**, *446–447*, 1421–1424.
30. Luo, Y.; Xu, K.; Huang, J.; Li, X.; Liu, T.; Qu, D.; Chen, P. Impact analysis of pressure-relief blasting on roadway stability in a deep mining area under high stress. *Tunn. Undergr. Space Technol.* **2021**, *110*, 103781. [CrossRef]
31. Cheng, S.; Ma, Z.; Gong, P.; Li, K.; Li, N.; Wang, T. Controlling the deformation of a small coal pillar retaining roadway by non-penetrating directional pre-splitting blasting with a deep hole: A case study in Wangzhuang coal mine. *Energies* **2020**, *13*, 3084. [CrossRef]
32. Chen, J.; Yang, C.; Zhu, Y.; He, F. Distressing maintenance analysis on roadway of large deformation in soft rocks with pillarless fully-mechanized sublevel caving system and its application. *Chin. J. Rock Mech. Eng.* **2002**, *21*, 5.
33. Zhang, X.; Pak, R.Y.; Gao, Y.; Liu, C.; Zhang, C.; Yang, J.; He, M. Field experiment on directional roof presplitting for pressure relief of retained roadways. *Int. J. Rock Mech. Min. Sci.* **2020**, *134*, 104436. [CrossRef]
34. Yin, D.; Chen, S.; Liu, S.; Ma, H. Effect of joint angle in coal on failure mechanical behavior of roof rock-coal combined body. *Q. J. Eng. Geol. Hydrogeol.* **2018**, *51*, 202–209. [CrossRef]
35. Chen, Y.; Zuo, J.; Liu, D.; Li, Y.; Wang, Z. Experimental and numerical study of coal-rock bimaterial composite bodies under triaxial compression. *Int. J. Coal Sci. Technol.* **2021**, *8*, 908–924. [CrossRef]
36. Wang, W.; Wang, G.; Zhao, W.; Wang, L.; Feng, Z.; Cui, R.; Du, F. Numerical assessment of the pressure relief effect of the hydraulic punching cavitation technique in a soft coal seam. *Geomech. Geophys. Geo-Energy Geo-Resour.* **2022**, *8*, 30. [CrossRef]

energies

Article

Study on Source Identification of Mixed Gas Emission and Law of Gas Emission Based on Isotope Method

Gang Xu *, Yaping Hou, Hongwei Jin and Zhongwei Wang

College of Safety Science and Engineering, Xi'an University of Science and Technology, Xi'an 710054, China
* Correspondence: xugang25193@xust.edu.cn

Abstract: It is of great significance to obtain the source of mixed gas emission from the working face and the law of gas emission from each coal seam for the targeted implementation of gas control measures. Based on the principle that the hydrocarbon isotope values of gas in different coal seams have significant variability, a hydrocarbon isotope method for identifying the source of gas emission is proposed. Taking Pingmei No. 6 Coal Mine as the study area, the distribution characteristics of each value were obtained by testing the values of carbon and hydrogen isotopes in the gas of mined coal seams and adjacent coal seams; by testing the hydrocarbon isotope value of CH_4 in the mixed gas of coal seam, the proportion of gas emission in each coal seam is determined and the law of gas emission in each coal seam is studied. The results show that the variation law of the proportion of gas emission in each coal seam can be divided into three stages: the dominant stage of gas emission in the mining layer (stage I), the stage of gas emission in the long-distance adjacent coal seam (stage II), and the dynamic equilibrium stage of gas emission in each coal seam (stage III). In the process of working face mining, the amount of gas emission in the mining layer remains in a small fluctuation state, and the proportion of gas emission decreases rapidly in stage I and stage II, and remains stable in stage III; the amount of gas emission and the proportion of gas emission in adjacent coal seams increase rapidly in stage I and stage II, and remain stable in stage III; the mixed gas emission of the working face increases rapidly in stage I and stage II, and remains stable in stage III. The calculation formula of the gas emission rate of the adjacent coal seam is established; during the development of the height of the mining fractured zone, the gas emission rate of the adjacent coal seam increases exponentially, and the gas emission ratio and gas emission amount of the adjacent coal seam increase; after the height of mining fracture zone tends to be stable, the gas emission rate, the proportion of gas emission, and the amount of gas emission remain of adjacent coal seams remain in a small fluctuation state.

Keywords: multi-seam mining; gas emission of working face; hydrocarbon isotope; gas emission rate of adjacent coal seam; mining fracture zone height

Citation: Xu, G.; Hou, Y.; Jin, H.; Wang, Z. Study on Source Identification of Mixed Gas Emission and Law of Gas Emission Based on Isotope Method. *Energies* **2023**, *16*, 1225. https://doi.org/10.3390/en16031225

Academic Editor: Sergey Zhironkin

Received: 9 January 2023
Revised: 19 January 2023
Accepted: 20 January 2023
Published: 23 January 2023

1. Introduction

Gas emission in the working face has the characteristics of high prediction difficulty, large damage range, and serious consequences, so it has always been the focus of mine disaster prevention and control. In particular, under the condition of multi-coal seam mining, when the layer spacing of each coal seam is small, the gas of the adjacent coal seam will pour into the mining coal seam, resulting in a relatively high gas emission of the mining coal seam, which increases the risk of the working face and the difficulty of gas control [1,2]. Gas extraction is the fundamental measure of gas control in order to reasonably and efficiently formulate the gas extraction measures of the working face, identify the source of gas emission from the working face, and analyze the law of gas emission from the working face [3,4].

The traditional prediction methods of gas emission mainly include the source prediction method [5] and the statistical analysis method [6]. The gas geology research group of Jiaozuo Mining Institute uses the statistical analysis method to analyze the geological factors affecting gas emission and predict the amount of gas emission, the influence of geological factors on gas emission is deeply studied from qualitative analysis to quantitative research [7]. However, the mathematical model established by this method fails to consider the influence of mine pressure and adjacent coal seam on gas emission during mining. Lunarzewski [8] used the geomechanics of "Floor gas" and "Roof gas" and the gas emission model to calculate the contribution ratio of different gas sources to the total gas content, but this method cannot realize the dynamic prediction of gas emission in the working face. Zhang et al. [9] used statistical analysis and source prediction methods to predict the gas emission of mining faces and analyzed the influencing factors and sources of gas emission in mining faces; however, this study failed to obtain the dynamic change characteristics of gas emission during mining. Whittles et al. [10] studied the influence of geological factors on the gas flow in the goaf of the longwall coal mining face in the United Kingdom by numerical simulation; however, this method failed to obtain the influence of working face mining on gas emission and the source of gas emission in the working face.

With the development of artificial intelligence technology, some mathematical models and methods have also been introduced into the prediction of gas emission. Gu and Zhang et al. [11,12] established a new prediction model of gas emission by combining gray theory and a wavelet neural network. According to the measured data of gas emission and related geological factors in the mining area of the mine, Zhang et al. [13] established a multi-factor mathematical geological model for predicting gas emission by using quantitative theory [14], considering various influencing factors including mining depth. Xiao and Zhu et al. [15,16] established a BP neural network source prediction model by combining the source prediction method with the neural network prediction technology. The above gas emission prediction method can obtain higher accuracy for short-term prediction in the case of existing gas emission data, but there is a large error for long-term prediction, and it is also difficult to obtain the source of gas emission in the working face.

The research on the law of gas emission in the process of working face mining has also achieved more results. Xu and Li et al. [17,18] studied the gas emission characteristics and gas distribution law of fully mechanized mining face through theoretical analysis and field measurement. Gao et al. [19] believed that when the mine pressure appeared, the absolute gas emission of the working face increased obviously, and the mining intensity was consistent with the change trend of gas concentration in the upper corner of the working face. Zhang et al. [20] believed that in the early stage of mining, the absolute amount of gas emission increased continuously, reached the peak when the initial pressure came, and then showed a wave-like downward trend. Cui et al. [21] considered that the absolute gas emission of the working face was positively correlated with daily output, and the relative gas emission was negatively correlated with output. Yuan and Dai et al. [22,23] mainly used the source prediction method to study the gas emission law of the protective layer working face and then took the corresponding gas control measures.

From the above, researchers have carried out a lot of research on gas emission prediction methods and emission laws, but mainly based on the constructed model to predict gas emission. The disadvantage of this method is that the applicability of the model is poor, and it is difficult to reflect the change in geological conditions over time. In addition, the research on the law of gas emission under the condition of multi-coal seam mining mainly adopts the source prediction method, which cannot realize the dynamic prediction of the proportion of gas emission and the amount of gas emission in each coal seam at the initial stage of mining. In view of the above problems, the hydrocarbon isotope method is used to dynamically predict the mixed gas in the working face, and the proportion of gas emission in each coal seam is quantitatively obtained. Then, the gas emission rate of adjacent coal seams and the variation law of gas emission in each coal seam during the mining process of the working face are studied, and the calculation formula of gas emission rate in adjacent layers is proposed, which provides a theoretical basis for formulating gas control measures in the working face under multi-coal seam mining conditions.

2. Materials and Methods

2.1. Determination Method of Mixed Gas Emission Source in the Working Face

2.1.1. Theoretical Foundation

Coal seam gas is generated with the formation of coal, the main gas component of coal seam gas is CH_4. During the peatification phase, i.e., the biochemical gas generation period, anaerobic microorganisms decompose organic matter to produce a large amount of CH_4; in the period of coalification metamorphism, under the action of high temperature and pressure, the volatile of organic matter decreases and the fixed carbon increases, and a large amount of CH_4 will be generated. Due to the different strengths of biodegradation or pyrolysis during the generation of CH_4, the effect of isotope fractionation is also different, resulting in obvious differences in the carbon and hydrogen isotope values of CH_4 in each coal seam gas, the difference in carbon and hydrogen isotope values is the theoretical basis for studying the source of mixed gas emission in the working face [24,25].

2.1.2. Calculation Model of Mixed Gas Emission Source in the Working Face

The mixed gas emission from the working face during multi-coal seam mining comes from different coal seams. The gas is only a simple physical mixture, and its chemical properties have not changed, but the carbon and hydrogen isotopes of the gas from different coal seams in the mixed gas are different, which provides the possibility for calculating the composition of gas emission from the working face. Based on this, the calculation formula of hydrocarbon isotope value in the mixed gas is derived according to the principle of mass conservation [26,27]:

$$\delta_{mix} = \frac{V_A\delta_A + V_B\delta_B}{V_A + V_B} \quad (1)$$

Among them, δ_{mix} is the measured value of CH_4 hydrocarbon isotope in mixed gas, ‰; δ_A is the hydrocarbon isotope value of CH_4 in the first coal seam, ‰; V_A is the volume of CH_4 in the first coal seam, m^3; δ_B is the hydrocarbon isotope value of CH_4 in the second coal seam, ‰; V_B is the volume of CH_4 in the second coal seam, m^3; δ_{mix}, δ_A, and δ_B can be directly measured by an isotope mass spectrometer.

For the unit volume of mixed gas, $V_A = aZ_A$, $V_B = bZ_B$, where a is the gas emission proportion of the first coal seam in the mixed gas, %; b is the gas emission proportion of the second coal seam in the mixed gas, %, $a + b = 1$; Z_A is the gas component of CH_4 in the first coal seam, %; Z_B is the gas component of CH_4 in the second coal seam, %; Z_A and Z_B can be tested by a gas chromatograph for the gas composition of each coal seam.

Similarly, when the composition source of mixed gas has n endmembers, the calculation formula of hydrocarbon isotope value in mixed gas can be expressed as:

$$\begin{cases} \delta_{mix} = aZ_A\delta_A + bZ_B\delta_B + \cdots + nZ_N\delta_N \\ a + b + \cdots + n = 1 \end{cases} \quad (2)$$

The gas emission ratio a and $b \cdots n$ of each coal seam in Equation (2) can be solved by the software MATLAB.

2.2. Test Method for Hydrocarbon Isotope Value

2.2.1. Field Test Background and Conditions

Pingdingshan Tian'an Coal Co., Ltd.'s sixth mine is affiliated with China Pingdingshan Shenma Group, the administrative division is under the jurisdiction of Pingdingshan City and Baofeng County, Henan Province. The approved production capacity of the mine is 3.2 million t/a, the mine adopts the multi-level development mode of vertical shaft and inclined shaft, adopts the long wall retreating mining technology, and manages the roof by all caving methods. It is a coal and gas outburst mine [28].

The main mining method of the Ding5−6 coal seam, the Wu8 coal seam, and the Wu9−10 coal seam in the minefield is multi-coal seam mining. The Ding5−6 coal seam belongs to the most unstable minable coal seam, the Ding5−6 coal seam is divided into two layers: Ding5 and Ding6, the direct roof of the coal seam is sandstone and sandy mudstone, the main roof is sandy mudstone, and the floor is sandy mudstone or mudstone. The coefficient of variation of coal thickness of the Wu8 coal seam is 39.88%, the mining index is 0.91, the elevation of the coal seam is 60 to −1000 m, the burial depth is 150 to 1100 m, and the coal-bearing area is 36.3 km^2. In terms of the characteristics of the roof and floor, the direct roof is sandy mudstone and mudstone, and the main roof is fine-grained sandstone; the floor is dark gray mudstone and sandy mudstone, and the old bottom is medium-grained sandstone, which belongs to the more stable roof and floor. The coefficient of variation of coal thickness of the Wu9−10 coal seam is 38.37%, the recoverability index is 0.95, the elevation of the coal seam is 60 to −1000 m, the burial depth is 150 to 1100 m, and the coal-bearing area is 31.3 km^2. It is a more stable area-wide recoverable coal seam, the characteristics of the top and bottom plate are: the direct top plate is mainly medium-grained sandstone or sandy mudstone, the old top is sandy mudstone, and the sandy mudstone pseudo-top can be seen locally. The direct bottom plate is mudstone, and the old bottom is sandy mudstone, which is a more stable top and bottom plate. The Wu8 −32010 working face is located in the third-level second mining area, and the working face elevation is −570 to −660 m. The strike length of the working face is 2300 m, the dip length is 220 m, and the mining height of the working face is 3.7 m. The long wall retreating mining technology is adopted, and the roof is managed by all caving methods, the working face strike section is shown in Figure 1. In order to reduce the impact of the risk of coal and gas outbursts, they first mine the Wu8 coal seam as a protective layer and then mine the Ding5−6 coal seam and the Wu9−10 coal seam. However, the average distance between the Wu8 coal seam and the upper Ding5−6 coal seam is 71 m, and the average distance between the Wu8 coal seam and the lower Wu9−10 coal seam is 13 m. During the mining process of the Wu8 coal seam, a large amount of gas from adjacent coal seams will enter the working face, which will bring potential safety hazards to the mining of the Wu8 coal seam. Therefore, it is of great practical significance to obtain the proportion of gas emission and the law of gas emission in each coal seam during the mining process of the Wu8 coal seam to ensure the safe and efficient mining of the working face.

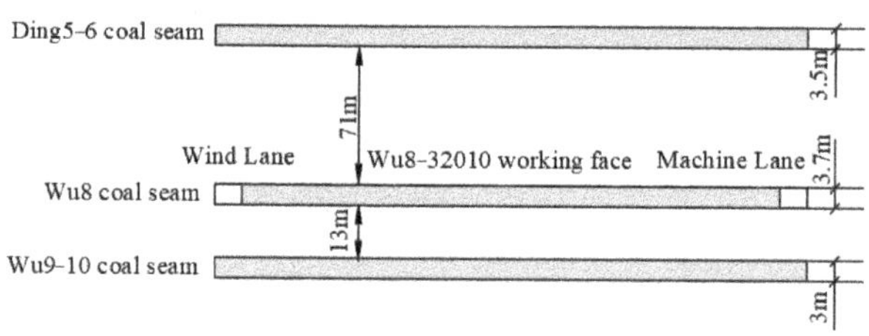

Figure 1. Wu8 −32010 working face strike section.

2.2.2. Testing of Hydrocarbon Isotope Value of Gas in Each Coal Seam

According to the standard AQ1018−2006 "mine gas emission prediction method" middle spacing and adjacent coal seam emission rate relationship curve [29], the Ding5−6 coal seam and Wu9−10 coal seam gas will pour into the Wu8 coal seam, other coal seam gas will not pour into the Wu8 coal seam. Therefore, desorption gas is collected in the Ding5−6 coal seam, the Wu8 coal seam, and the Wu9−10 coal seam, respectively, to determine the gas composition and hydrocarbon isotope values, including ^{13}C (CH_4), ^{13}C (CO_2), ^{13}C (C_2H_6), and ^{2}H (CH_4). Four samples are collected from each coal seam, and a total of twelve samples are collected.

The specific test method is as follows: first, six sealed tanks with a capacity of 1 L should be prepared, the sealed tank should be washed and dried before use, and the air tightness of the sealed tank should be ensured to be intact, so there is no air leakage at 300 to 400 kPa; then the coal sample containing gas is drilled by the special drilling rig for coal core in the selected place of the coal mine, and the coal sample is put into the prepared sealed tank; after leaving the well, the sealed tank is connected with the experimental equipment for gas component test and hydrocarbon isotope value test. The instrument used for the gas component test is the GC−2000 TCD gas chromatograph, and the instrument used for the hydrocarbon isotope value test is the Delta V stable isotope mass spectrometer. The experimental instruments are shown in Figure 2.

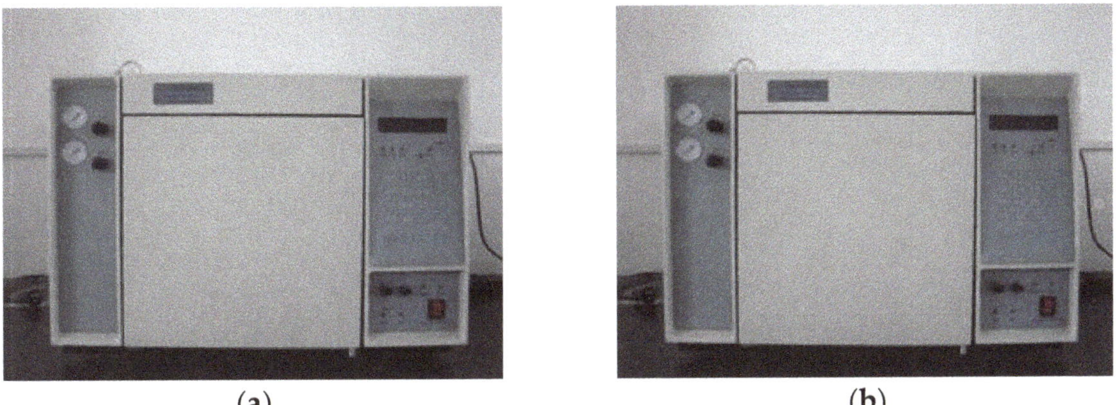

Figure 2. Experimental instrument diagram. (**a**) GC−2000TCD gas chromatograph. (**b**) Delta V stable isotope mass spectrometer.

2.2.3. Testing of CH_4 Hydrocarbon Isotope Value of Mixed Gas in the Working Face

(1) Working face mixed gas CH_4 hydrocarbon isotope value determination scheme

The two areas of the upper corner and goaf of the Wu8−32010 working face are selected as the sampling sites of mixed gas samples, in which the sampling point of the goaf is 20 m deep into the goaf. The specific sampling site is shown in Figure 3. The sampling time is calculated from the beginning of the working face, and the sampling time is one month, that is, from 7 June 2018 to 6 July 2018. The samples are collected once a day after the samples are collected, and the samples are sent to the laboratory for mixed gas component test and CH_4 hydrocarbon isotope value test. During this period, a total of 30 groups of mixed gas samples are collected.

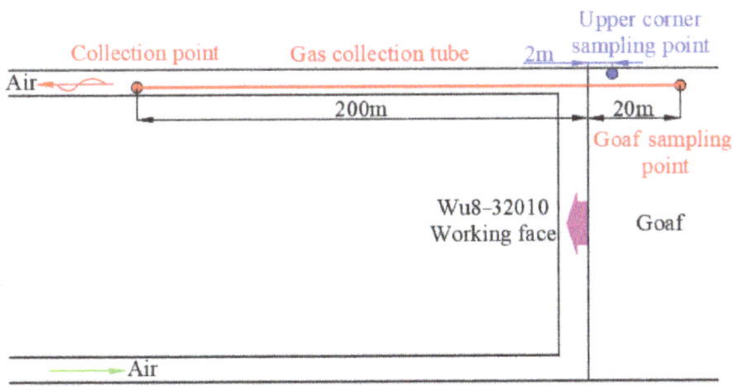

Figure 3. Mixed gas sample collection location diagram.

(2) Sample collection method of mixed gas

The collection method of mixed gas samples in the upper corner of the working face is as follows: the sampling personnel stands on the footplate of the end support of the return air side and extends the expansion rod to a distance of 1.5 to 2 m from the support and a distance of 300 mm from the top side; the other end of the telescopic rod is connected with a high negative pressure suction tube, the gas in the sampling tube and the airbag is discharged, and the sampling bag is connected to start sampling. It is strictly prohibited to extend the head into the windshield during sampling.

The collection method of mixed gas samples in goaf is as follows: the sampling tube with a diameter of 20 mm is arranged in the return airway; one end of the sampling tube is 20 m deep into the goaf and a sampling device is arranged; the other end is connected to the extraction system in the return air trough, and the valve is set when connected to facilitate sampling.

3. Results

3.1. Test Results of Hydrocarbon Isotope Value of Gas in Each Coal Seam

The test results of gas composition and hydrocarbon isotope value of each coal seam are shown in Table 1. From Table 1, it can be seen that CH_4 is the main gas component of each coal seam, among which the average proportion of CH_4 gas component in the Ding5−6 coal seam is 71.563%, the average proportion of CH_4 gas component in Wu8 coal seam is 87.340%, and the average proportion of CH_4 gas component in the Wu9−10 coal seam is 88.642%, which indicates that there are some differences in CH_4 gas components in each coal seam. In addition to CH_4, the gas composition of each coal seam also includes N_2, CO_2, and C_2H_6.

Table 1. Test results of gas components and hydrocarbon isotope values of each coal seam.

Coal Seam	Sample Number	Gas Component/%				Carbon Isotope Values/‰			Hydrogen Isotope Values/‰
		CH_4	C_2H_6	N_2	CO_2	^{13}C (CH_4)	^{13}C (CO_2)	^{13}C (C_2H_6)	^2H (CH_4)
Ding5−6 coal seam	1	73.269	0.015	17.26	9.456	−32.335	−13.210	−8.152	−164.257
	2	70.456	0.01	20.493	9.041	−32.864	−10.548	−12.754	−165.952
	3	74.387	0.014	18.031	7.568	−29.543	−9.526	−11.853	−172.346
	4	68.14	0.009	21.752	10.099	−30.790	−13.284	−8.265	−170.765
	Average value	71.563	0.012	19.384	9.041	−31.383	−11.642	−10.256	−168.33
Wu8 coal seam	1	85.562	0.021	12.06	2.357	−28.250	−9.518	−8.265	−158.359
	2	89.843	0.024	8.541	1.592	−28.652	−10.475	−8.351	−155.458
	3	89.425	0.016	8.695	1.864	−28.225	−8.865	−9.026	−155.269
	4	84.53	0.011	13.444	2.015	−28.613	−9.642	−8.938	−156.314
	Average value	87.340	0.018.	10.685	1.957	−28.435	−9.625	−8.645	−156.35
Wu9−10 coal seam	1	90.365	0.02	8.268	1.347	−26.158	−7.854	−6.892	−155.128
	2	86.525	0.024	8.469	4.982	−28.045	−8.526	−7.264	−153.284
	3	88.942	0.028	9.251	1.779	−27.816	−10.365	−9.350	−153.649
	4	88.736	0.02	9.296	1.948	−26.629	−10.263	−9.742	−155.059
	Average value	88.642	0.023	8.821	2.514	−27.162	−9.252	−8.312	−154.28

In order to more intuitively show the distribution characteristics of hydrocarbon isotope values of gas in each coal seam, the tested hydrocarbon isotope values are presented in a box diagram (Figure 4), the upper and lower bounds of the box represent 75% and 25% quantiles of the data, respectively, and the hollow point in the middle of the box represents the average value of the data, while the upper and lower bounds of the vertical lines represent the maximum and minimum values of the data, respectively.

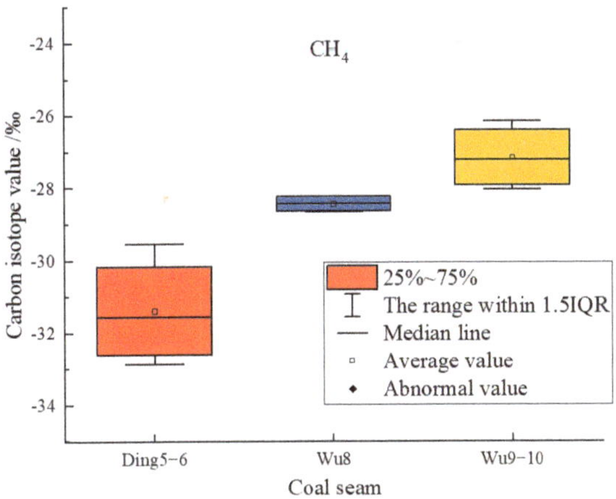

(**a**) Carbon isotope values of CH_4

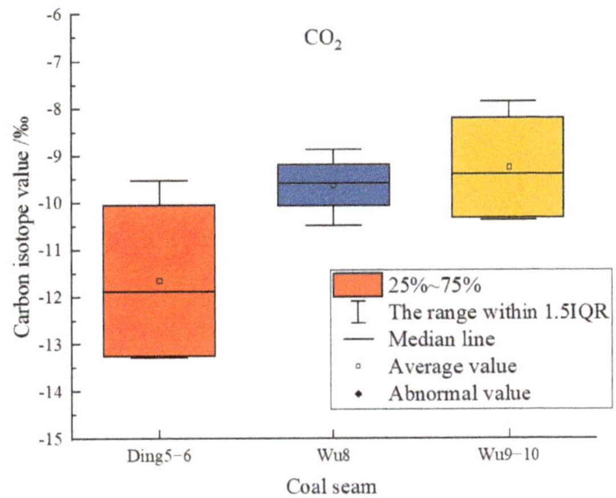

(**b**) Carbon isotope values of CO_2

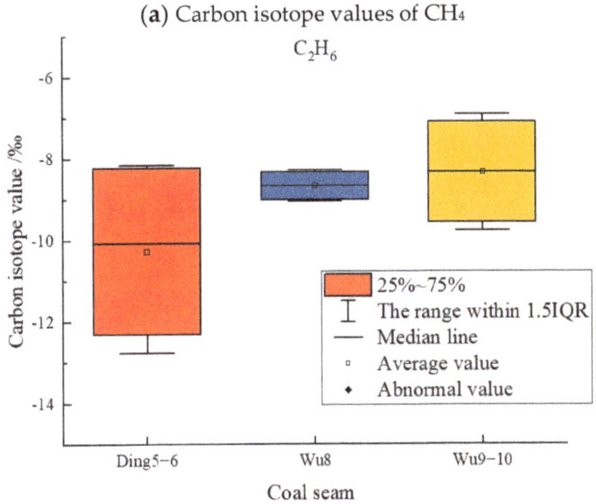

(**c**) Carbon isotope values of C_2H_6

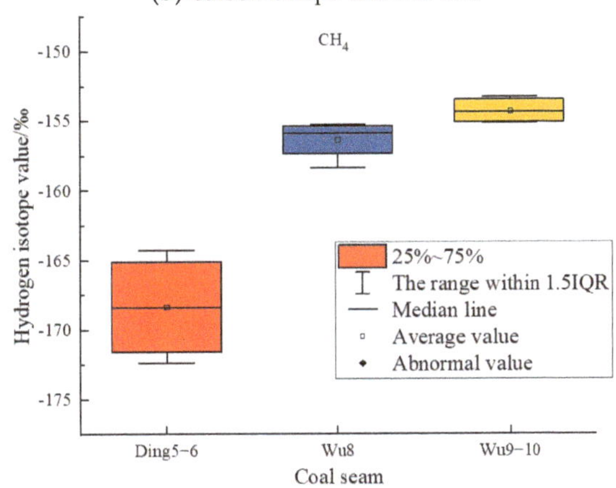

(**d**) The hydrogen isotope value of CH_4

Figure 4. Box diagram of hydrocarbon isotope value distribution of gas in each coal seam.

It can be seen from Figure 4a that the distribution range (25% to 75%) of the carbon isotope values of CH_4 gas in each coal seam does not overlap, so the carbon isotope value of CH_4 gas has the condition to identify the source of mixed gas emission. It can be seen from Figure 4b that the distribution range (25% to 75%) of the carbon isotope values of CO_2 gas in the Wu8 coal seam and the Wu9−10 coal seam overlaps, so the carbon isotope value of CO_2 gas does not have the condition to identify the source of mixed gas emission. It can be seen from Figure 4c that the distribution range (25% to 75%) of carbon isotope values of C_2H_6 gas in the Ding5−6 coal seam, the Wu8 coal seam, and the Wu9−10 coal seam overlaps, so the carbon isotope values of C_2H_6 gas does not have the condition to identify the source of mixed gas emission. It can be seen from Figure 4d that the distribution range (25% to 75%) of hydrogen isotope values of CH_4 gas in each coal seam does not overlap, so the hydrogen isotope value of CH_4 gas has the condition to identify the source of mixed gas emission. Based on the above analysis, the hydrocarbon isotope value of CH_4 gas can be selected to identify the source of mixed gas emission.

3.2. Test Results of Mixed Gas Emission Source in the Working Face

The working face mixed gas CH_4 hydrocarbon isotope value test results are shown in Table 2.

Table 2. Test results of CH_4 hydrocarbon isotope value in the mixed gas of the working face.

Measurement Locations	Isotope Category	7 June	8 June	9 June	10 June	11 June	12 June	13 June	14 June	15 June	16 June
Upper corner	$^{13}C(CH_4)$/‰	−24.778	−24.756	−24.742	−24.676	−24.610	−24.579	−24.557	−24.502	−24.482	−24.425
	$^{2}H(CH_4)$/‰	−136.302	−136.315	−136.326	−136.162	−136.019	−136.031	−136.041	−135.872	−135.896	−135.407
Goaf	$^{13}C(CH_4)$/‰	−24.769	−24.753	−24.733	−24.705	−24.659	−24.620	−24.570	−24.529	−24.509	−24.449
	$^{2}H(CH_4)$/‰	−136.263	−136.253	−136.282	−136.306	−136.321	−136.328	−136.152	−136.021	−135.996	−135.537
Measurement locations	Isotope category	17 June	18 June	19 June	20 June	21 June	22 June	23 June	24 June	25 June	26 June
Upper corner	$^{13}C(CH_4)$/‰	−24.430	−24.366	−24.343	−24.270	−24.227	−24.161	−24.200	−24.243	−24.273	−24.245
	$^{2}H(CH_4)$/‰	−135.405	−134.906	−134.908	−134.257	−133.958	−133.460	−133.773	−134.121	−134.276	−134.087
Goaf	$^{13}C(CH_4)$/‰	−24.405	−24.339	−24.318	−24.278	−24.253	−24.243	−24.179	−24.235	−24.197	−24.202
	$^{2}H(CH_4)$/‰	−135.187	−134.688	−134.543	−134.219	−134.057	−134.042	−133.557	−133.989	−133.718	−133.706
Measurement locations	Isotope category	27 June	28 June	29 June	30 June	1 July	2 July	3 July	4 July	5 July	6 July
Upper corner	$^{13}C(CH_4)$/‰	−24.209	−24.187	−24.202	−24.282	−24.178	−24.251	−24.273	−24.203	−24.161	−24.199
	$^{2}H(CH_4)$/‰	−133.788	−133.631	−133.793	−134.281	−133.631	−134.077	−134.266	−133.800	−133.466	−133.777
Goaf	$^{13}C(CH_4)$/‰	−24.213	−24.171	−24.196	−24.205	−24.237	−24.217	−24.200	−24.156	−24.168	−24.173
	$^{2}H(CH_4)$/‰	−133.881	−133.548	−133.718	−133.728	−134.042	−133.867	−133.744	−133.394	−133.575	−133.557

We substitute the test results of CH_4 hydrocarbon isotope values in Tables 1 and 2 into Equation (2), and use MATLAB software to obtain the proportion of gas emission from each coal seam in the upper corner and goaf during the mining process of the working face (Figures 5 and 6). From Figures 5 and 6, it can be seen that the change trend of gas emission proportion of each coal seam in the upper corner and goaf during the mining process of the working face is basically the same. The gas emission proportion of the Wu8 coal seam shows a change rule of decreasing first and then stabilizing, and the gas emission proportion of the Wu9−10 coal seam and the Ding5−6 coal seam shows a change rule of increasing first and then stabilizing.

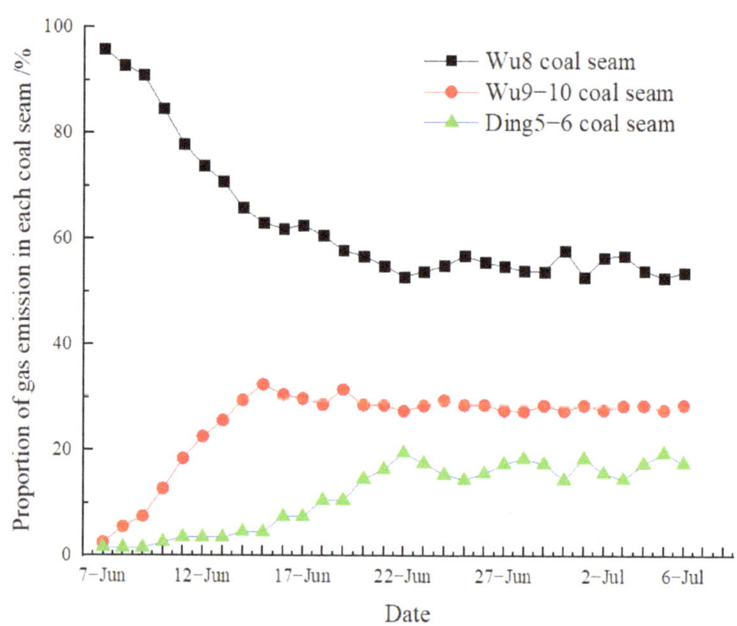

Figure 5. Gas emission proportion of each coal seam in the upper corner.

Further analysis of Figures 5 and 6 shows that the change law of the proportion of gas emission in each coal seam can be divided into three stages: the dominant stage of gas emission in the mining layer (stage I), the stage of gas emission in long-distance adjacent coal seams (stage II), and the dynamic equilibrium stage of gas emission in each coal seam (stage III).

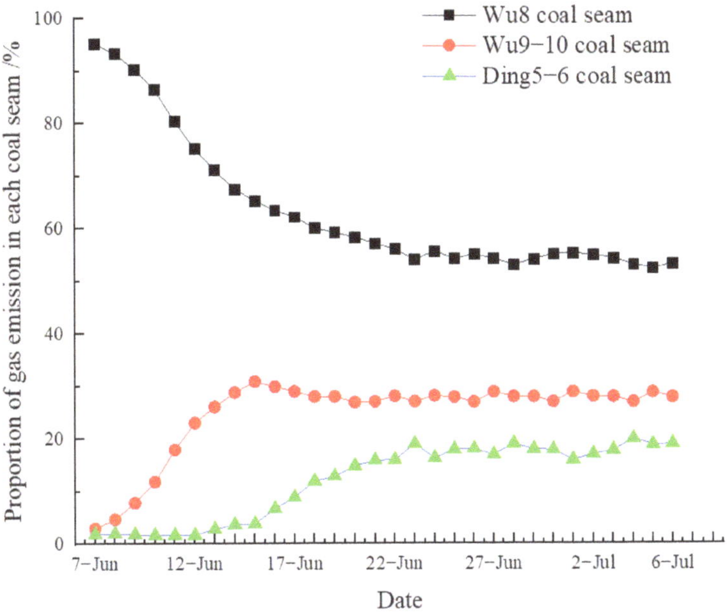

Figure 6. Gas emission proportion of each coal seam in goaf.

The mining time of the stage I working face is from 7 June to 15 June, and the advancing distance of the working face is 54 m. The proportion of gas emission in the Wu8 coal seam shows a trend of rapid decline, and the proportion of gas emission decreases from 95% at the beginning of mining to 60%; the proportion of gas emission in the Wu9−10 coal seam shows a trend of rapid increase, and the proportion of gas emission increases from 3% at the beginning of mining to 30%; the proportion of gas emission in the Ding5−6 coal seam shows a trend of slow increase, and the proportion of gas emission increases from 1% at the beginning of mining to 5%.

The mining time of the stage II working face is from 16 June to 22 June, and the advancing distance of the working face is 42 m. The proportion of gas emission in the Wu8 coal seam shows a trend of slow decline, and the proportion of gas emission decreases from 60% to 53%; the proportion of gas emission in the Wu9−10 coal seam shows a slow downward trend, and the proportion of gas emission decreases from 30% to 27%; the proportion of gas emission in the Ding5−6 coal seam increases rapidly from 5% to 18%.

The mining time of the stage III working face is from 23 June to 6 July, and the advance distance of the working face is 84 m. The proportion of gas emission in each coal seam enters the dynamic equilibrium stage. The proportion of gas emission in the Wu8 coal seam is stable at about 54%, the proportion of gas emission in the Wu9−10 coal seam is stable at about 28%, and the proportion of gas emission in the Ding5−6 coal seam is stable at about 18%.

4. Discussion

4.1. Gas Emission Law of the Working Face and Each Coal Seam

While testing the CH_4 hydrocarbon isotope value of the working face, the gas concentration value in the return airflow of the working face is recorded. Combining this with the test results of the air volume of the working face and the gas emission ratio of each coal seam in the upper corner, the gas emission of each coal seam can be obtained (Table 3).

Table 3. Gas emission test results of each coal seam.

Category / Date	7 June	8 June	9 June	10 June	11 June	12 June	13 June	14 June	15 June	16 June
Gas concentration/%	0.41	0.43	0.46	0.51	0.56	0.62	0.66	0.69	0.72	0.74
Wu8 coal seam gas emission/m^3t^{-1}	4.72	4.79	5.02	5.18	5.24	5.49	5.61	5.46	5.45	5.51
Wu9−10 coal seam gas emission/m^3t^{-1}	0.12	0.29	0.42	0.78	1.24	1.68	2.03	2.44	2.80	2.71
Ding5−6 coal seam gas emission/m^3t^{-1}	0.08	0.08	0.08	0.16	0.24	0.26	0.28	0.38	0.39	0.67
Working face gas emission/m^3t^{-1}	4.92	5.16	5.52	6.12	6.72	7.44	7.92	8.28	8.64	8.88

Category / Date	17 June	18 June	19 June	20 June	21 June	22 June	23 June	24 June	25 June	26 June
Gas concentration/%	0.75	0.77	0.78	0.79	0.81	0.82	0.81	0.82	0.83	0.81
Wu8 coal seam gas emission/m^3t^{-1}	5.64	5.61	5.41	5.38	5.34	5.21	5.24	5.41	5.66	5.41
Wu9−10 coal seam gas emission/m^3t^{-1}	2.68	2.65	2.95	2.71	2.77	2.71	2.77	2.91	2.85	2.78
Ding5−6 coal seam gas emission/m^3t^{-1}	0.68	0.98	1.00	1.39	1.60	1.93	1.72	1.53	1.45	1.53
Working face gas emission/m^3t^{-1}	9.00	9.24	9.36	9.48	9.72	9.84	9.72	9.84	9.96	9.72

Category / Date	27 June	28 June	29 June	30 June	1 July	2 July	3 July	4 July	5 July	6 July
Gas concentration/%	0.82	0.81	0.81	0.82	0.83	0.81	0.81	0.83	0.82	0.81
Wu8 coal seam gas emission/m^3t^{-1}	5.40	5.25	5.24	5.71	5.27	5.50	5.54	5.38	5.19	5.23
Wu9−10 coal seam gas emission/m^3t^{-1}	2.71	2.67	2.77	2.71	2.84	2.69	2.77	2.84	2.72	2.78
Ding5−6 coal seam gas emission/m^3t^{-1}	1.73	1.80	1.70	1.43	1.85	1.53	1.42	1.74	1.92	1.71
Working face gas emission/m^3t^{-1}	9.84	9.72	9.72	9.84	9.96	9.72	9.72	9.96	9.84	9.72

Based on the basic parameters of the working face, the "mine gas emission prediction method" (AQ1018−2006) is used to calculate the gas emission of the working face. Combining this with the gas emission test results of each coal seam in Table 3, the change trend of the measured value and the calculated value of the gas emission can be obtained (Figure 7).

The measured value in Figure 7 is the test result using the method in this paper, and the calculated value is the calculated result using the "mine gas emission prediction method" (AQ1018−2006). It can be seen from Figure 7 that the gas emission of each coal seam and working face has different variation rules. During the mining process of the working face, the gas emission of the Wu8 coal seam is maintained in a small fluctuation state, and the difference between the measured value and the calculated value is small. From the beginning of mining to 15 June, the gas emission of the Wu9-10 coal seam increases rapidly, after 15 June, the gas emission of the Wu9-10 coal seam remains in a small fluctuation state when it reaches the calculated value. From the beginning of mining to 15 June, the gas emission of the Ding5-6 coal seam increases slowly, from 15 June to 22 June, the gas emission of the Ding5-6 coal seam increases rapidly, and remains in a slight fluctuation state after reaching the calculated value. The mixed gas emission of the working face increases rapidly before 22 June and remains in a small fluctuation state after reaching the calculated value.

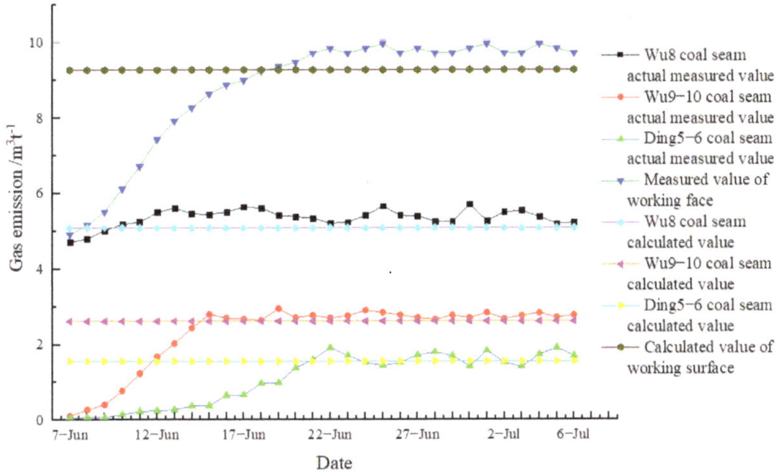

Figure 7. Variation trend of measured and calculated values of gas emission.

In summary, it can be seen that the gas emission amount of the working face and the measured value of the gas emission amount of each coal seam show an increasing trend in the early stage of the working face mining. The change range and change time of the gas emission amount are related to the geological conditions of the working face mining. The existing gas emission prediction method [30,31] makes it difficult to dynamically display and describe this change law; when the working face advances for a certain distance, the measured values of the gas emission amount of the working face and the gas emission amount of each coal seam will remain in a small fluctuation state. This small fluctuation state actually reflects the dynamic changes in coal seam mining conditions and geological conditions. The existing gas emission prediction method [32,33] usually cannot show this small fluctuation state. From the test results, it can be seen that the measured gas emission in this paper shows a small fluctuation state, which indicates that the mining conditions and geological conditions of the working face in this paper change little; however, in some cases, the changes of coal seam mining conditions and geological conditions are more intense, at this time, the gas emission will also fluctuate greatly, which will bring serious hidden dangers to the safe production of coal mines. Through the test method in this paper, the dynamic change law of gas emission in each coal seam can be obtained, so as to formulate and implement gas control measures in a targeted manner and avoid the occurrence of coal mine safety production accidents.

4.2. Analysis of Influencing Factors of Gas Emission

4.2.1. Analysis of Influencing Factors of Gas Emission in the Mining Layer

According to the standard "mine gas emission prediction method" (AQ1018 –2006), the calculation method of the mining layer gas emission is as follows:

$$Q_1 = K_1 \cdot K_2 \cdot K_3 \cdot \frac{m}{M} \cdot (W_0 - W_c) \tag{3}$$

where Q_1 is the relative gas emission of the mining coal seam (including surrounding rock), m^3/t; K_1 is the gas emission coefficient of the surrounding rock, and the value range is 1.1 to 1.3. When the roof is managed by all caving methods, 1.3 is taken; K_2 is the coal loss coefficient of the working face, which is the reciprocal of the recovery rate of the working face; K_3 is the influence coefficient of roadway pre-drainage gas on gas emission in mining layer; m is the thickness of mining layer, m; M is the mining height of the working face, m; W_0 is the original gas content of coal seam, m^3/t; W_c is the residual gas content of coal seam, m^3/t.

The calculation method of K_3 is as follows:

$$K_3 = \frac{L - 2h}{L} \quad (4)$$

where L is the length of the mining face, m; h is the width of the coal seam gas emission zone in the roadway, m, which is considered according to the average exposure time of 200 days in the roadway.

It can be seen from Equation (3) that the gas emission of the mining layer is related to the gas emission coefficient of the surrounding rock, the coal loss coefficient of the working face, the influence coefficient of the pre −drainage gas of the preparation roadway, and the original gas content of the coal seam. In the test of this paper, the basic parameters such as the gas emission coefficient of surrounding rock, the coal loss coefficient of the working face, the influence coefficient of pre-discharge gas in preparation roadway, and the original gas content of coal seam do not change, so the gas emission of mining layer remains in a small fluctuation state and the difference between the measured value and the calculated value is small (Figure 7).

4.2.2. Analysis of Influencing Factors of Gas Emission in Adjacent Coal Seam

The standard "mine gas emission prediction method" (AQ1018 –2006) also gives the adjacent coal seam gas emission calculation method as follows:

$$Q_2 = \sum_{i=1}^{n} (W_{0i} - W_{ci}) \cdot \frac{m_i}{M} \cdot \eta_i \quad (5)$$

where Q_2 is the relative gas emission of the adjacent coal seam, m^3/t; m_i is the thickness of the ith adjacent coal seam, m; M is the mining thickness of the mining layer, m; W_{0i} is the original gas content of the ith adjacent coal seam, m^3/t; W_{ci} is the residual gas content of the ith adjacent coal seam, m^3/t; η_i is the ith adjacent coal seam gas emission rate, %, which can be seen in Figure 8.

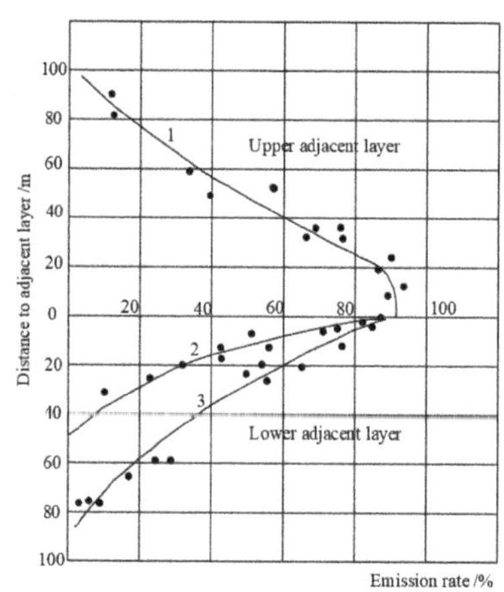

1: Upper adjacent coal seam. 2: Adjacent coal seam under gently inclined coal seam.
3: Adjacent coal seam under inclined and steeply inclined coal seams

Figure 8. Relationship curve between gas emission rate of adjacent coal seam and layer spacing.

Equation (5) is a common method for calculating the gas emission of adjacent coal seams [34]. According to Equation (5), the gas emission of adjacent coal seams is positively correlated with the original gas content of adjacent coal seams, the thickness of adjacent coal seams, and the gas emission rate of adjacent coal seams. In the process of working face mining, the original gas content of the adjacent coal seam, the thickness of the adjacent coal seam, and other parameters remain unchanged, they have no effect on the change of gas emission of the adjacent coal seam, so the gas emission rate of the adjacent coal seam is the main parameter affecting the change of gas emission of the adjacent coal seam. When using Equation (5) to calculate the gas emission of adjacent coal seams, the gas emission rate of adjacent coal seams is usually selected from Figure 8; however, the gas emission rate of adjacent coal seams in Figure 8 is the gas emission rate when the dynamic equilibrium stage is reached after the full mining of the working face. Before the dynamic equilibrium stage is reached, the calculation of the gas emission rate of adjacent coal seams in Figure 8 will inevitably produce large errors, which is the problem with using Equation (5) to calculate the gas emission of adjacent coal seams.

The gas emission rate of adjacent coal seams is related to the gas flow characteristics of adjacent coal seams. The principle of adjacent coal seam gas flowing to the working face is (Figure 9): after the mining of the mining layer, a certain mining space will be formed, the coal rock layer around the coal seam will move to the mining space, and the original equilibrium relationship of the coal rock mass will be disturbed and destroyed, so that the original stress–strain state of the coal rock mass will change. The results of the change will lead to the release of pressure and elastic potential energy in the coal rock around the coal seam, thus forming mining fractures in the coal rock around the coal seam. As an associated gas of the coal seam, the gas itself has fluidity, and the generation of mining cracks will provide a channel for the seepage movement of gas in the coal seam. During the mining process of the mining layer, the pressure-relief gas of the adjacent coal seam flows into the mining face through the mining-induced cracks, which increases the gas emission of the mining face [35].

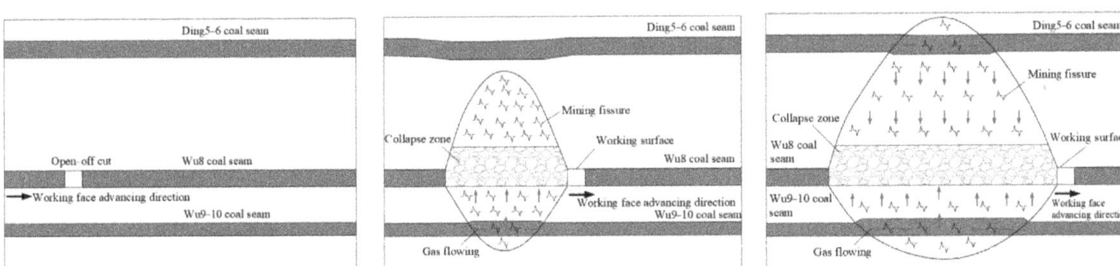

(a) (b) (c)

Figure 9. Gas flow characteristics of adjacent coal seams during advancing of the working face. (**a**) Unmined coal seams. (**b**) Stage I. (**c**) Stage II.

In order to further analyze the influence of the gas emission rate of the adjacent coal seam on the gas emission of the adjacent coal seam, based on the geological conditions of this paper, FLAC3D software is used to simulate the development characteristics of the fracture zone in the process of the working face advancing. According to the numerical simulation results, the change trend of the development height of the mining fracture zone with the advancing distance of the working face can be obtained. Then the adjacent coal seam gas emission rate is calculated according to Equation (5) by using the measured data of the adjacent coal seam gas emission (Figure 10).

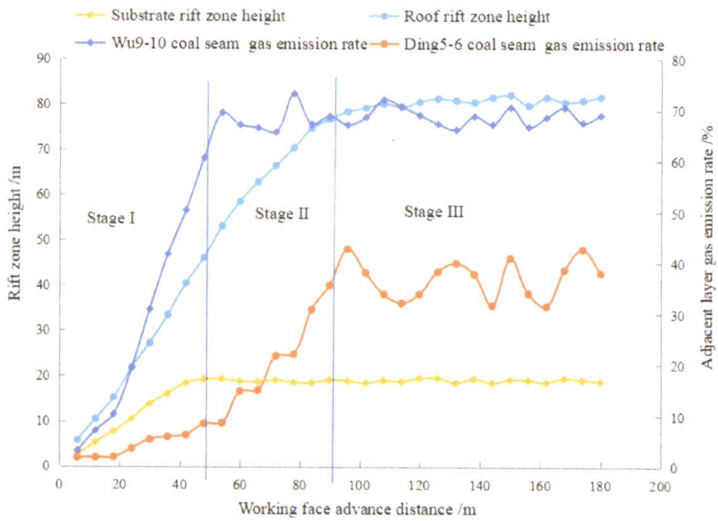

Figure 10. Variation of mining fracture zone height and gas emission rate of adjacent coal seams with the advancing distance of the working face.

It can be seen from Figure 10 that the height of the mining fracture zone increases with the advance of the working face. The height of the fracture zone increases rapidly in the early stage of mining, and the growth rate slows down after full mining; when the working face advances to a certain distance, the height of fracture zone does not increase anymore. It can also be seen in Figure 10 that the adjacent coal seam gas emission rate is a dynamic change parameter in the working face mining process.

Based on the test results of this paper, combined with Figures 9 and 10, the influence of the gas emission rate of adjacent coal seams on gas emission of adjacent coal seams can be analyzed as follows:

(1) In the process of advancing the working face, the height of the mining-induced fracture zone of the coal seam roof and the coal seam floor in stage I shows an increasing trend. Due to the small interlayer spacing between the Wu8 coal seam and the Wu9–10 coal seam, the Wu9–10 coal seam quickly enters the mining-induced fracture zone of the coal seam floor, so that a large amount of gas in the Wu9–10 coal seam enters the working face through the mining-induced fracture zone of the coal seam floor, resulting in a rapid increase in the gas emission rate of the Wu9–10 coal seam. Although the height of the mining fracture zone of the coal seam roof increases rapidly, due to the large interlayer spacing between the Wu8 coal seam and the Ding5–6 coal seam, the Ding5–6 coal seam has not yet entered the mining fracture zone of the coal seam roof, so that only a small amount of the Ding5–6 coal seam gas enters the working face, so the gas emission rate of the Ding5–6 coal seam increases slowly.

(2) In the process of advancing the working face, the height of the mining-induced fracture zone of the coal seam floor in stage II tends to be stable. Because the coal seam of Wu9–10 has entered the mining-induced fracture zone of the coal seam floor in stage I, the gas of the Wu9–10 coal seam entering the working face in stage II will remain in a small fluctuation state, resulting in the gas emission rate of the Wu9–10 coal seam also remaining in a small fluctuation state. In stage II, the height of the mining fracture zone of the coal seam roof is still increasing. With the increase of the height of the mining fracture zone of the coal seam roof, the Ding5–6 coal seam will enter the mining fracture zone of the coal seam roof, so that the gas of the Ding5–6 coal seam enters the working face through the mining fracture zone of the coal seam

roof, resulting in the rapid increase of the gas emission rate of the Ding5−6 coal seam in stage II.

(3) With the advance of the working face, in stage III, the mining-induced fracture zone height of the coal seam roof and floor tends to be stable. The Ding5−6 coal seam and the Wu9−10 coal seam also enter into the mining-induced fracture zone of the coal seam roof and floor, respectively, the gas emission rate of adjacent coal seams and the gas emission of adjacent coal seams will remain in a small fluctuation state until the end of working face mining.

According to the above analysis, the gas emission rate of adjacent coal seams in the process of the working face shows a phased change characteristic, which is the result of the combined effect of the working face mining method and geological conditions. Therefore, the gas emission of adjacent coal seams is the result of a combination of multiple factors, it is difficult to predict the gas emission of adjacent coal seams only by establishing a mathematical model or calculation formula [36–38], and there will be obvious defects in accuracy and timeliness.

4.3. Gas Emission Rate of Adjacent Coal Seams

In Section 4.2., the influencing factors and changing trends of the gas emission rate of adjacent coal seams have been analyzed. Because the gas emission rate of adjacent coal seams shows a phased change during the mining process of the working face, the calculation formula for the gas emission rate of adjacent coal seams will also be established in stages.

Firstly, according to Figure 11, the calculation formula of the height of the fracture zone of the coal seam roof with the advancing distance of the working face is obtained by fitting:

$$h_1 = -0.004l^2 + 1.337l - 5.884 \quad (R^2 = 0.994) \quad l \le l_1 \tag{6}$$

where l is the advancing distance of the working face, m; h_1 is the height of the coal seam roof fracture zone, m; l_1 is the advancing distance of the working face when the height of the fracture zone of the coal seam roof reaches the upper adjacent coal seam, m.

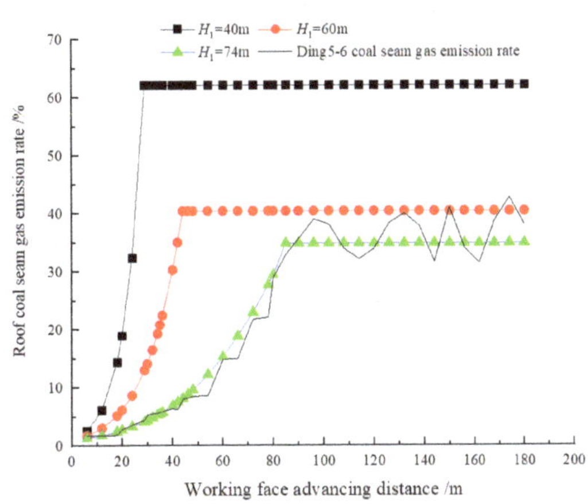

Figure 11. Variation of the gas emission rate of lower adjacent coal seams with the advancing distance of the working face.

Similarly, according to Figure 11, the calculation formula of the height of the fracture zone in the coal seam floor with the advancing distance of the working face is obtained by fitting:

$$h_2 = -0.002l^2 + 0.572l - 1.055 \quad (R^2 = 0.994) \quad l \le l_2 \tag{7}$$

where l is the advancing distance of the working face, m; h_2 is the height of the coal seam floor fracture zone, m; l_2 is the advancing distance of the working face when the height of the fracture zone of the coal seam floor reaches the lower adjacent coal seam, m.

Then, according to Figure 11, the calculation formula of the gas emission rate of the upper adjacent coal seam with the height of the coal seam roof fracture zone is obtained by fitting:

$$P_1 = 1.236e^{0.042h_1} \quad (R^2 = 0.98) \quad l \leq l_1 \tag{8}$$

where P_1 is the upper adjacent coal seam gas emission rate, %.

Similarly, according to Figure 11, the calculation formula of the gas emission rate of the lower adjacent coal seam with the height of the fracture zone of the coal seam floor is obtained by fitting:

$$P_2 = 2.609e^{0.168h_2} \quad (R^2 = 0.976) \quad l \leq l_2 \tag{9}$$

where P_2 is the gas emission rate of the lower adjacent coal seam, %.

According to the gas flow characteristics of the adjacent coal seam during the advancing process of the working face, the higher the height of the mining fracture zone, the higher the gas emission rate of the adjacent coal seam, and the gas emission rate of the adjacent coal seam and the height of the mining fracture zone are exponentially related. From the layer spacing of adjacent coal seams, the smaller the layer spacing of adjacent coal seams, the higher the gas emission rate of adjacent coal seams, and the layer spacing of adjacent coal seams is inversely proportional to the gas emission rate of adjacent coal seams. In addition, it can be seen from Equations (8) and (9) that the gas emission rates of upper adjacent coal seams and lower adjacent coal seams have similar variation trends with the height of the fracture zone, and the gas emission rates of upper adjacent coal seams and lower adjacent coal seams can be combined for study.

Based on the above analysis, combined with Equation (6)–(9) and Figure 8, the calculation formula of the gas emission rate of the upper adjacent coal seam with the advancing distance of the working face can be obtained by introducing the interlayer spacing H_1 of the upper adjacent coal seam and the interlayer spacing H_2 of the lower adjacent coal seam:

$$\begin{cases} h_1 = -0.004l^2 + 1.337l - 5.884 \quad (R^2 = 0.994) \quad l \leq l_1 \\ P_1 = (-0.024)H_1 - 16(+2.609)e^{(-0.00218(H_1-16)+0.168)h_1} \quad l \leq l_1 \\ P_1 = (115 - H_1)/115 \quad l > l_1 \end{cases} \tag{10}$$

where H_1 is the distance between the upper adjacent coal seam and the mining layer, m.

Similarly, the calculation formula of the gas emission rate of the lower adjacent coal seam with the advancing distance of the working face can be obtained:

$$\begin{cases} h_2 = -0.002l^2 + 0.572l - 1.055 \quad (R^2 = 0.994) \quad l \leq l_2 \\ P_2 = (-0.024)H_2 - 16(+2.609)e^{(-0.00218(H_2-16)+0.168)h_2} \quad l \leq l_2 \\ P_2 = (50 - H_2)/50 \quad l > l_2 \end{cases} \tag{11}$$

where H_2 is the distance between the lower adjacent coal seam and the mining layer, m.

According to Equations (10) and (11), the variation trend of the gas emission rate of adjacent coal seams with the advancing distance of the working face can be drawn under different interlayer spacings (Figures 11 and 12). From Figures 11 and 12 it can be seen that as the working face advances, the gas emission rate of the upper adjacent coal seam and the lower adjacent coal seam has the same change trend. In the early stage of working face mining, the gas emission rate of adjacent coal seams increases exponentially with the advancing distance of the working face. After the working face is fully mined, the gas emission rate of adjacent coal seams remains stable; the larger the layer spacing with the mining coal seam is, the smaller the gas emission rate of the adjacent coal seam is and the longer the distance for the working face to reach the dynamic equilibrium state. In Figures 11 and 12, the curve of seam gas emission rate of adjacent coal seam with an interlayer spacing of 74 m and the curve of seam gas emission rate of adjacent coal seam of the Ding5 –6 coal seam basically coincide, and the curve of seam gas emission rate of adjacent coal seam with an interlayer spacing of 16 m and the curve of seam gas emission rate of adjacent coal seam of Wu9-10 coal seam basically coincide. This shows that the error between the calculated value and the test result is small, and the accuracy of the gas emission rate of adjacent coal seam calculated by Equations (10) and (11) is high, which can meet the needs of gas emission prediction of the adjacent coal seam in production practice.

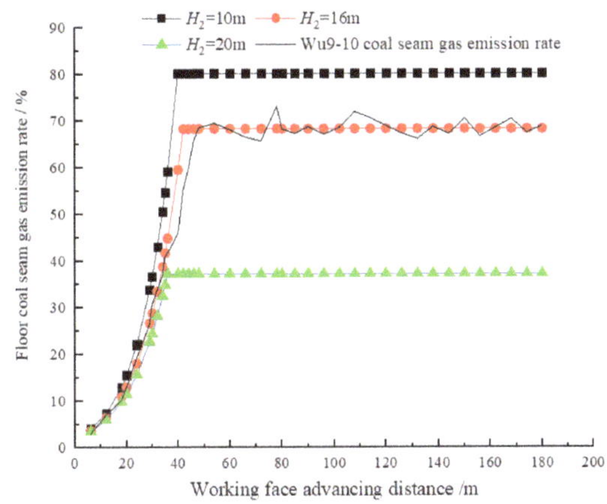

Figure 12. Variation of the gas emission rate of upper adjacent coal seams with the advancing distance of the working face.

The test method of gas emission in adjacent coal seams and the calculation formula of the gas emission rate in adjacent coal seams proposed in this paper can make up for the shortcomings of the existing methods, and have good applicability under complex geological conditions. Its popularization and application under the complex geological conditions of multiple coal seams can obtain considerable safety and economic benefits.

5. Conclusions

1. The CH_4 hydrocarbon isotope values in the gas of the Wu8 coal seam, the Wu9−10 coal seam, and the Ding5−6 coal seam are obviously different, which has the conditions to identify the source of gas emission in the working face by using the hydrocarbon isotope method.
2. The variation law of the proportion of gas emission in each coal seam during the mining process of the working face can be divided into the dominant stage of gas emission in the mining layer (stage I), the stage of gas emission in the long-distance adjacent coal seam (stage II), and the dynamic equilibrium stage of gas emission in each coal seam (stage III). The proportion of gas emission in the Wu8 coal seam decreases from 95% to 60% in stage I, from 60% to 53% in stage II, and stabilizes at about 54% in stage III. The proportion of gas emission in the Wu9−10 coal seam increases from 3% to 30% in stage I, decreases from 30% to 27% in stage II, and stabilizes at about 28% in stage III. The proportion of gas emission in the Ding5-6 coal seam increases from 1% to 5% in stage I, from 5% to 18% in stage II, and stabilizes at about 18% in stage III.
3. During the mining process of the working face, the gas emission amount of the Wu8 coal seam remains in a stable state. The gas emission amount of the Wu9−10 coal seam increases rapidly in stage I and remains stable in stage II and stage III. The gas emission amount of the Ding5−6 coal seam increases slowly in stage I, increases rapidly in stage II, and remains stable in stage III. The mixed gas emission of the working face increases rapidly in stage I and stage II and remains stable in stage III.

4. The adjacent coal seam gas emission rate is the main influencing factor of the adjacent coal seam gas emission, and the adjacent coal seam gas emission rate is related to the development process of the mining fracture zone. When the height of the mining fracture zone increases, the gas emission rate of the adjacent coal seam increases exponentially, and the gas emission of the working face increases. After the height of the mining fracture zone tends to be stable, the gas emission rate of adjacent coal seams and the gas emission amount of each coal seam remain in a small fluctuation state.
5. The calculation formula of the gas emission rate of the adjacent coal seam is established. Compared with the test results, the error of the calculated value is small, which can meet the needs of gas emission prediction of adjacent coal seams in production practice.

Author Contributions: Conceptualization, G.X. and Y.H.; Data curation, G.X., Y.H. and Z.W.; Formal analysis, G.X., Y.H. and H.J.; Funding acquisition, G.X., Z.W. and H.J.; Methodology, G.X. and Y.H.; Project administration, H.J.; Software, G.X., Y.H. and Z.W.; Supervision, H.J.; Writing—original draft, G.X. and Y.H.; Writing—review and editing, G.X., H.J., Y.H. and Z.W. All authors have read and agreed to the published version of the manuscript.

Funding: All of these are gratefully acknowledged. This work was financially supported by the Natural Science Foundation of China (51904231), the National Key Research and Development Program (2018YFC0807805), and the Natural Science Basic Research Program of Shaanxi (2019JM-072).

Institutional Review Board Statement: Not applicable.

Informed Consent Statement: Not applicable.

Data Availability Statement: Not applicable.

Acknowledgments: Many thanks to all the contributions and support given by the authors in preparing the writing of this article. Moreover, thanks to the Sixth Mine of Pingdingshan Tian'an Coal Co., Ltd. for providing the field test for this study.

Conflicts of Interest: The authors declare no conflict of interest.

References

1. Kurnia, J.C.; Xu, P.; Sasmito, A.P. A novel concept of enhanced gas recovery strategy from ventilation air methane in underground coal mines –A computational investigation. *J. Nat. Gas Sci. Eng.* **2016**, *35*, 661–672. [CrossRef]
2. Mishra, D.P.; Panigrahi, D.D.; Kumar, P. Computational investigation on effects of geo-mining parameters on layering and dispersion of methane in underground coal mines –A case study of Moonidih Colliery. *J. Nat. Gas Sci. Eng.* **2018**, *56*, 110–124. [CrossRef]
3. Karacan, C.Ö.; Diamond, W.P.; Schatzel, S.J. Numerical analysis of the influence of in-seam horizontal methane drainage boreholes on longwall face emission rates. *Int. J. Coal Geol.* **2007**, *72*, 15–32. [CrossRef]
4. Sander, R.; Connell, L.D. A probabilistic assessment of enhanced coal mine methane drainage (ECMM) as a fugitive emission reduction strategy for open cut coal mines. *Int. J. Coal Geol.* **2014**, *131*, 288–303. [CrossRef]
5. Qi, Q.J.; Xia, S.Y. Construction of sharing platform for the gas emission rate prediction based on source prediction. *Min. Saf. Environ. Prot.* **2018**, *45*, 59–64
6. Cang, Y.X.; Li, W.G.; Shu, Z.C.; Yan, C.Y. Statistical analysis on coal mine gas explosion and optimization of safety input. *Adv. Mater. Resea.* **2011**, *1043*, 1316–1321.
7. Gas Geology Group of Jiaozuo Mining Institute. Attempt of statistical analysis in prediction of gas outburst. *Coalfield Geol. Explo.* **1983**, 26–30+8.
8. Lunarzewski, W.L. Gas emission prediction and recovery in underground coal mines. *Int. J. Coal Geol.* **1998**, *35*, 117–145. [CrossRef]
9. Zhang, Y.P.; Ye, Q.; Jia, Z.Z.; Jiang, W.W. The analysis and forecast of gas emission in workface. *China Min. J.* **2007**, *123*, 46–49+52.
10. Whittles, D.N.; Lowndes, I.S.; Kingman, S.W.; Yates, C.; Jobling, S. Influence of geotechnical factors on gas flow experienced in a UK longwall coal mine panel. *Int. J. Rock Mech. Min.* **2006**, *43*, 369–387. [CrossRef]
11. Gu, S.; Cui, H.Q.; Feng, W.L. Mine gas gushing forecasting based on grey model and wavelet neural network. *J. China Coal Soc.* **2007**, *156*, 964–966.
12. Zhang, S.R.; Wang, B.T.; Li, X.E.; Chen, H. Research and application of improved gas concentration prediction model based on grey theory and BP neural network in digital mine. *Procedia CIRP* **2016**, *56*, 471–475. [CrossRef]

13. Zhang, X.L.; Shan, J.P.; Peng, S.P. Mathematical geology technique and method for prediction of gas content and emission. *J. China Coal Soc.* **2009**, *34*, 350–354.
14. Ock, J.H. Activity duration quantification under uncertainty: Fuzzy set theory application. *Cost Eng.* **1996**, *38*, 26.
15. Xiao, J.P.; Dai, G.L. Study on different-source prediction of gas emission in fully mechanized coal face based on BP neural network. *J. Anhui Univ. Sci. and Techno.* **2011**, *31*, 51–55.
16. Zhu, H.Q.; Chang, W.J.; Zhang, B. Different-source gas emission prediction model of working face based on BP artificial neural network and its application. *J. China Coal Soc.* **2007**, *152*, 504–508.
17. Xu, Q.Y.; Zhang, L.; Li, Y.M. Fully mechanized working face gas distribution law and constitution characteristics in BU Ertai coal mine. *J. Liaoning Techni. Univ.* **2014**, 33, 887–891.
18. Li, S.G.; Ding, Y.; An, Z.F.; Wei, W.B. Research on gas emission and distribution characteristics of fully Mechanized Coal Face in High Methane and Thick Seam. *Coal Techno.* **2015**, *34*, 113–116.
19. Gao, L.; Li, X.J.; Pan, J.C. Main controlling factors and control measures of gas emission in mining face of Buertai mine. *J. Saf. Sci. Techno.* **2019**, *15*, 130–135.
20. Zhang, P.; Wang, W.C. Research on gas emission and distribution of the first face in high gas and thick seam. *Sci. Techno. Eng.* **2017**, *17*, 176–180.
21. Cui, H.L.; Wang, Y.; Zhao, H.B.; Zou, D.L. Distribution of gas emission at fully mechanized top coal caving face in high gas and super thick coal seam in xiagou mine. *Saf. Coal Min.* **2016**, *47*, 178–181.
22. Yuan, L. Gas distribution of the mined-outside and extraction technology of first mined key seam relief-mining in gassy multi-seams of low permeability. *J. China Coal Soc.* **2008**, 33, 1362–1367.
23. Dai, G.L.; Wang, Y.Q.; Zhang, C.R.; Li, Q.M.; Shao, G.Y. Forecast of the gas effused form the face in protective seam. *J. China Coal Soc.* **2007**, *32*, 382–385.
24. Wu, R.X.; Xiao, Y.L.; Fang, L.C. Geophysical and geochemical prospecting for hard heading gas abnormal places between high gas coal seams. *Coal Geol. China.* **2014**, *26*, 74–77.
25. Huang, H.; Jiang, F.W.; Han, B.W.; Zhang, P.S. Comprehensive detection analysis on the reason of abnormal gas blow-out from the drilling hole through the floor limestone of A group coal seam in Huainan mining area. *J. China Coal Soc.* **2013**, *38*, 1988–1992.
26. Liang, W.X.; Li, J.T.; Fu, W.; Zhang, Y. Research and application of mixed gas source identification technology based on stable isotope. *Min. Saf. Environ. Prot.* **2022**, *49*, 56–61.
27. Zhou, W.; Yuan, L.; Zhang, G.L.; Dou, H.L.; Xue, S.; He, G.H.; Han, Y.C. A new method for determining the individual sources of goaf gas emissions: A case study in Sihe Coal Mine. *J. China Coal Soc.* **2018**, *43*, 1016–1023.
28. Yang, W.; Lin, B.Q.; Gao, Y.B.; Lv, Y.; Wang, Y.; Mao, X.; Wang, N.; Wang, D.; Wang, Y. Optimal coal discharge of hydraulic cutting inside coal seams for stimulating gas production: A case study in Pingmei coalfield. *J. Nat. Gas Sci. Eng.* **2016**, *28*, 379–388. [CrossRef]
29. AQ1018 –2006. Prediction method of mine gas emission. *Syst. Sci. Control. Eng.* **2006**, *6*, 85–91.
30. Qin, Y.J.; Su, W.W.; Jiang, W.Z.; Chen, Y.P. Research progress and development direction of mine gas emission forecast technology in china. *Saf. Coal Min.* **2020**, *51*, 52–59.
31. Wang, L.; Li, J.H.; Zhang, W.B.; Li, Y. Research on the gas emission quantity prediction model of improved artificial bee colony algorithm and weighted least squares support vector machine (IABC-WLSSVM). *Appl. Bionics Biomech.* **2022**, *2022*, 4792988. [CrossRef]
32. Xu, G.; Wang, L.; Jin, H.W.; Liu, P.D. Gas emission prediction in mining face by Factor Analysis and BP neural network coupling model. *J. Xi'an Univ. Sci. Techno.* **2019**, *39*, 965–971.
33. Liu, C.; Li, S.; Yang, S. Gas emission quantity prediction and drainage technology of steeply inclined and extremely thick coal seams. *Int. J. Min. Sci. Techno.* **2018**, *28*, 415–422.
34. Li, D.Y. Discussion on gas emission prediction method in adjacent layer of fully mechanized caving face. *Coal Min. Saf.* **2014**, *45*, 169–171.
35. Wang, H.F.; Fang, L.; Cheng, Y.P.; Zhou, H.R. Pressure-relief gas extraction of lower adjacent coal seam based on strata movement and its application. *J. Min. Saf. Eng.* **2013**, *30*, 128–131.
36. Liu, G.F.; Wang, H.X.; Song, Z.M. Numerical simulation study on gas emission rate owing to mining influence of adjacent coal seam. *J. Heinan Polytechnic Univ.* **2015**, *34*, 445–450.
37. Wang, W.; Peng, L.; Wang, X. Prediction of coal mine gas emission quantity based on grey-gas geologic method. *Math. Ppobl. Eng.* **2018**, *2018 Pt 17*, 4397237.
38. Qu, Q.D.; Balusu, R.; Belle, B. Specific gas emissions in Bowen Basin longwall mines, Australia. *Int. J. Coal Geol.* **2022**, *261*, 104076. [CrossRef]

energies MDPI

Review

Application of Gob-Side Entry Driving in Fully Mechanized Caving Mining: A Review of Theory and Technology

Dongdong Chen, Jingkun Zhu, Qiucheng Ye, Xiang Ma, Shengrong Xie *, Wenke Guo, Zijian Li, Zhiqiang Wang, Shaohua Feng and Xiangxiang Yan

School of Energy and Mining Engineering, China University of Mining and Technology-Beijing, Beijing 100083, China; chendongbcg@163.com (D.C.); zjk17860715825@163.com (J.Z.); yeqiucheng0314@163.com (Q.Y.); max_cumtb@163.com (X.M.); gwk0129@163.com (W.G.); lzj_te_amo_j@163.com (Z.L.); wang18336811606@163.com (Z.W.); fshahsx@163.com (S.F.); yxxcumtb@163.com (X.Y.)

* Correspondence: xsrxcq@163.com

Abstract: China has abundant coal resources, and the distribution of coal seams is complex. Thick coal seams account for more than 45% of all coal seams. Fully mechanized top coal caving mining has the advantages of large production, high efficiency, and low cost. In fully mechanized caving mining, especially in fully mechanized caving mining of extra-thick coal seams, the mining space is ample, the mine pressure is severe, and the roadway maintenance is complex. As a result, it is necessary to summarize and discuss the gob-side entry driving of fully mechanized caving in theory and technology, which will help to promote the further development of fully mechanized caving gob-side entry driving technology. First, in recent years, the research hotspots of gob-side entry driving have focused on the deformation mechanism and the control method of the roadway surrounding rock. Secondly, this paper discusses the theoretical models of the "triangle-block" and "beam" for the activity law of the overlying strata in gob-side entry driving, including the lateral breaking "large structure" model, compound key triangle block structure model in the middle and low position, the high and low right angle key block stability mechanics model, elastic foundation beam model, low-level combined cantilever beam + high-level multilayer masonry beam structure model, and the vertical triangular slip zone structure model. It introduces the "internal and external stress field theory" and the "stress limit equilibrium zone model". Thirdly, it summarizes several numerical simulation analysis methods in different conditions or research focuses and selects appropriate constitutive models and simulation software. Finally, it introduces surrounding rock control technology, including two ribs, the roof, and under challenging conditions. It provides a method reference for support in similar projects.

Keywords: fully mechanized top coal caving mining; gob-side entry driving; triangle-block structure; beam structure; cable truss; support

Citation: Chen, D.; Zhu, J.; Ye, Q.; Ma, X.; Xie, S.; Guo, W.; Li, Z.; Wang, Z.; Feng, S.; Yan, X. Application of Gob-Side Entry Driving in Fully Mechanized Caving Mining: A Review of Theory and Technology. *Energies* **2023**, *16*, 2691. https://doi.org/10.3390/en16062691

Academic Editor: Krzysztof Skrzypkowski

Received: 6 February 2023
Revised: 9 March 2023
Accepted: 10 March 2023
Published: 13 March 2023

1. Introduction

The recoverable reserves of thick coal seams in China account for about 45% of the total reserves of production mines [1]. The thick coal seam mining method has always been an important research topic in the coal industry [2–4]. The most common method of mining thick coal seams in China is fully mechanized top coal caving. In mining, most of the working face roadways will leave a certain width of a section coal pillar to protect the roadway, which is responsible for supporting the overlying strata and isolating the goaf water and harmful gas [5–8]. The width of the coal pillar increases as the mining intensity and geological conditions become more complex, resulting in a significant loss of coal resources [9,10]. Nonpillar or narrow coal pillar mining methods have been proposed to improve the recovery rate of coal resources. Gob-side entry driving is an important method of nonpillar mining in mining roadways [11]. Maintaining narrow coal pillars not only puts the roadway in the stress reduction zone [12,13] but also reduces coal pillar loss and improves the resource recovery rate. As a result, gob-side entry driving mining has been extensively used in thick coal seam mining.

Scholars from both home and abroad have researched the related problems of gob-side entry driving [14–16]. In terms of the activity law of the overlying strata [17] in gob-side entry driving, they put forward various theories based on the mechanical model of the "triangle-block" [18–21] and "beam". In fully mechanized caving section roadway surrounding rock control theory, they put forward the "internal and external stress field theory" and the "stress limit equilibrium zone model". The proposal of these theoretical models has extensively promoted the development and application of gob-side entry driving in fully mechanized caving mining. Based on these theories, we can make correct guidance for production practices. However, the current theory has limitations, and theoretical research must be strengthened. The numerical simulation [22] of gob-side entry driving in fully mechanized caving found that the strain-softening constitutive model is mainly used to study the reasonable width of the coal pillar [23–26]. Considering the compaction effect of gangue in goaf, the double-yield model is often used to study its influence on stress redistribution [27]. Numerical simulation is an indispensable technical means in mining engineering. Although the numerical simulation results sometimes do not reflect the actual situation well, they can also provide some reference. To control the surrounding rock of gob-side entry driving in fully mechanized caving, with the development of coal mine support technology [28–30], this method gradually changed from an initial shotcrete wall [31,32] and shed support [33] to a high prestressed anchor bolt and anchor cable support. In addition, for the asymmetric deformation problem of gob-side entry, they proposed an asymmetric cable truss [34–37] and step-bundled anchor cable [31,32] support way and achieved an excellent supporting effect. In order to ensure efficient and safe production, roadway surrounding rock control has always been an important research topic in the mining field. The mine pressure is severe in a fully mechanized caving roadway, and roadway maintenance is complex. Therefore, a safe and efficient support method is needed.

Gob-side entry driving is usually arranged at the edge of the goaf of the previous working face. After the overlying rock in the goaf has collapsed and become stable, retaining the smaller width coal pillar (generally 5–8 m) plays an isolation role. It drives the roadway along the edge of the goaf [38]. The popularization and application of gob-side entry driving mining are conducive to promoting the development of mining roadway support theory, effectively improving the recovery rate of coal resources and having obvious social and economic benefits. As shown in Figure 1, according to the mining geological conditions, gob-side entry driving in fully mechanized caving mining is mainly applied to incline extra-thick coal seam caving mining, three soft coal seam caving mining, large mining height caving mining, thick and hard basic roof caving mining, island working face caving mining, deep well caving mining, and large section caving mining [39–41].

With theoretical research and mine equipment development, gob-side entry driving has gradually become an essential means in fully mechanized caving mining [42]. Therefore, the coal recovery rate increases, the surrounding rock of the roadway is effectively controlled, and the incidence of accident disasters is reduced, promoting high-quality and efficient coal mining. The research hotspot map was obtained by analyzing the research status of gob-side entry driving in fully mechanized caving mining in recent years, as shown in Figure 2.

It can be seen from Figure 2 that the research hotspots of gob-side entry driving in fully mechanized caving mainly focus on the following five aspects: ① stability of the surrounding rock; ② deformation mechanism of the surrounding rock; ③ reasonable roadway positions and the coal pillar width; ④ numerical simulation analysis of gob-side entry driving; and ⑤ roadway surrounding rock control technology.

By searching the keywords "gob-side entry driving" in CNKI, more than 170 related pieces of literature in recent years were obtained. Figure 3 depicts the subject words of these research papers: gob-side entry driving, extra-thick coal seam mining, fully mechanized top coal caving mining, reasonable width of coal pillar, surrounding rock control [43], large mining height mining, research and application, large section, numerical simulation, and island working face. Among them, research on the appropriate section coal pillar width [44,45] of gob-side entry driving in fully mechanized caving is a hot topic, accounting for 25%.

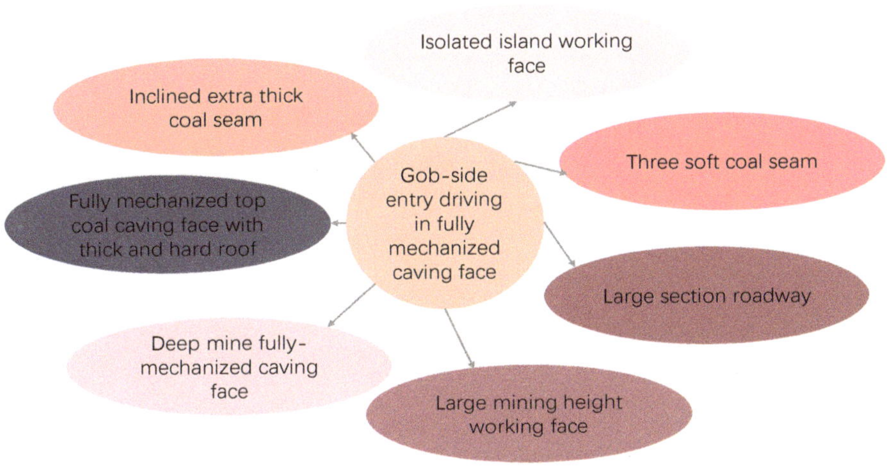

Figure 1. Some research directions of gob-side entry driving in fully mechanized caving mining.

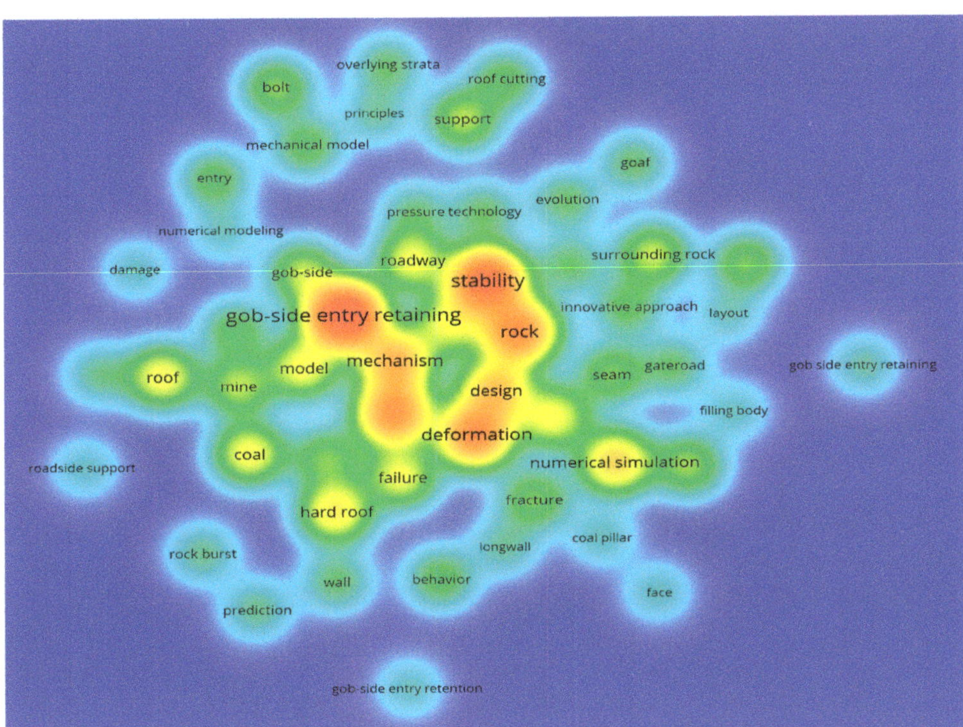

Figure 2. Research hotspot map of gob-side entry driving in fully mechanized caving.

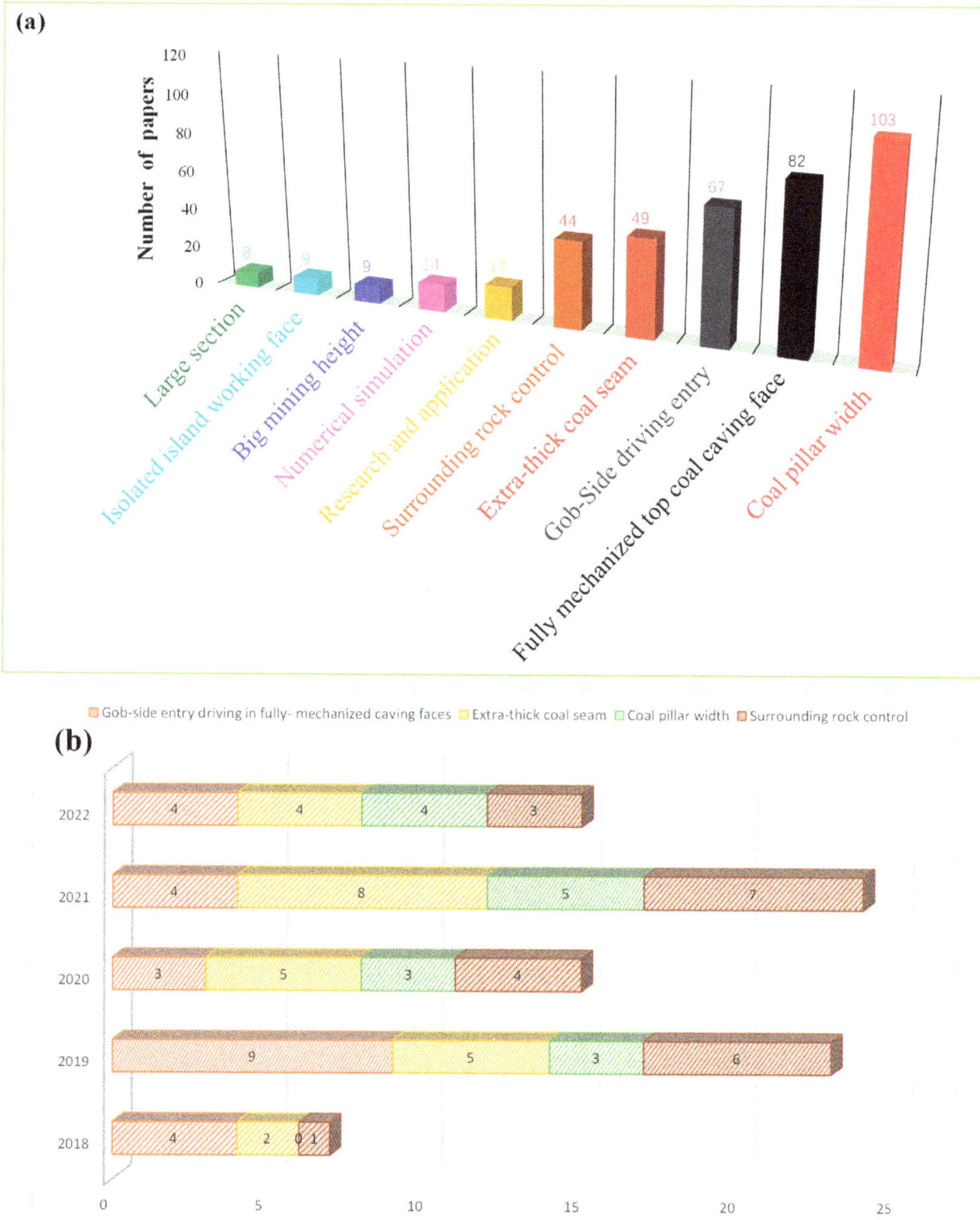

Figure 3. Research on gob-side entry driving technology in fully mechanized caving mining in China: (**a**) 10 research hotspots and (**b**) research trends in recent 5 years.

In this paper, the technology of gob-side entry driving with fully mechanized caving in recent years will be summarized in an all-around way. The related theories, numerical simulation constitutive models, research methods, and surrounding rock control technology will be systematically expounded. With proposals such as "carbon neutralization" and "carbon peak", coal mining is increasingly advocating for the development of safe mining technology with a high recovery rate. Fully mechanized caving along the goaf is widely used to realize a coal pillar reduction or no coal pillar. Therefore, this paper takes this as the research object. We analyze the application status and development prospects of gob-side entry driving in fully mechanized caving. When some experts or scholars need to preliminarily understand the relevant knowledge on gob-side entry driving, this paper can provide some guidance.

2. Research on the Theoretical Model of Gob-Side Entry Driving in Fully Mechanized Caving Mining

Fully mechanized top coal caving mining is a necessary technical means for high yield and high efficiency of coal mines in China. In terms of the movement law of the overlying strata in gob-side entry driving, scholars have constructed a mechanical model of the "triangle-block" and "beam" of the overlying rock, including the lateral breaking "large structure" model, compound key triangle block structure model in the middle and low position, the "high and low right angle key block stability mechanics model", elastic foundation beam model, low-level combined cantilever beam + high-level multilayer masonry beam structure model, and the vertical triangular slip zone structure model. In fully mechanized caving section roadway surrounding rock control theory, they put forward the "internal and external stress field theory" and the "stress limit equilibrium zone model".

2.1. Activity Law of Overlying Strata in Gob-Side Entry Driving

2.1.1. Mechanical Model of the "Triangle-Block"

(1) The lateral breaking "large structure" model

When the working face is left with narrow coal pillars, mining will significantly affect the surrounding rock, and the movement of the overlying strata will be violent, which will cause potential safety hazards. By analyzing the characteristics of the surrounding rock, scholars put forward the stability control principle of the "large structure" of the lateral breaking of the overlying strata in the adjacent goaf and the "small structure" of the overlying roof of the roadway surrounding rock. They established the large structure model of the overlying rock mass of the roadway driving along the goaf, as shown in Figure 4 [46]. By analyzing the stability of the triangular arc block of a basic roof, as shown in Figure 5, scholars proposed a surrounding rock control mechanism. Among them, F_Z is the resultant force of the self-weight of key block B, F_R is the resultant force of the self-weight of the upper weak overburden rock, the vertical force and horizontal force of rock block A to key block B are R_A and F_A, respectively, and the vertical shear force and horizontal thrust of structural block C to key block B are R_B and F_B, respectively, the supporting force of the gangue, end coal, and lateral coal in the goaf to key block B are F_G, F_{SM}, and F_M.

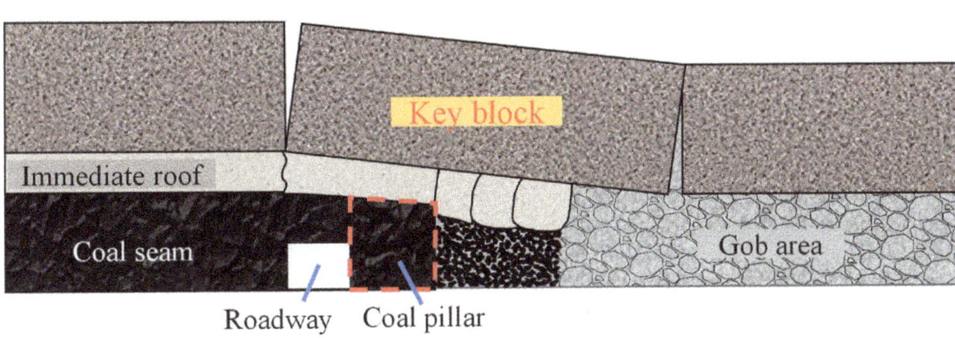

Figure 4. The lateral breaking structure model [46].

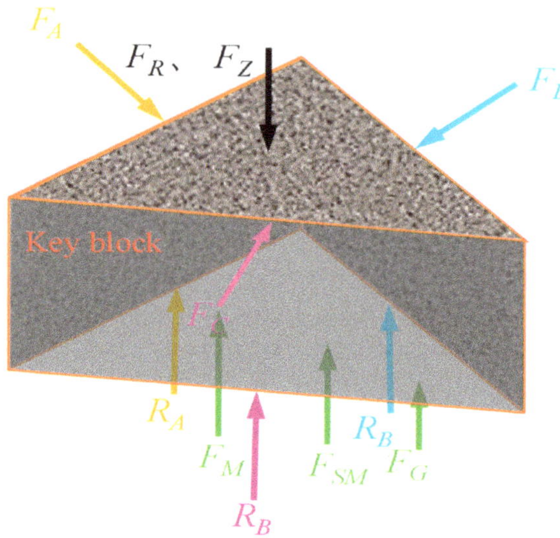

Figure 5. Mechanical analysis of the lateral breaking "large structure".

(2) Compound key triangle block structure model in the middle and low position

The construction of a mechanical model of a middle and low composite key triangular plate structure with synchronous and asynchronous migration of a basic roof (low position) at the end of the adjacent goaf of a fully mechanized caving face and the adjacent key hard rock layer (middle position) is shown in Figures 6 and 7. The stability characteristics and engineering disaster conditions of the key triangular plate structure in three-time states (before the formation of the section coal lane, after the formation of the section coal lane, and at the time of mining) were explored. The fracture position of the main roof of fully mechanized caving and its influence were obtained [47]. The same symbols in Figures 5 and 7 also have the same meaning. In addition, F'_Z is the resultant force of the self-weight of key block B (mesoposition), and F'_R is the resultant force of the self-weight of the upper weak overburden (mesoposition). The movement of weak rock strata above key block B separates from the hard rock strata above it and loses the transmission of force, namely, $F_X = 0$.

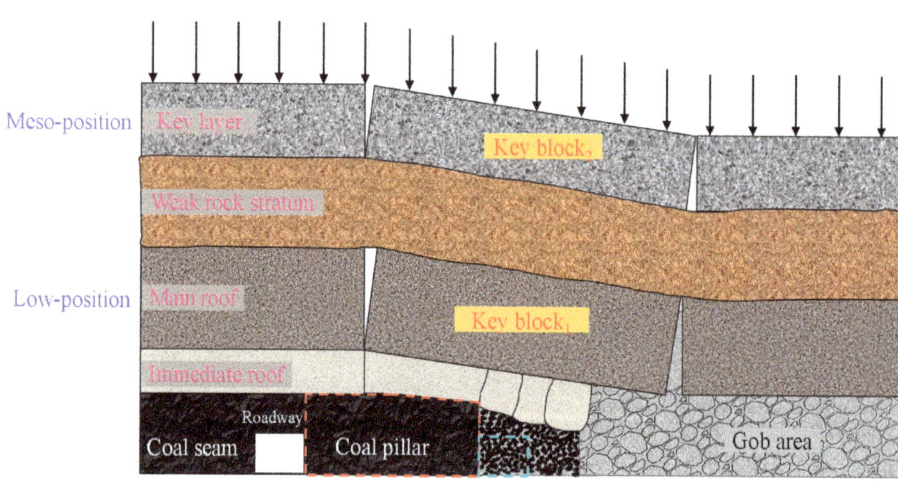

Figure 6. Compound triangle block structure model [47].

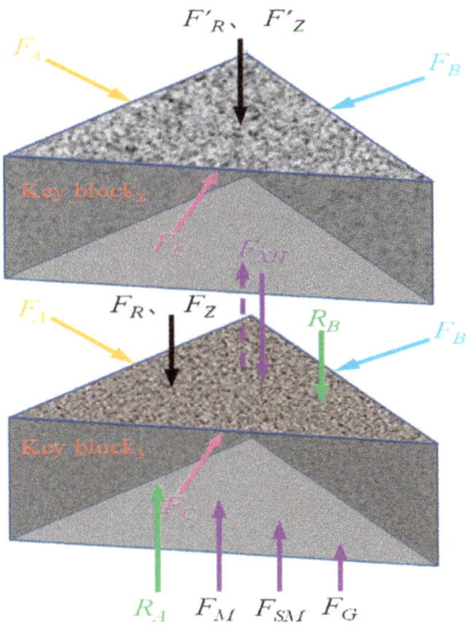

Figure 7. Mechanical analysis of the key triangular plate structure [47].

(3) The high and low right angle key block stability mechanics model

In fully mechanized caving mining of extra-thick coal seam with a hard and thick main roof, the mining space is large, and the strength is high, which causes the overlying strata to fracture, move, and collapse, and the influence is vast. Therefore, the low-key and high-key strata play a key role in the stability of the surrounding rock of the gob-side entry. Based on the theory of internal and external stress field and limit equilibrium theory, as shown in Figures 8 and 9, scholars have deduced a mechanical model for the stability of high and low right-angle key blocks with periodic breaking [48] and analyzed the joint stability of high and low key blocks.

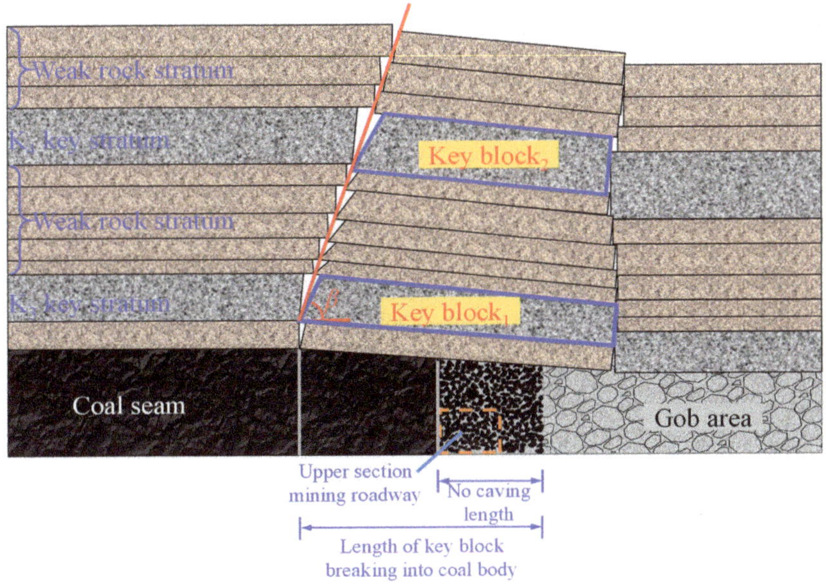

Figure 8. Fracture morphology of key blocks [48].

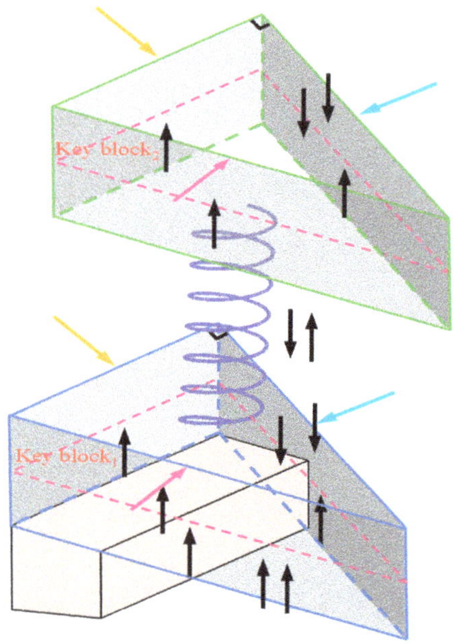

Figure 9. Mechanical analysis of high and low right angle key blocks [48].

2.1.2. Mechanical Model of the "Beam"

(1) Elastic foundation beam model

This model was based on the theory of masonry beams with lateral roof fracture, considering the deformation characteristics of coal seam and immediate roof. Scholars have established the basic roof elastic foundation beam model, as shown in Figure 10. They deduced the expressions of the lateral roof bending moment and displacement and then obtained the basic roof breaking position; the influence of the main roof, immediate roof, and coal seam thickness and elastic modulus on the lateral fracture position of the roof was explored [49]. Figure 11 shows the overburdened structure after the basic roof is broken.

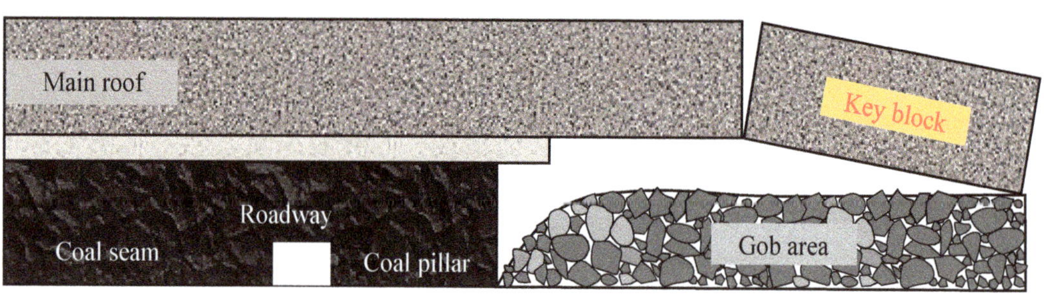

Figure 10. Structural model of elastic foundation beam [49].

(2) Low-level combined cantilever beam + high-level multilayer masonry beam structure model

The mining space formed by the mining of the fully mechanized caving face in the extra thick coal seam is large, the roof breaking and migration range are wide, and the development height of the overburdened caving zone and the fracture zone is significantly increased. When there are multiple layers of hard rock strata in the roof, the overlying hard key stratum breaks not only one layer but multiple layers. Therefore, for the mining of extra-thick coal seam with a hard roof, scholars have proposed a lateral overburden structure model of "low-level combined cantilever beam + high-level multi-layer masonry beam" in the roadway of extra-thick coal seam with a hard roof, as shown in Figure 12 [50].

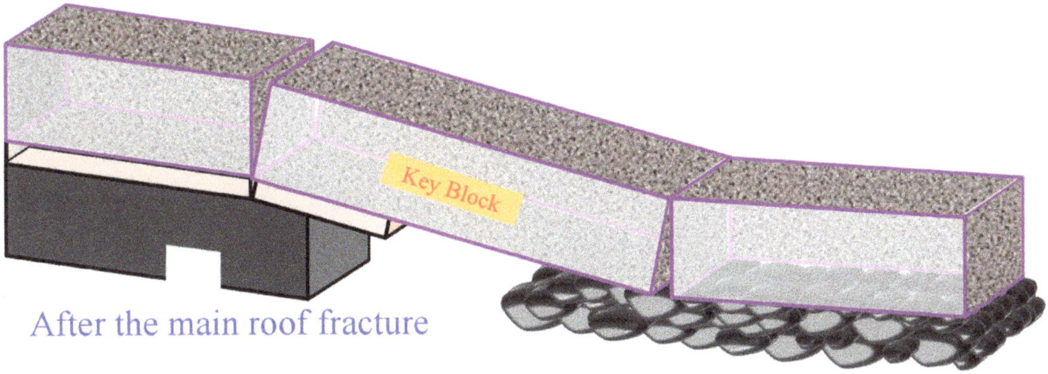

Figure 11. Overburden rock structure.

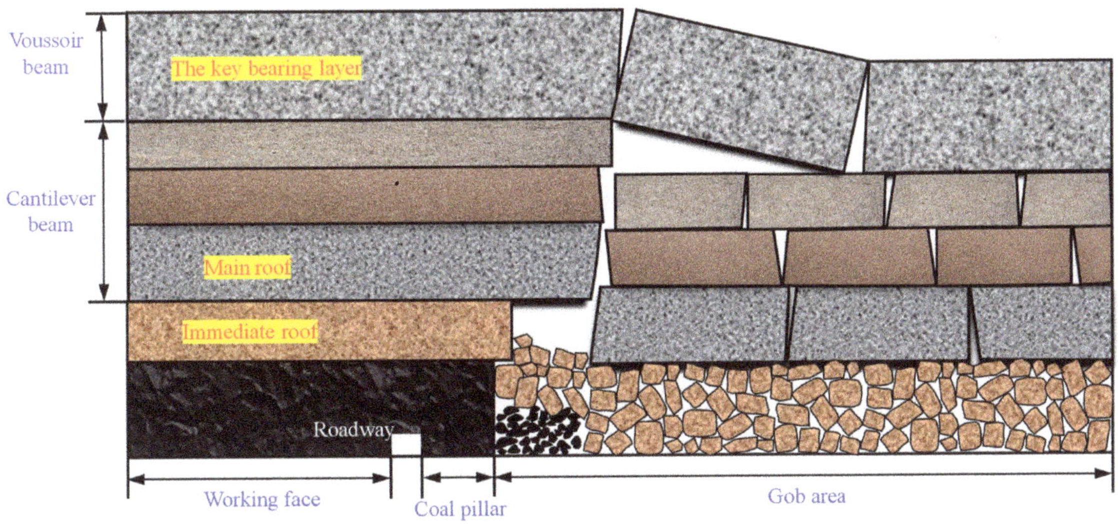

Figure 12. Low-level combined cantilever beam + high-level multilayer masonry beam structure model [50].

(3) The vertical triangular slip zone structure model

Scholars took a fully mechanized caving face with a large mining height in an extra-thick coal seam as their research object and analyzed the activity range, fracture field distribution, motion characteristics, and structural characteristics of the overlying strata at the end of the working face with a large mining height in the extra-thick coal seam. It was proposed that there is a stable stress reduction zone with a triangular slip zone structure at the end of the goaf. This is conducive to the layout of gob-side entry driving with a small coal pillar and the maintenance of a small coal pillar roadway. According to the movement characteristics of the overlying strata in the goaf and the time–space relationship, the reasonable position and time of the small coal pillar driving along the goaf were determined, as shown in Figure 13 [51].

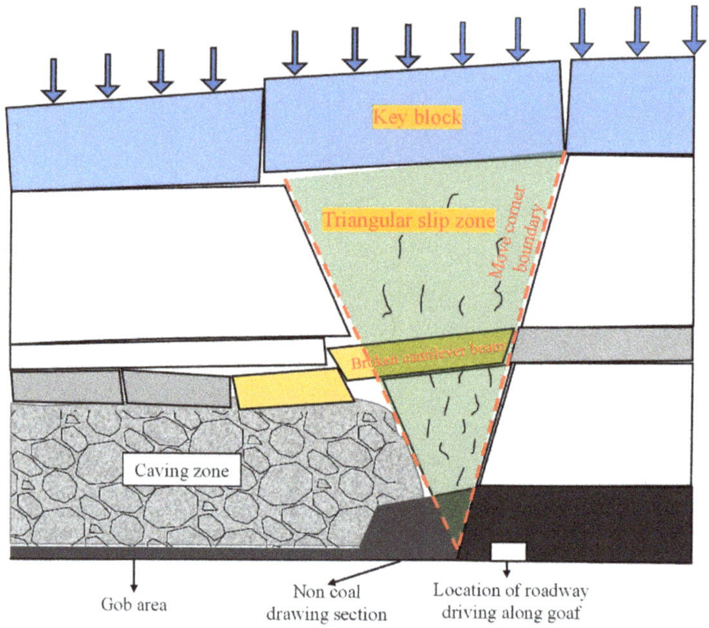

Figure 13. The vertical triangular slip zone structure model [51].

2.2. Surrounding Rock Control of Fully Mechanized Caving Section Roadway

2.2.1. The "Internal and External Stress Field Theory"

Scholars have established the structural mechanics model of gob-side entry driving by the theoretical analysis method. They established the expression of "internal stress field" width and determined the reasonable position of gob-side entry driving and the reasonable width of the coal pillar. They predicted the deformation of the surrounding rock of gob-side entry driving, as shown in Figure 14 [52]. The movement law and deformation failure characteristics of the surrounding rock of the roadway and the curved triangular block of the end basic roof in different stages were studied. The overall mechanical environment of the fully mechanized caving along the goaf was analyzed, and stability control theory was preliminarily formed. In Figures 15–17 [53], scholars also established a mechanical model of the roadway surrounding rock structure. The relationship between the key block's rotation angle and the coal pillar's overlying load was obtained, and the coal pillar's width was calculated to determine the fracture position of the basic roof. In Figure 14, the interval calibrated by S_1 is called the "internal stress field", and the interval calibrated by S_2 is called the "external stress field". σ_y is the lateral support pressure; K is the stress concentration factor; and γ is the average bulk density of the overlying strata. Moreover, H is the buried depth of the roadway.

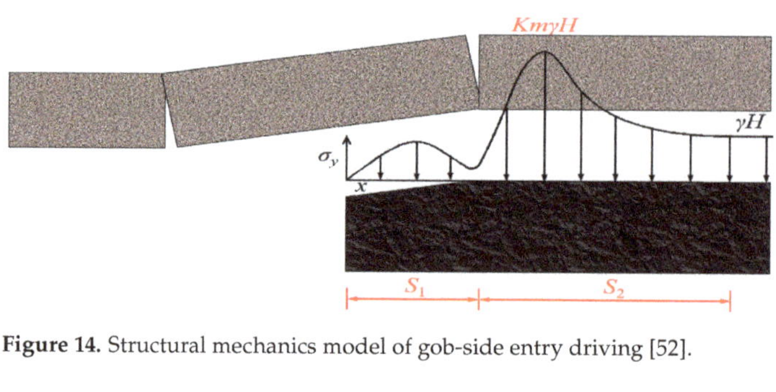

Figure 14. Structural mechanics model of gob-side entry driving [52].

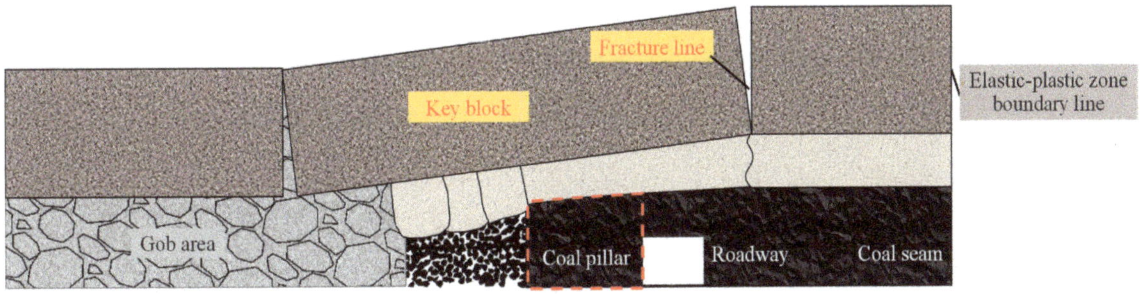

Figure 15. Structure model with main roof fracture line above the solid coal [53].

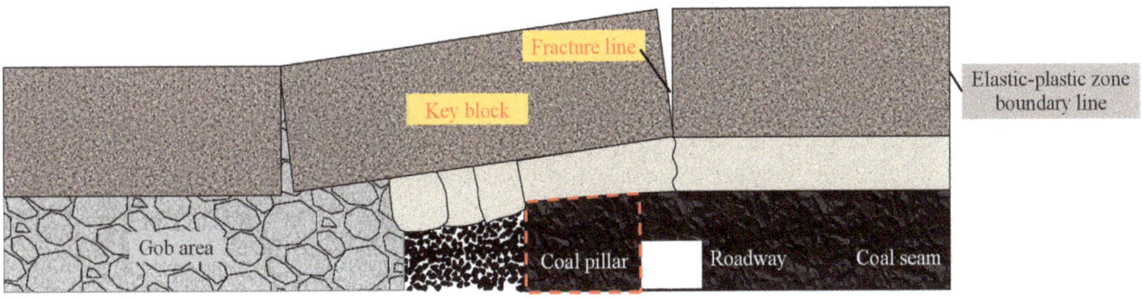

Figure 16. Structure model with main roof fracture line above the roadway [53].

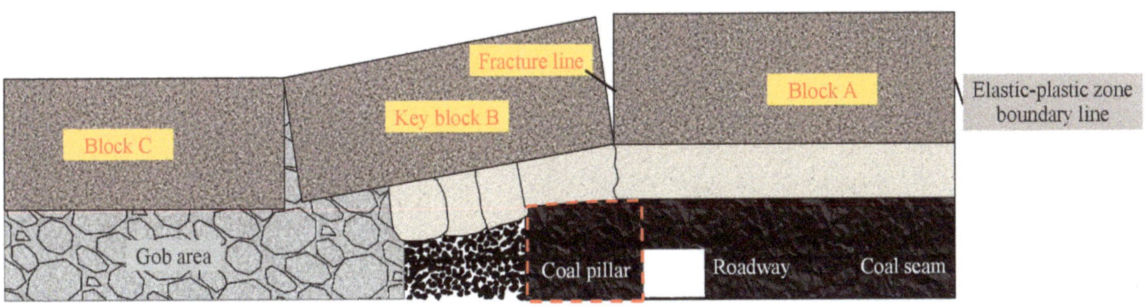

Figure 17. Structure model with main roof fracture line above the coal pillar [53].

2.2.2. The "Stress Limit Equilibrium Zone Model"

In Figures 18 and 19, based on the stress characteristics of the surrounding rock of gob-side entry driving in deep-well fully mechanized caving, considering the strength-softening characteristics of coal and rock mass at the interface between the roadway side and roof and floor, scholars have established the mechanical analysis model of two ribs and deduced the theoretical calculation of the limit equilibrium zone width and the coal stress displacement of two ribs [54]. Based on the distribution characteristics of the inclined abutment pressure of the coal body on the goaf side and the limit equilibrium theory of a coal pillar in roadway protection, the analytical expressions of the upper and lower limits of a reasonable width of a narrow coal pillar in roadway protection were determined [55]. Based on analyzing the stress environment of gob-side entry driving, the principle of damage mechanics was used to analyze the abutment pressure distribution of the solid coal side of gob-side entry driving under the given deformation, and the relationship between the abutment pressure distribution and parameters such as the coal rock thickness and elastic modulus was discussed. It is of great significance to the upkeep of gob-side entry as well as the study of the floor heave mechanism and control [56]. In Figure 18, the shear stress at the interface between the coal seam and the roof and floor is τ_{xy}, the vertical pressure is σ_y, the horizontal stress inside the coal rock mass is σ_x, and the roadway rib support resistance is f_i. In Figure 19, f_s is the support resistance of the general goaf side, f_z is the support resistance of the roadway side, and the peak stress in the limit equilibrium zone of the coal pillar side should be σ_{ym}.

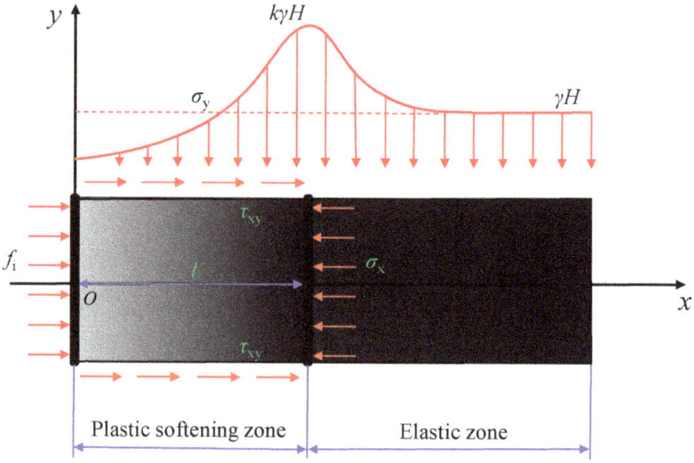

Figure 18. The mechanical model of solid coal rib [54].

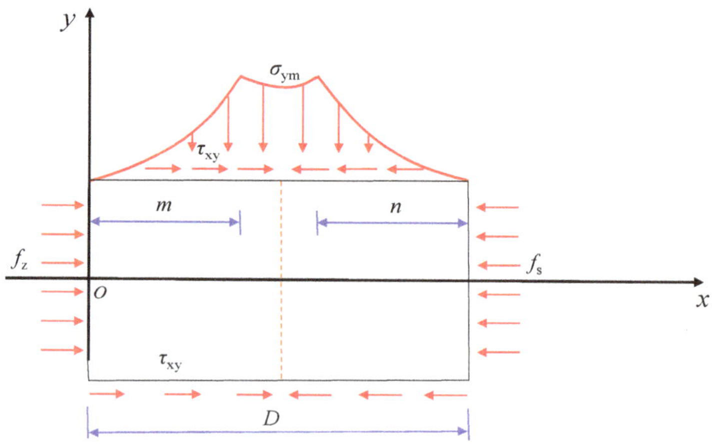

Figure 19. The mechanical model of coal pillar rib [54].

Among the above many theoretical models, the rock beam model belongs to a relatively simplified model, and the triangular block structure is a theoretical model that is more recognized by scholars at present. It is also more often used in the related research of gob-side entry driving.

3. Numerical Analysis Method of Gob-Side Entry Driving in Fully Mechanized Caving Mining

Numerical simulation is an indispensable technical means in mining engineering. At present, the most commonly used numerical analysis methods are the finite element method, boundary element method, finite difference method, weighted residual method, discrete element method, rigid body element method, discontinuous deformation analysis

(DDA) method, manifold element method [57], etc. In the study of fully mechanized caving gob-side entry driving engineering problems, scholars often use FLAC^{3D}, UDEC, PFC, and other numerical software according to different working conditions and select the appropriate software analysis [58–61]. As shown in Figure 20, it shows the constitutive model used in FLAC^{3D}.

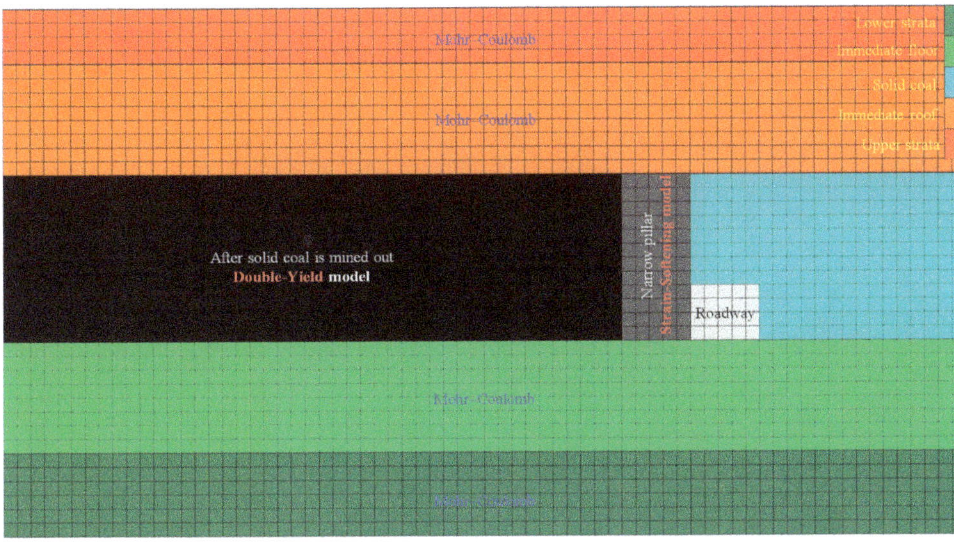

Figure 20. Constitutive model of FLAC^{3D} used gob-side entry driving in fully mechanized caving.

3.1. *Constitutive Model of FLAC^{3D}*

3.1.1. Strain-Softening Model of Yielding Coal Pillar

In studying the surrounding rock control of gob-side entry driving in fully mechanized caving, many scholars mostly choose the strain-softening constitutive model when studying reasonable coal pillar width. Compared with the Mohr–Coulomb constitutive model, it can more truly reflect the yield of the small coal pillar, especially when retaining a small coal pillar. It can also provide an accurate and reliable basis for formulating a sensible coal pillar width and support [23–26]. The strain-softening model reflects the real failure properties of coal pillars as follows: the elastic stage is consistent with the Mohr–Coulomb model. After entering the plasticity, the cohesion and friction will gradually decrease with the plastic strain [30,62,63].

FLAC3D finite difference software has a strain-softening constitutive model [60]. Scholars have established a standard specimen model for a uniaxial compression 1:1 simulation and carried out the corresponding parameter iterative inversion through indoor test and numerical simulation then fitted the standard specimen parameters for numerical simulation [64–66].

3.1.2. Double-Yield Model of Goaf

Due to the compaction of gangue in the goaf, the stress state of the surrounding rock in the gob-side entry will be affected [67]. Therefore, the double-yield model can well-simulate the influence of gangue on stress redistribution [27]. In the gob-side entry driving process, the coal pillar's bearing capacity needs to be considered, and the influence on the goaf cannot be ignored [68–73]. In the numerical simulation, the "cap pressure" is the parameter that mainly determines the compaction characteristics of the goaf material in the simulation, which is controlled by Table [74,75]. According to the classical theoretical formula of Salamon, the corresponding parameter inversion is carried out by establishing a 1 m × 1 m × 1 m model in the numerical simulation so that the parameters can precisely reflect the actual condition [76,77]. The specific parameter inversion process in the numerical model is shown in Figure 21.

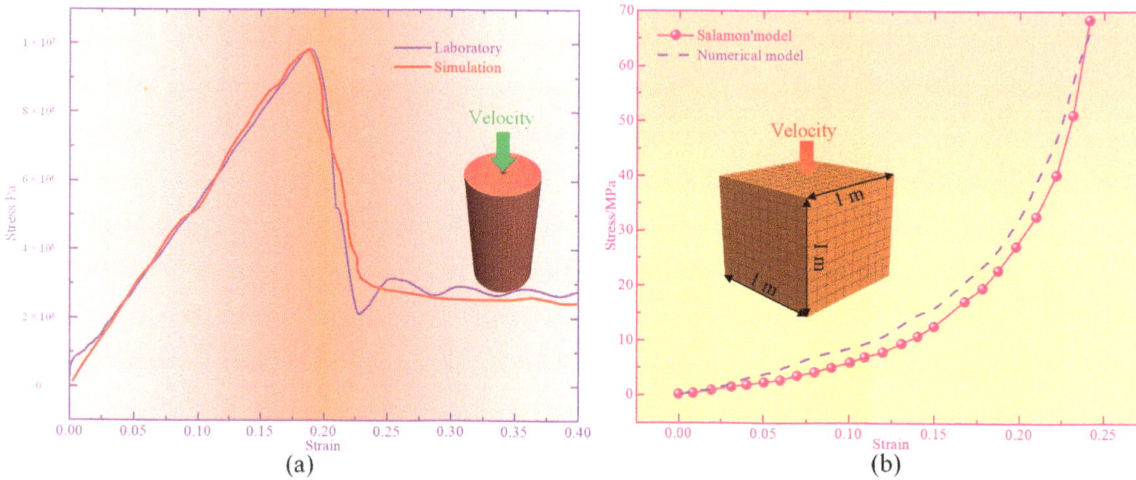

Figure 21. Inversion fitting process of constitutive model parameters in numerical simulation: (**a**) strain-softening model and (**b**) double-yield model.

3.2. UDEC Simulation of Coal Pillar Fracture

Through the UDEC (a discrete element software), many scholars can intuitively see the fracture development and damage degree in the coal pillar of gob-side entry driving [78–81]. By analyzing the crack propagation morphology and plastic state in the coal pillar, they can guide the reasonable setting position of the gob-side entry and coal pillar width to minimize the influence of the coal pillar on the stability of the roadway [82–84]. As shown in Figure 22, it indicates the development state of coal pillar cracks in gob-side entry driving.

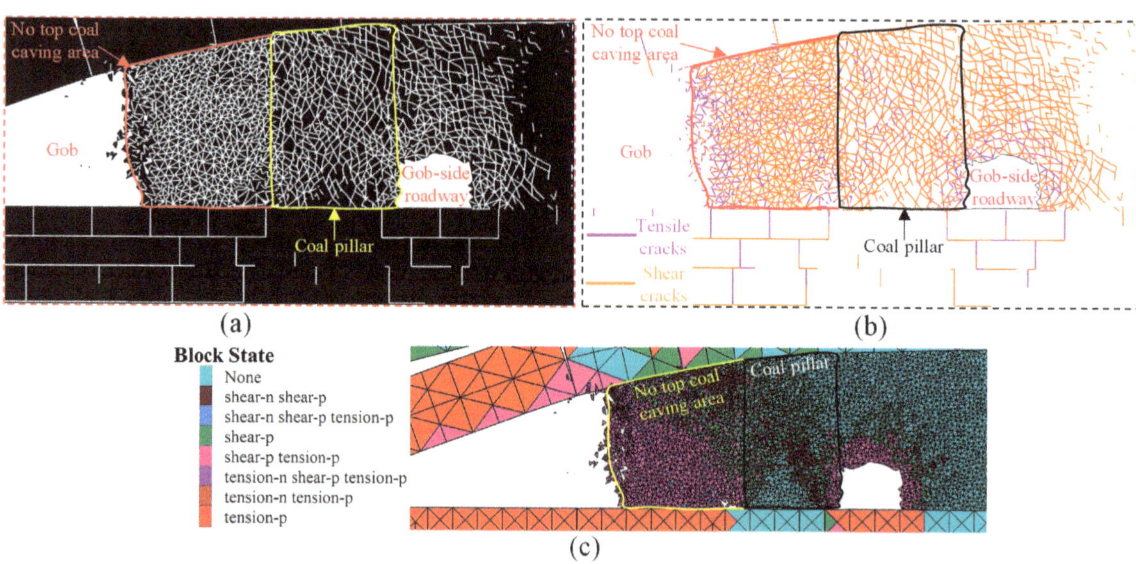

Figure 22. Development state of coal pillar fracture of gob-side entry driving in fully mechanized caving: (**a**) fracture pattern, (**b**) cracks, and (**c**) state.

3.3. Other Numerical Analysis Methods

Many scholars have carried out uniaxial and triaxial compression simulation tests by sampling coal on-site and combining discrete element software PFC^{2D}/PFC^{3D} to explore the crack propagation law of coal specimens under the condition of prefabricated cracks [85–87] to guide the setting of coal pillars in gob-side entry driving to ensure the optimal stress field environment of roadway surrounding rock. Some scholars have studied the dynamic, progressive failure process of coal rock samples through CDEM to analyze the influence of cracks on the stability of coal rock columns [88,89].

4. Surrounding Rock Control Technology of Gob-Side Entry Driving in Fully Mechanized Caving Mining

The structural characteristics of overburdened roadway rock differ in different mining stages of gob-side entry driving, which has a significant impact on the surrounding rock support [90]. It is essential for the stability of the surrounding rock to optimize the support parameters according to the structural characteristics of the overlying rock [91]. Scholars have put forward various supporting technologies for roadway driving along the goaf. This paper will introduce the surrounding rock control technology from three aspects: two ribs, the roof, and other complex conditions.

4.1. Two-Rib Support of the Roadway

4.1.1. Support of Coal Pillar Rib

In gob-side entry of fully mechanized caving, the coal pillar rib usually adopts general support forms, such as a bolt + ladder beam of steel (W, JW steel strip) + mesh and anchor cable support. However, with the progress of fully mechanized caving mining, the fracture development of the coal pillar is obvious, and the stress concentration at the end of fracture is obvious, which leads to the weakening of the bearing capacity and an antideformation and failure ability of the coal pillar [92,93]. The linkage between the coal pillar and top coal is large. The failure of the coal pillar reduces the stability of the top coal, increasing the deformation and pressure of the roadway and increasing the difficulty of support. Using an ordinary bolt and cable support on the coal pillar's rib to maintain stability is challenging. Roadside support can assist the coal pillar in bearing roof pressure and improve the bearing capacity of the coal pillar.

The roadside support of gob-side entry driving is mainly divided into concrete wall support on the side of the coal pillar roadway and filling support on the side of the coal pillar goaf, as shown in Figure 23. The pouring concrete wall support on the side of the coal pillar roadway refers to establishing a certain width of the reinforced concrete wall in the roadway. The supporting wall is connected with the roof, floor, and the coal pillar rib through the preset high-strength bolt. Therefore, the reinforced concrete wall and the surrounding rock are coordinated [31,32]. The filling support of the goaf side of the coal pillar refers to the injection of foam, fly ash material, high water material, paste material, or cement slurry near the goaf side of the coal pillar. It can replace part of the falling gangue or directly fill it, thereby reducing the roof activity space [94,95].

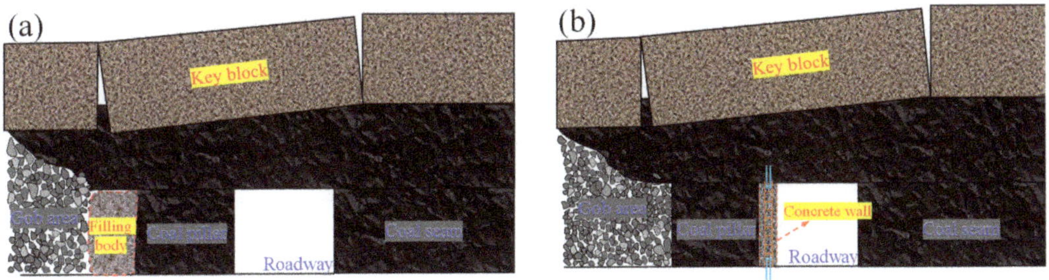

Figure 23. *Cont.*

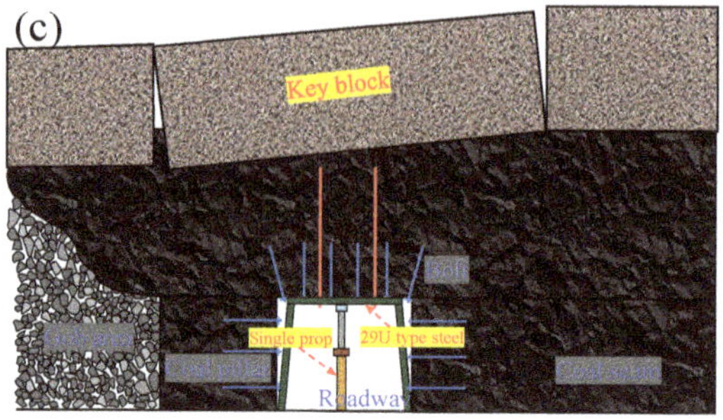

Figure 23. Support diagram of coal pillar rib: (**a**) filling support on the side of the goaf, (**b**) concrete wall support on the side of the roadway, and (**c**) general bolt and anchor cable support. (**a**,**b**) [31,32] and (**c**) [94,95].

4.1.2. Support of Virgin Coal Rib

The degree of mine pressure on the virgin coal rib is small, and the damage to the surrounding rock of the roadway is also tiny. As a result, the support method is simple. Bolt + ladder beam of steel (W, JW steel strip) + mesh is often used, and sometimes a single anchor cable is also used for reinforcement support.

4.2. Roof Support of Gob-Side Entry Driving

For a long time, scholars have conducted much research on the problem of roof control in gob-side entry driving. The roof is usually controlled with a combination of various support methods, among which, the most commonly used is bolt support [44]. The arrangement of bolt support in various fully mechanized caving roadways is the same, so this paper focuses on roof control technology with an anchor cable as the core.

4.2.1. Single Anchor Cable Support or Anchor Cable + Steel Strip Support

When the coal seam thickness is less than six m, two or three independent single anchor cables are arranged on the roof for support, as shown in Figure 24a,b, and the anchor cable can be anchored to the stable rock stratum. W steel strips are also commonly used to connect the anchor cables, as shown in Figure 24c,d.

4.2.2. Anchor Cable Truss Support

The anchor cable truss comprises a long anchor cable and a special connecting lock device [96]. Figure 24e,f represents symmetric and asymmetric layouts, respectively. The special connecting lock device connecting the long anchor cable is shown in Figure 24g.

Figure 24h indicates its control principle [97,98]: The cable truss system gradually locks during roof rock deformation, increasing the compression value of shallow surrounding rock and preventing excessive deformation of roadway surrounding rock; the anchor cable truss has a long length and solid shear resistance. It crosses the greatest shear stress area at the coal pillar–roof junction obliquely, enhancing the surrounding rock's shear resistance and maintaining its stability in the coal pillar's corner area.

4.2.3. Cable Beam Truss Support

The cable beam truss structure comprises a long anchor cable, channel steel support beam, steel support beam, and lock. The single anchor cable is first connected with a high-strength steel support beam, and the anchor cable near the side of the coal pillar is connected with a channel steel support beam for a secondary connection. The support structure is arranged near the side of the coal pillar rib, as shown in Figure 24i,j. Figure 24k is the on-site support diagram, and Figure 24l is the supporting principle diagram.

The control principle is as follows: After applying a high pretightening force, the anchor cable, steel (channel steel) support beam, and coal–rock mass form an inverted trapezoidal bearing structure. When subjected to unbalanced abutment pressure, the inverted trapezoidal structure jams the two corners. The greater the load, the greater the force of the anchor cable and the formation of a stress arch with a base point at the two corners. The formation of the stress arch weakens the transfer of external pressure to the interior, reducing the asymmetric subsidence of the roof and horizontal extrusion deformation [99]. The anchor cables are connected by the high-strength steel support beam, which is more flexible to adapt to extrusion deformation and can prevent the connection structure from failing due to horizontal dislocation of surrounding rock. Increasing the internal hole size of the channel steel can reserve the deformation space for horizontal movement and avoid the stress concentration between the channel steel and the anchor cable due to the horizontal movement of the rock stratum [100].

4.2.4. Anchor Cable + Channel Steel Support

In mining hard and thick main roof coal seams, there are some control problems, such as severe overburden activity and asymmetric deformation in the roof. Scholars have proposed the asymmetric combined control technology of the roof with anchor cable–channel steel combination [101]. Figure 24m,n shows that each row contains four or five anchor cables. The anchor cables near the two ribs are deflected outward by 15°, and the anchor cables in the middle position are arranged perpendicular to the roof. Figure 24p is the supporting principle diagram.

4.2.5. Step Bundled Anchor Cable Support

For the roof support of ultra-thick coal seams (up to 15 m), some scholars have proposed the supporting technology of step-bundled anchor cables, as shown in Figure 24q,r. Moreover, the step bundle anchor cable comprises 5 anchor cables and a bundle anchor cable tray arranged in a "2-1-2" manner. The center is a 22 mm × 10,300 mm anchor cable surrounded by two 22 mm × 6300 mm and two 22 mm × 8300 mm anchor cables. The anchor cables are arranged diagonally and connected by a porous tray [32], as shown in Figure 24s [32].

4.3. *Support under Difficult Conditions such as a Broken Roof*

Figure 25 depicts an early support form of roadway driving along the goaf, primarily shed support, including I-steel and U-steel support. Secondary or multiple mining may influence the roads during layered mining and the mining of coal seam groups, and the deformation and failure of the surrounding rock are severe. The shed support and steel mesh combination can effectively limit the roadway's severe deformation. At the same time, it can improve the stress environment and the mechanical properties of the surrounding rock with a high-strength grouting anchor cable, indirectly improve the majestic residual strength and self-bearing capacity, significantly slow down the large deformation of the surrounding rock, and ensure the stability of the surrounding rock [33].

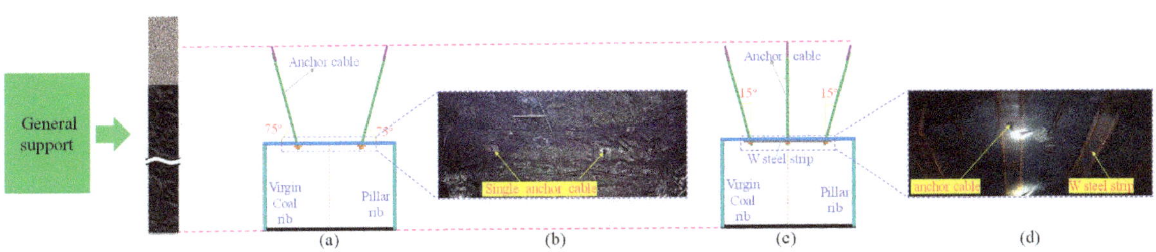

Figure 24. *Cont.*

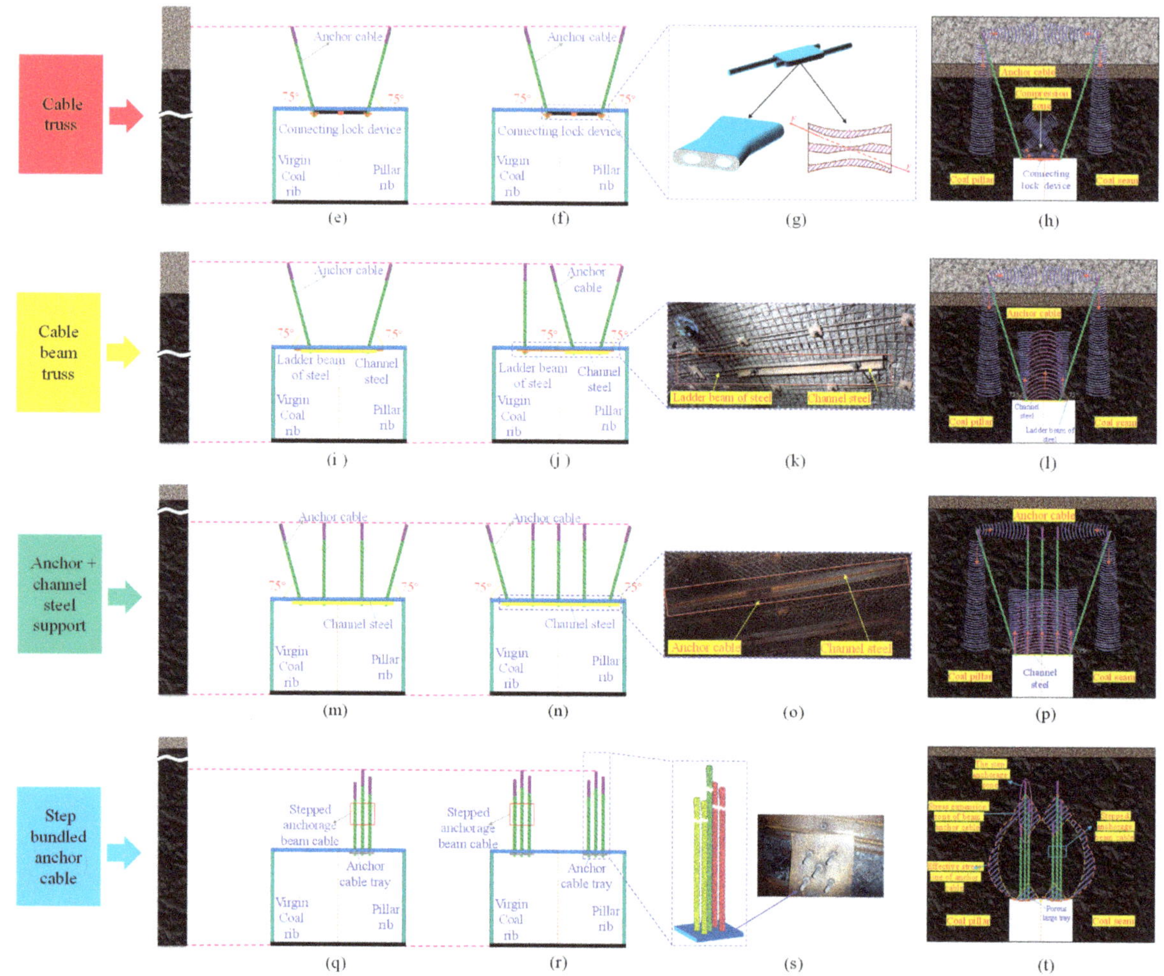

Figure 24. The roof support of gob-side entry driving: (**a**) single anchor cable support, (**b**) on-site support diagram, (**c**) anchor cable + W steel strip support, (**d**) on-site support diagram, (**e**) symmetric cable truss support, (**f**) asymmetric cable truss support, (**g**) connecting lock device, (**h**) supporting principle diagram, (**i**) asymmetric double anchor cable support, (**j**) cable truss + single anchor cable support, (**k**) on-site support diagram, (**l**) supporting principle diagram, (**m**) channel steel + four anchor cables support, (**n**) channel steel + five anchor cables support, (**o**) on site support diagram, (**p**) supporting principle diagram, (**q**) single-step bundled anchor cable support, (**r**) step bundled anchor cable support, (**s**) stepped anchorage beam cable, and (**t**) supporting principle diagram. (**f**) [96], (**h**) [97,98], and (**m**,**n**) [101].

A single support method in the gob-side entry of fully mechanized caving often cannot meet the support requirements. Therefore, the combined support form with an anchor cable support as the core and other support methods (bolt, metal mesh, etc.) is often used to achieve adequate control of the surrounding rock, as shown in Table 1. The '$\surd$' in Table 1 represents the support form used in a certain spatial orientation of the roadway.

Table 1. Roof combined support table of gob-side entry driving in fully mechanized caving.

Support Position		Support Pattern	Bolt + Ladder Beam (W, JW Steel Strip) + Net	Single Anchor Cable	Anchor Cable + W, JW Steel Strip	Grouting	Roadway Side Concreate Wall	Gob Side Filling	Shelf	Anchor + Connecting Lock Device	Anchor + Channel Steel	Step Bundled Anchor Cable + Porous Large Tray
Pillar rib		I	√									
		II	√									
		III	√			√	√					
		IV	√	√				√				
		V	√						√			
		…										
Virgin coal rib		I	√									
		II	√	√								
		III	√						√			
		…										
Roadway roof	General support		√									
	Anchor truss	I	√							√		
		II	√							√	√	
	Step bundled anchor cable		√		√							√
	…											
Special condition		I	√	√					√			
		…										

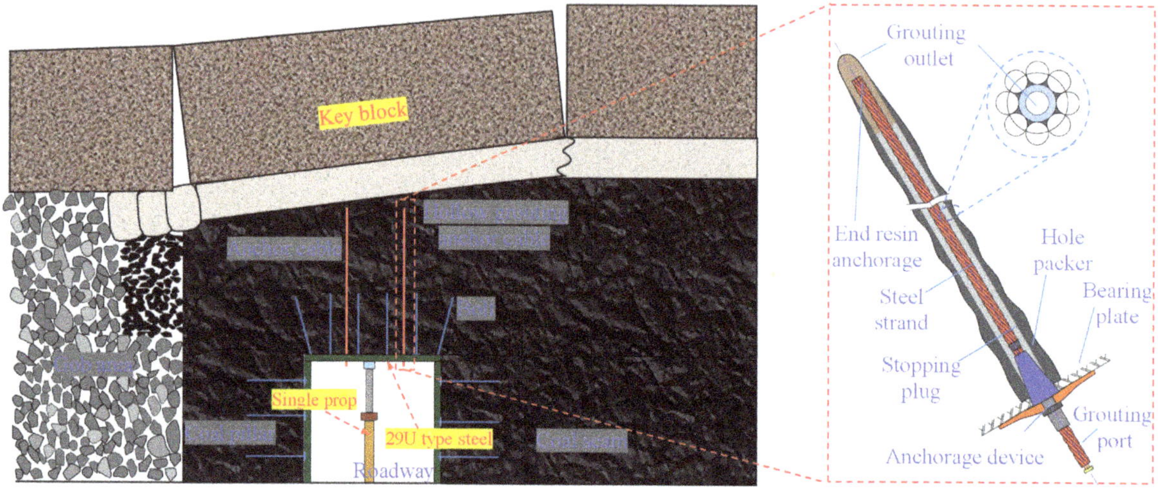

Figure 25. Shed support-grouting anchor cable cooperative support diagram [33].

5. Engineering Monitoring

In addition to the above research, other scholars also analyzed the stress and deformation laws of surrounding rock in gob-side entry driving using field engineering monitoring.

The lateral roof structure type and lateral abutment pressure distribution characteristics were determined using microseismic monitoring and stress dynamic monitoring [44,102]. As a result, the deformation, failure mechanism, and control of fully mechanized caving roadway along the goaf were studied. The borehole peeping method was used to measure the main roof's fracture position and the roof's two-way movement characteristics. Combined with the CT identification of the asymmetric evolution process of the microcracks in the roof coal, comprehensive support technology was proposed [37,103]. In addition, to evaluate the feasibility of the support scheme and understand the working state of the support scheme in detail, the surface displacement of the roadway was monitored by arranging the measuring station; using the steel ruler and the measuring line, the separation sensor monitored the roof separation; the stress of the coal pillar was monitored with the borehole stress meter; and the bolt cable dynamometer monitored the working resistance of the bolt cable [104,105]. As shown in Figure 26, represents a variety of monitoring instruments.

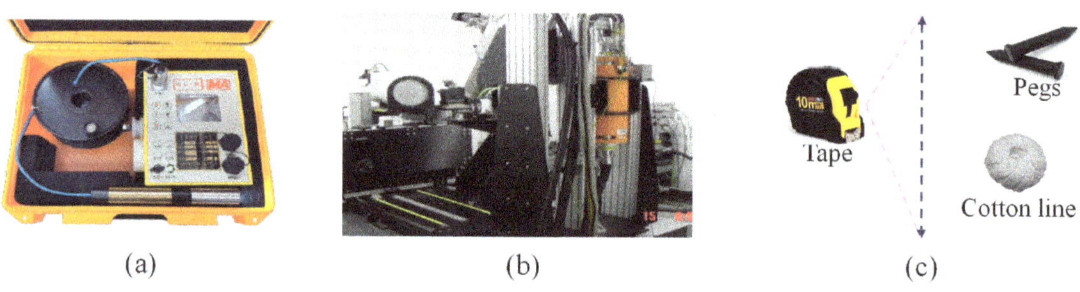

Figure 26. *Cont.*

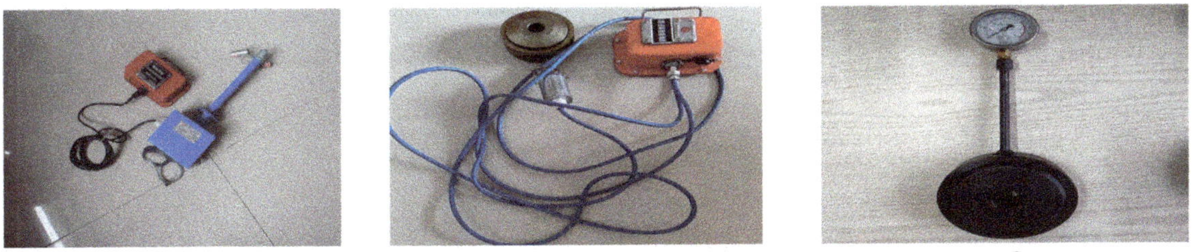

(d) (e) (f)

Figure 26. Engineering monitoring instruments: (**a**) detection equipment for internal fracture of the coal mass, (**b**) industrial CT scanning system, (**c**) roadway surface displacement monitoring equipment, (**d**) roof abscission layer instrument, (**e**) borehole stress gauge, and (**f**) anchor cable dynamometer.

6. Discussion

Gob-side entry driving is usually arranged at the edge of the goaf of the prior working face, retaining a certain width coal pillar (generally 5–8 m).

This paper summarized seven theoretical models of the overlying strata activity law and surrounding rock control in fully mechanized caving gob-side entry driving (the lateral breaking "large structure" model, compound key triangle block structure model in the middle and low position, the high and low right angle key block stability mechanics model, elastic foundation beam model, low-level combined cantilever beam + high-level multilayer masonry beam structure model, the vertical triangular slip zone structure model, the "internal and external stress field theory" and the "stress limit equilibrium zone model"). Three kinds of constitutive models (strain-softening, Mohr–Coulomb, double-yield) and numerical simulation methods were discussed. The support methods of coal pillar ribs, virgin coal ribs, and roofs of gob-side entry driving in a fully mechanized caving face are summarized. The following conclusions and prospects have been reached:

(1) With the wide application of gob-side entry driving without a coal pillar or with a narrow coal pillar, theoretical research on the activity law of the overlying strata and the stability control of the surrounding rock is gradually improved and developed. It lays a solid theoretical foundation for further promotion and application.
(2) Numerical simulation is an important technical means to study the problem of gob-side entry driving in fully mechanized caving. By selecting the appropriate constitutive model and numerical simulation software, the crack propagation morphology and plastic state in the coal pillar can be simulated, which can guide the reasonable location of gob-side entry driving and the design width of the coal pillar.
(3) The stability of the surrounding rock in the gob-side entry of fully mechanized caving is essential. On the side of the virgin coal rib, bolt cable support is often used with a ladder beam of steel (W, JW steel strip) and mesh. On the side of the coal pillar rib, it is mainly divided into three categories: anchor cable support, concrete wall support on the side of the roadway, and filling support on the side of the goaf; among them, the process of pouring a concrete wall and filling support is cumbersome and costly, which is only used under some special conditions. In the roof of the roadway, five kinds of support forms, such as anchor cable support, anchor cable truss support, anchor beam truss support, anchor cable + channel steel support, and step bundle anchor cable support, are often used. The roof often adopts a combination of multiple support methods. Under complete conditions, such as a broken roof, the deformation of the roadway is controlled by shed support, anchor cable grouting composite support, and other forms.
(4) Mine pressure monitoring is a research method often used in engineering. Understanding the stress and deformation law of surrounding rock in gob-side entry driving

is essential in optimizing the support design scheme using microseismic monitoring, dynamic stress monitoring, and borehole peeping for on-site engineering monitoring.

(5) Gob-side entry driving in fully mechanized caving is an important method of thick coal seam mining. In the future, we still need to strengthen the research on the basic theory to help us gain a more in-depth understanding of the various problems in gob-side entry driving. Further, we need to explore the nonpillar mining technology and develop more effective surrounding rock control technology to improve the recovery rate of coal resources in fully mechanized caving mining. In addition, the future method of gob-side entry driving with fully mechanized caving will also take into account precision, automation, and greening to realize safe and efficient mining of coal mines.

In this paper, we have summarized the relevant theories and technologies of gob-side entry driving in fully mechanized caving face and put forward that its future development direction should focus on theoretical research, nonpillar mining, and efficient surrounding rock control, which is helpful to promote the further development of gob-side entry driving technology in fully mechanized caving face.

Author Contributions: Conceptualization, D.C. and S.X.; data curation, J.Z.; formal analysis, Q.Y. and X.M.; funding acquisition, D.C.; investigation, W.G., Z.L. and Z.W.; methodology, D.C., J.Z. and Q.Y.; project administration, D.C. and S.X.; software, D.C., X.M., S.F. and X.Y.; supervision, S.X.; validation, X.M. and W.G.; writing–original draft, D.C., J.Z. and Q.Y. writing–review and editing, D.C., J.Z. and Q.Y. All authors have read and agreed to the published version of the manuscript.

Funding: This work was financially supported by the National Natural Science Foundation of China (Grant No. 52004286), the Fundamental Research Funds for the Central Universities (Grant No. 2022XJNY02), the China Postdoctoral Science Foundation (Grant No. 2020T130701, 2019M650895), all of which were gratefully acknowledged.

Conflicts of Interest: The authors declare no conflict of interest.

References

1. Zhang, P.; Zhai, E.; Cheng, A.; Xu, S.; Sun, B. Optical Fiber Monitoring Study on Characteristics of Deformation in Floor of Deep and Thick Coal Seam During Mining. *Chin. J. Undergr. Space Eng.* **2019**, *15*, 1197–1203+1211.
2. Xie, S.; Wu, Y.; Ma, X.; Chen, D.; Guo, F.; Jiang, Z.; Li, H.; Zou, H.; Liu, R.; Zhang, X. Reasonable Stopping Method and Retracement Channel Support at Fully Mechanized Top Coal Caving Working Face of 15 m Extra-Thick Coal Seam: A Case Study. *Energy Sci. Eng.* **2022**, *10*, 4336–4357. [CrossRef]
3. Klishin, V.I. Innovative Technologies for Thick Coal Seam Mining on the Basis of Powered Roof Support with Controlled Coal Discharge. *IOP Conf. Ser. Earth Environ. Sci.* **2019**, *262*, 12027. [CrossRef]
4. Kurlenya, M.V.; Klishin, V.I.; Kubanychbek. Pressure Control and Top Coal Drawing by Powered Machine Complexes in Mining at Thick Coal Seams. *Gospod. Surowcami Miner.* **2008**, *24*, 25–35.
5. Ju, W. Study on Reasons of Rock Burst in Roadway and Prevention Technology of Steeply-Inclined and Extremely Thick Coal Seam with Horizontally Slicing Way. Ph.D. Thesis, Beijing Jiaotong University, Beijing, China, 2009.
6. Song, G.; Stankus, J. Control Mechanism of a Tensioned Bolt System in the Laminated Roof with a Large Horizontal Stress. In Proceedings of the 16th International Conference on Ground Control in Mining, Morgantown, WV, USA, 5–7 August 1997.
7. Sellers, E.J.; Klerck, P. Modeling of the Effect of Discontinuities on the Extent of the Fracture Zone Surrounding Deep Tunnels. *Tunn. Undergr. Space Technol.* **2000**, *15*, 463–469. [CrossRef]
8. Fairhurst, C. Deformation, Yield, Rupture and Stability of Excavations at Great Depth. *Int. Soc. Rock Mech. Rock Eng.* **1989**, *98*, 1103–1114.
9. Mei, X. Surrounding Rock Stability and Asymmetric Support of Large Section Gob-Side Entry Driving in Fully Mechanized Caving. Ph.D. Thesis, China University of Mining and Technology, Beijing, China, 2016.
10. Yang, K.; Xie, G. Study on Asymmetrical Bolting Support of Retreating Entry in Large Dip Coal Seam and It's Practice. *Chin. J. Undergr. Space Eng.* **2013**, *9*, 924–927.
11. Kang, H.; Zhang, X.; Wang, D.; Tian, J.; Yi, Z.; Jiang, W. Strata Control Technology and Applications of Non-Pillar Coal Mining. *J. China Coal Soc.* **2022**, *47*, 16–44.
12. Chen, D.; Guo, F.; Li, Z.; Ma, X.; Xie, S.; Wu, Y.; Wang, Z. Study on the Influence and Control of Stress Direction Partial-Stress Boosting of Main Deflection and Roadways Surrounding Rock and under the Influence of Multi-Seam Mining. *Energies* **2022**, *15*, 8257. [CrossRef]

13. Xie, S.; Wu, Y.; Guo, F.; Zou, H.; Chen, D.; Zhang, X.; Ma, X.; Liu, R.; Wu, C. Application of Pre-Splitting and Roof-Cutting Control Technology in Coal Mining: A Review of Technology. *Energies* **2022**, *15*, 6489. [CrossRef]
14. Dychkovskyi, R.; Shavarskyi, I.; Saik, P.; Lozynskyi, V.; Falshtynskyi, V.; Cabana, E. Research into Stress-Strain State of the Rock Mass Condition in the Process of the Operation of Double-Unit Longwalls. *Min. Miner. Depos.* **2020**, *14*, 85–94. [CrossRef]
15. Vu, T.T. Solutions to Prevent Face Spall and Roof Falling in Fully Mechanized Longwall at Underground Mines, Vietnam. *Min. Miner. Depos.* **2022**, *16*, 127–134. [CrossRef]
16. Matayev, A.; Kainazarova, A.; Arystan, I.; Abeuov, Y.; Kainazarov, A.; Baizbayev, M.; Demin, V.; Sultanov, M. Research into Rock Mass Geomechanical Situation in the Zone of Stope Operations Influence at the 10th Anniversary of Kazakhstan's Independence Mine. *Min. Miner. Depos.* **2021**, *15*, 103–111. [CrossRef]
17. Kumar, R.; Mishra, A.K.; Kumar, A.; Singh, A.K.; Ram, S.; Singh, R. Importance of Fracturing Hard and Massive Overlying Strata for Complete Extraction of Thick Coal Seam—Case Studies. *J. Geol. Soc. India* **2022**, *98*, 203–210. [CrossRef]
18. Xie, S.; Chen, D.; Zeng, J.; Gao, M.; Zhang, T.; Liu, J. First Fracturing Characteristics of Main Roof Plate Structure Considering Coal Pillars and Elastic Coal Foundation on Both Sides of Stope. *J. China Coal Soc.* **2019**, *44*, 115–126.
19. He, F.; He, W.; Chen, D.; Xie, S.; Li, H.; He, C. First Fracture Structure Characteristics of Main Roof Plate Considering Elastic-Plastic Deformation of Coal. *J. China Coal Soc.* **2020**, *45*, 2704–2717.
20. Chen, D.; Wu, Y.; Xie, S.; He, F.; Sun, Y.; Shi, S.; Jing, Z. Study on the First Fracture of the Main Roof Plate Structure with One Side Goaf and Elastic-Plastic Foundation Boundary. *J. China Coal Soc.* **2021**, *46*, 3090–3105.
21. Chen, D.; Guo, F.; Wu, Y.; Xie, S.; He, F.; Liu, R.; Wu, C. Study on the First Fracture of The Main Roof Plate Structure with the Long Side Goaf and Elastic-Plastic Softening Foundation Boundary. *J. China Coal Soc.* **2022**, *47*, 1473–1489.
22. Castelli, F.; Grasso, S.; Lentini, V.; Sammito, M.S.V. Effects of Soil-Foundation-Interaction on the Seismic Response of a Cooling Tower by 3D-FEM Analysis. *Geosciences* **2021**, *11*, 200. [CrossRef]
23. Zhang, G.; He, F.; Jia, H.; Lai, Y. Analysis of Gateroad Stability in Relation to Yield Pillar Size: A Case Study. *Rock Mech. Rock Eng.* **2017**, *50*, 1263–1278. [CrossRef]
24. Li, W.; Bai, J.; Peng, S.; Wang, X.; Xu, Y. Numerical Modeling for Yield Pillar Design: A Case Study. *Rock Mech. Rock Eng.* **2015**, *48*, 305–318. [CrossRef]
25. Jiang, L.; Zhang, P.; Chen, L.; Hao, Z.; Sainoki, A.; Mitri, H.S.; Wang, Q. Numerical Approach for Goaf-Side Entry Layout and Yield Pillar Design in Fractured Ground Conditions. *Rock Mech. Rock Eng.* **2017**, *50*, 3049–3071. [CrossRef]
26. Wang, E.; Xie, S. Determination of Coal Pillar Width for Gob-Side Entry Driving in Isolated Coal Face and Its Control in Deep Soft-broken Coal Seam: A Case Study. *Energy Sci. Eng.* **2022**, *10*, 2305–2316. [CrossRef]
27. Shen, W.; Xiao, T.; Wang, M.; Bai, J.; Wang, X. Numerical Modeling of Entry Position Design: A Field Case. *Int. J. Min. Sci. Technol.* **2018**, *28*, 985–990. [CrossRef]
28. Xie, S.; Wang, E.; Chen, D.; Li, H.; Jiang, Z.; Yang, H. Stability Analysis and Control Technology of Gob-Side Entry Retaining with Double Roadways by Filling with High-Water Material in Gently Inclined Coal Seam. *Int. J. Min. Sci. Technol.* **2022**, *9*, 52. [CrossRef]
29. Chen, D.; Wu, Y.; Xie, S.; Guo, F.; He, F.; Liu, R. Reasonable Location of Stopping Line in Close-Distance Underlying Coal Seam and Partition Support of Large Cross-Section Roadway. *Int. J. Min. Sci. Technol.* **2022**, *9*, 52. [CrossRef]
30. Xie, S.; Wu, Y.; Chen, D.; Liu, R.; Han, X.; Ye, Q. Failure Analysis and Control Technology of Intersections of Large-Scale Variable Cross-Section Roadways in Deep Soft Rock. *Int. J. Min. Sci. Technol.* **2022**, *9*, 19. [CrossRef]
31. Yue, S.; Xie, S.; Chen, D.; Gao, M.; Zhang, Q.; Zhang, T. Research on Surrounding Rocks Control of Narrow Pillar with High-Intensity Fully-Mechanized Top Coal Caving Mining in 15 m Ultra-Thick Coal Seam. *J. Min. Saf. Eng.* **2017**, *34*, 905–913.
32. Guo, J.; Wang, W.; Yue, S.; He, F.; Gao, M.; Xie, S. Surrounding Rock Control Mechanism and Its Application of Gob-Side Driving Entry in Extra Thick Coal Seam. *J. China Coal Soc.* **2017**, *42*, 825–832.
33. Xie, S.; Wu, X.; Chen, D.; Sun, Y.; Wang, E.; Wu, X.; Duan, X.; Jiang, Z.; Qi, P.; Shi, S. Failure Mechanism and Control Technology of Thick and Soft Coal Fully Mechanized Caving Roadway under Double Gobs in Close Coal Seams. *Shock Vib.* **2020**, *2020*, 8846014. [CrossRef]
34. Zhang, G.; He, F. Asymmetric Failure and Control Measures of Large Cross-Section Entry Roof with Strong Mining Disturbance and Fully-Mechanized Caving Mining. *Chin. J. Rock Mech. Eng.* **2016**, *35*, 806–818.
35. Guo, J.; Wang, W.; He, F.; Zhang, G. Main Roof Break Structure and Surrounding Stability Analysis in Gob-Side Entry with Fully-Mechanized Caving Mining. *J. Min. Saf. Eng.* **2019**, *36*, 446–454+464.
36. Wei, Z.; He, F.; Zhang, G.; Gao, Y.; Lai, Y. Failure Mechanism and Cable Truss Control of Large-Scale Section Gob-Side Entry Roof with Fully-Mechanized Caving. *J. Min. Saf. Eng.* **2017**, *34*, 1.
37. Zhang, G.; He, F. Pillar Width Determination and Surrounding Rocks Control of Gob-Side Entry with Large Cross-Section and Fully-Mechanized Mining. *Rock Soil Mech.* **2016**, *37*, 1721–1728+1736.
38. Jiang, W. Deformation Mechanism and Stability Control of Roadway Along Goaf in Fully Mechanized Top Coal Caving Face with Thick and Hard Roof. Ph.D. Thesis, China University of Mining and Technology, Beijing, China, 2021.
39. Xie, S.; Jiang, Z.; Chen, D.; Wang, E.; Lv, F. A New Pressure Relief Technology by Internal Hole-Making to Protect Roadway in Two Sides of Deep Coal Roadway: A Case Study. *Rock Mech. Rock Eng.* **2023**, *56*, 1537–1561. [CrossRef]
40. Xie, S.; Li, H.; Chen, D.; Feng, S.; Ma, X.; Jiang, Z.; Cui, J. New Technology of Pressure Relief Control in Soft Coal Roadways with Deep, Violent Mining and Large Deformation: A Key Study. *Energies* **2022**, *15*, 9208. [CrossRef]

41. Xie, S.; Wang, E.; Chen, D.; Jiang, Z.; Li, H.; Liu, R. Collaborative Control Technology of External Anchor-Internal Unloading of Surrounding Rock in Deep Large-Section coal Roadway Under Strong Mining Influence. *J. China Coal Soc.* **2022**, *47*, 1946–1957.
42. Bo, J.; Wang, W.; Hou, Z.; Huang, H. Control Mechanism and Support Technique about Gateway Driven along Goaf in Fully Mechanized Top Coal Caving Face. *J. China Coal Soc.* **2000**, *5*, 478–481.
43. Xie, S.; Wu, Y.; Guo, F.; Chen, D.; Wang, E.; Zhang, X.; Zou, H.; Liu, R.; Ma, X.; Li, S. Interaction Mechanism of the Upper and Lower Main Roofs with Different Properties in Close Coal Seams: A Case Study. *Energies* **2022**, *15*, 5533. [CrossRef]
44. Wang, D.; Li, S.; Wang, Q.; Li, W.; Wang, F.; Wang, H.; Peng, P.; Ruan, G. Experimental Study of Reasonable Coal Pillar Width in Fully Mechanized Top Coal Caving Face of Deep Thick Coal Seam. *Chin. J. Rock Mech. Eng.* **2014**, *33*, 539–548.
45. Bo, J.; Hou, C.; Huang, H. Numerical Simulation Study on Stability of Narrow Coal Pillar of Roadway Driving Along Goaf. *Chin. J. Rock Mech. Eng.* **2004**, *20*, 3475–3479.
46. Hou, C.; Li, X. Stability Principle of Big and Small Structures of Rock Surrounding Roadway Driven Along Goaf in Fully Mechanized Top Coal Caving face. *J. China Coal Soc.* **2001**, *1*, 1–7.
47. Gao, S. Mechanism and Control of Surrounding Rock Failure of the Adjacent Goaf Broken Coal Roadway in Lower Position of Double Lane in Different Layers at Fully-Mechanized Coal Caving Sector. Ph.D. Thesis, China University of Mining and Technology, Beijing, China, 2021.
48. He, W. Study on Strata Structure Characteristic and Surrounding Rock Control of Gob-Side Entry with Narrow Coal Pillar for Longwall Caving Mining in Extra-Thick Coal Seam. Ph.D. Thesis, China University of Mining and Technology, Beijing, China, 2021.
49. Xie, F. Fracture Location and Structural Stability of Overlying Strata in Top-Coal Caving Gob-Side Entry. Ph.D. Thesis, China University of Mining and Technology, Beijing, China, 2017.
50. Du, B. Study on Strong Ground Pressure Mechanism of Gob-Side Roadway and its Control in Extra Thick Coal Seam with Hard Roof. Ph.D. Thesis, China University of Mining and Technology, Beijing, China, 2022.
51. Li, H.; Wang, Z. Research and Application of Small Coal Pillar Gob-Side Entry Driving in Extremely Thick Coal Seam with Large Mining Height and Fully Mechanized Caving Face. *Datong Coal Sci. Technol.* **2019**, *03*, 1–8+53.
52. Li, L.; Bo, J.; Wang, X. Rational Position and Control Technique of Roadway Driving Along Next Goaf in Fully Mechanized Top Coal Caving Face. *J. China Coal Soc.* **2012**, *37*, 1564–1569.
53. Cha, W.; Li, X.; Hua, X.; Wu, T.; Yin, S. Impact and Application on Narrow Coal Pillar for Roadway Protecting from Fracture Position of Upper Roof. *J. China Coal Soc.* **2014**, *39*, 332–338.
54. Wang, D.; Wang, H.; Li, S.; Wang, Q.; Song, Y. Stress and Deformation Analysis of the Side Wall in a Roadway Driven Along Goaf Incorporating the Strength Softening Characteristics of Coal Mass. *J. China Univ. Min. Technol.* **2019**, *48*, 295–304.
55. Qi, F.; Zhou, Y.; Cao, Z.; Zhang, Q.; Li, N. Width Optimization of Narrow Coal Pillar of Roadway Driving Along Goaf in Fully Mechanized Top Coal Caving Face. *J. Min. Saf. Eng.* **2016**, *33*, 475–480.
56. Wang, W.; Huang, C.; Hou, C.; Peng, G. Mechanical Deformation Analysis of Floor Strata of Roadway Driving Along Next Goaf in Fully Mechanized Sub-Level Caving Face. *J. China Coal Soc.* **2002**, *1*, 26–30.
57. Zhu, W.; Guan, K.; Liu, H.; Niu, L.; Liu, X.; Hou, C.; Deng, W. *Applied Rock Mechanics*; Metallurgical Industry Press: Beijing, China, 2022.
58. Shi, C.; Zhang, Q.; Wang, S. *Numerical Simulation Technology and Application with Particle Flow Code (PFC5.0)*; China Building Industry Press: Beijing, China, 2018.
59. Itasca Consulting Group Inc. *UDEC (Universal Distinct Element Code)User's Manual Version 6.0*; ITASCA Consulting Group Inc.: Minneapolis, MN, USA, 2014.
60. Itasca Consulting Group Inc. *FLAC 3D (Version 5.0) User's Manual*; ITASCA Consulting Group Inc.: Minneapolis, MN, USA, 1986.
61. Wang, T.; Han, X.; Zhao, X. *FLAC3D Numerical Simulation Method and Engineering Application—In-Depth Analysis of FLAC3D5.0*; China Building Industry Press: Beijing, China, 2015.
62. Chen, D.; Wang, E.; Xie, S.; He, F.; Wang, L.; Zhang, Q.; Wu, X.; Jiang, Z.; Li, Y.; Shi, S. Roadway Surrounding Rock under Multi-Coal-Seam Mining: Deviatoric Stress Evolution and Control Technology. *Adv. Civ. Eng.* **2020**, *2020*, 9891825. [CrossRef]
63. Chen, D.; Guo, F.; Xie, S.; Wang, E.; Wu, Y.; Jiang, Z.; Wang, L.; Cui, J.; Zhang, X.; Liu, R. Mining-induced Failure Characteristics and Surrounding Rock Control of Gob-side Entry Driving Adjacent to Filling Working Face in the Deep Coal Mine. *Energy Sci. Eng.* **2022**, *10*, 2593–2611. [CrossRef]
64. Lu, Y.; Wang, L.; Yang, F.; Li, Y.; Chen, H. Post-Peak Strain Softening Mechanical Properties of Weak Rock. *Chin. J. Rock Mech. Eng.* **2010**, *29*, 640–648.
65. Zhang, F.; Sheng, Q.; Zhu, Z.; Zhang, Y. Study on Post-Peak Mechanical Behaviour and Strain-Softening Model of Three Gorges Granite. *Chin. J. Rock Mech. Eng.* **2008**, *27*, 2651–2655.
66. Medhurst, T.P.; Brown, E.T. A Study of the Mechanical Behaviour of Coal for Pillar Design. *Int. J. Rock Mech. Min. Sci.* **1998**, *35*, 1087–1105. [CrossRef]
67. Yavuz, H. An Estimation Method for Cover Pressure Re-Establishment Distance and Pressure Distribution in the Goaf of Longwall Coal Mines. *Int. J. Rock Mech. Min. Sci.* **2004**, *41*, 193–205. [CrossRef]
68. Guo, Z.; Zhang, L.; Ma, Z.; Zhong, F.; Yu, J.; Wang, S. Numerical Investigation of the Influence of Roof Fracturing Angle on the Stability of Gob-Side Entry Subjected to Dynamic Loading. *Shock Vib.* **2019**, *2019*, 1434135. [CrossRef]
69. Zhang, G.; Liang, S.; Tan, Y.; Xie, F.; Chen, S.; Jia, H. Numerical Modeling for Longwall Pillar Design: A Case Study from a Typical Longwall Panel in China. *J. Geophys. Eng.* **2018**, *15*, 121–134. [CrossRef]

70. Wang, J.; Qiu, P.; Ning, J.; Zhuang, L.; Yang, S. A Numerical Study of the Mining-induced Energy Redistribution in a Coal Seam Adjacent to an Extracted Coal Panel during Longwall Face Mining: A Case Study. *Energy Sci. Eng.* **2020**, *8*, 817–835. [CrossRef]
71. Yang, R.; Zhu, Y.; Li, Y.; Li, W.; Lin, H. Coal Pillar Size Design and Surrounding Rock Control Techniques in Deep Longwall Entry. *Arab. J. Geosci.* **2020**, *13*, 453. [CrossRef]
72. Li, M.; Zhang, J.; Huang, Y.; Zhou, N. Effects of Particle Size of Crushed Gangue Backfill Materials on Surface Subsidence and Its Application under Buildings. *Environ. Earth Sci.* **2017**, *76*, 603. [CrossRef]
73. Yadav, A.; Behera, B.; Sahoo, S.K.; Singh, G.S.P.; Sharma, S.K. An Approach for Numerical Modeling of Gob Compaction Process in Longwall Mining. *Min. Metall. Explor.* **2020**, *37*, 631–649. [CrossRef]
74. Chen, Y.; Xu, D. *FLAC/FLAC3D Foundation and Engineering Examples*; China Water & Power Press: Beijing, China, 2013.
75. Peng, W. *FLAC3D Practical Tutorial, Version 2*; China Machine Press: Beijing, China, 2020.
76. Salamon, M.D.G. Stability, Instability and Design of Pillar Workings. *Int. J. Rock Mech. Min. Sci. Geomech. Abstr.* **1970**, *7*, 613–631. [CrossRef]
77. Salamon, M.D.G. Mechanism of Caving in Longwall Coal Mining. In Proceedings of the 31st US Symposium on Rock Mechanics, Colorado School of Mines, Golden, CO, USA, 18–20 June 1990.
78. Gao, F.; Stead, D.; Kang, H. Simulation of Roof Shear Failure in Coal Mine Roadways Using an Innovative UDEC Trigon Approach. *Comput. Geotech.* **2014**, *61*, 33–41. [CrossRef]
79. Gao, F.; Stead, D.; Kang, H. Numerical Simulation of Squeezing Failure in a Coal Mine Roadway Due to Mining-Induced Stresses. *Rock Mech. Rock Eng.* **2015**, *48*, 1635–1645. [CrossRef]
80. Wu, W.; Bai, J.; Wang, X.; Yan, S.; Wu, S. Numerical Study of Failure Mechanisms and Control Techniques for a Gob-Side Yield Pillar in the Sijiazhuang Coal Mine, China. *Rock Mech. Rock Eng.* **2019**, *52*, 1231–1245. [CrossRef]
81. Gao, F.; Stead, D.; Kang, H.; Wu, Y. Discrete Element Modelling of Deformation and Damage of a Roadway Driven along an Unstable Goaf—A Case Study. *Int. J. Coal Geol.* **2014**, *127*, 100–110. [CrossRef]
82. Vergara, M.R.; Van Sint Jan, M.; Lorig, L. Numerical Model for the Study of the Strength and Failure Modes of Rock Containing Non-Persistent Joints. *Rock Mech. Rock Eng.* **2016**, *49*, 1211–1226. [CrossRef]
83. Wu, B.; Wang, X.; Bai, J.; Wu, W.; Zhu, X.; Li, G. Study on Crack Evolution Mechanism of Roadside Backfill Body in Gob-Side Entry Retaining Based on UDEC Trigon Model. *Rock Mech. Rock Eng.* **2019**, *52*, 3385–3399. [CrossRef]
84. Gao, F.Q.; Kang, H.P. Effects of Pre-Existing Discontinuities on the Residual Strength of Rock Mass—Insight from a Discrete Element Method Simulation. *J. Struct. Geol.* **2016**, *85*, 40–50. [CrossRef]
85. Chen, X.; Xiao, T.; She, H. Triaxial Compression Condition Order Fracture Rock Fissure Extension Study. *Sci. Technol. Eng.* **2022**, *22*, 11567–11576.
86. Rong, H.; Li, G.; Liang, D.; Xu, J.; Hu, Y. Particle Flow Simulation of Mechanical Properties of High Stress Rock under the Influence of Stress Path. *J. Min. Saf. Eng.* **2022**, *39*, 163–173.
87. Wu, D.; Wei, L.; Su, H.; Wu, J.; Liu, R.; Zhou, J. Experimental Study and PFC3D Simulation on Crack Propagation of Fractured Rock-like Specimens with Bolts under Uniaxial Compression. *Rock Soil Mech.* **2021**, *42*, 1681–1692.
88. Hou, F.; Zhang, L.; Jiang, Q.; Feng, C.; Li, S. Hydraulic Fracturing Characteristics of Water Resisting Rock Mass with Layered Joints Based on CDEM. *Hazard Control Tunn. Undergr. Eng.* **2021**, *3*, 76–84.
89. Zhao, Y.; Sun, Z.; Song, H.; Zhao, S. Crack Propagation Law of Mode I Dynamic Fracture of Coal: Experiment and Numerical Simulation. *J. China Coal Soc.* **2020**, *45*, 3961–3972.
90. Zhang, W.; Zhang, D.; Chen, J.; Xu, M. Control of Surrounding Rock Deformation for Gob-Side Entry Driving in Narrow Coal Pillar of Island Coalface. *J. China Univ. Min. Technol.* **2014**, *43*, 36–42+55.
91. Zhao, Q.; Du, F.; Li, Q.; Qian, W.; Shan, Y. Surrounding Rock Control Technology of Gateway Driven Along Goaf in Fully-Mechanized Coal Mining Face. *Coal Sci. Technol.* **2015**, *43*, 23–28. [CrossRef]
92. Dong, H. Ground Control of Narrow Coal Pillar in Gob Side Entry Driving with Fully Mechanized Top Coal Caving Mining in Extra-Thick Coal Seam. *J. Min. Strat. Control Eng.* **2021**, *3*, 32–42.
93. Li, X.; Ju, M.; Jia, S.; Chong, Z. Study of Influential Factors on the Stability of Narrow Coal Pillar in Gob-Side Entry Driving and Its Engineering Application. *J. Min. Saf. Eng.* **2016**, *33*, 761–769.
94. Zhang, H.; Wan, Z.; Zhang, Y.; Ma, Z.; Zhang, J.; Liu, S.; Ge, L. Deformation Mechanism of Narrow Coal Pillar in the Fully-Mechanized Gob-Side Entry with Incompletely Stable Overlying Strata. *J. Min. Saf. Eng.* **2016**, 33, 692–698.
95. Zhang, H.; Wan, Z.; Zhang, Y. Mechanism of Grouted-Reinforcement in Last Roadway for Pillar in the Fully-Mechanized Gob-Side Entry. *J. Min. Saf. Eng.* **2018**, *35*, 489–495.
96. He, F.; Li, K. Arc-Connection Locker for Girder Anchor Cable. CN Patent 201236719[P], 13 March 2009.
97. Yan, H.; He, F.; Wang, S. Safety Control and Evaluation of Roadway with Super-Large Cross-Section and Soft-Weak Thick Coal Roof. *Chin. J. Rock Mech. Eng.* **2014**, *33*, 1014–1023.
98. Kang, R.; He, F.; Li, H. Application of Truss and Anchor to Support Seam Gateway with Broken Roof. *Coal Sci. Technol.* **2010**, *38*, 28–33.
99. Kang, H.; Wang, J.; Lin, J. Study and Applications of Roadway Support Techniques for Coalmines. *J. China Coal Soc.* **2012**, *35*, 1809–1814.
100. Su, X.; Song, X.; Li, H.; Yuan, H.; Li, M. Study on Coupled Arch-Beam Support Structure of Roadway with Extra-Thick Soft Compound Roof. *Chin. J. Rock Mech. Eng.* **2014**, *33*, 1828–1836.

101. He, W.; He, F.; Chen, D.; Chen, Q. Pillar Width and Surrounding Rock Control of Gob-Side Roadway with Mechanical Caved Mining in Extra-Thick Coal Seams under Hard-Thick Main Roof. *J. Min. Saf. Eng.* **2020**, *37*, 349–358, 365.
102. Cheng, Y.; Jiang, F.; Pang, J. Research on Lateral Strata Pressure Characteristic in Goaf of Top Coal Caving in Extra Thick Coal Seam and Its Application. *J. China Coal Soc.* **2012**, *37*, 1088–1093.
103. Yin, S.; Cheng, G.; He, F.; Xie, F.; Shan, Y. An Asymmetric Support Technique for Fully-Mechanized Coal Roadway Nearby Narrow Pillar Based on the Fracture Position Analysis in Basic Roof. *Chin. J. Rock Mech. Eng.* **2016**, *35*, 3162–3174.
104. Zhang, G. Asymmetric Failure Mechanism and Regulation System of Gob-Side Entry Roof with Fully-Mechanized Caving Mining and a Loose and Weak Coal Pillar. Ph.D. Thesis, China University of Mining and Technology, Beijing, China, 2017.
105. Hao, D.; Cui, Q.; He, J.; Guo, X.; Cheng, P. Deformation Characteristics and Separation Monitoring of Layered Roof Roadway Supported with Bolts and Cables. *J. China Coal Soc.* **2017**, *42*, 43–50.

energies

MDPI

Article

Mechanism and Control of Asymmetric Floor Heave in the Gob-Side Coal Roadway under Mining Pressure in Extra-Thick Coal Seams

Deqiu Wang [1,*], Yun Zheng [2,*], Fulian He [1], Jiayu Song [1], Jianlong Zhang [1], Yanhao Wu [1], Pengpeng Jia [1], Xiaohui Wang [1], Baoping Liu [1], Feifei Wang [1], Yajiang Zhang [1] and Kai Tao [1]

1 School of Energy & Mining Engineering, China University of Mining & Technology (Beijing), Beijing 100083, China; fulian202210@163.com (F.H.); 18611440178@163.com (J.S.); jianlong__zhang@163.com (J.Z.); wu_yhck@126.com (Y.W.); 15237935736@163.com (P.J.); wxhde2020@163.com (X.W.); 18325157157@163.com (B.L.); feifeiwang258@163.com (F.W.); yajiang_zhang@163.com (Y.Z.); tk18201271613@163.com (K.T.)

2 School of Safety Engineering, North China Institute of Science and Technology, Langfang 065201, China

* Correspondence: 15639178973@163.com (D.W.); zytcha@163.com (Y.Z.)

Abstract: Due to their tense mining succession relationship, gob-side roadways may undergo significant deformation under multi-mining pressure. In this article, many methods, such as on-site research, a theoretical analysis, a numerical simulation and an industrial experiment, are used to research the mechanism of asymmetric floor heave in a gob-side coal roadway affected by mining pressure during the mining of extra-thick coal seams. Our main research is as follows: (1) By monitoring the floor deformation in the roadway on site, it is concluded that the roadway floor shows asymmetry, indicating that the floor displacement near the coal pillar side is relatively large. (2) Based on a lateral overburden structure model of the roadway, the calculation formulas of the horizontal vertical stress caused by the roadway excavation and the excavation of the upper working face are derived separately, and the vertical stress coupling curves on both sides of the roadway during the mining of the upper working face are obtained through a numerical simulation. It is concluded that the cause of the asymmetric floor heave in the roadway is an uneven distribution of vertical stress. (3) The numerical simulation shows a symmetrical distribution of the floor displacement curve during the roadway excavation with a max. displacement of 49.5 mm. The floor displacement curve during the mining of the upper working face is asymmetric with a max. displacement of 873 mm at a distance of 1 m from the central axis near the coal pillar side. The range of the plastic zone in the roadway gradually expands with the mining of the upper working face, and the maximum depth of floor failure is 5.5 m. (4) According to the cooperative control principle of "roof + two sides + floor", an asymmetric floor heave joint control scheme of "floor leveling + anchor cable support + concrete hardening" is proposed. The floor deformation monitoring results indicate that the max. floor heave at the measurement point near the coal pillar in the roadway is 167 mm, and the floor heave is effectively controlled.

Keywords: mining pressures; gob-side coal roadway; asymmetric floor heave; failure mechanism; cooperative control

Citation: Wang, D.; Zheng, Y.; He, F.; Song, J.; Zhang, J.; Wu, Y.; Jia, P.; Wang, X.; Liu, B.; Wang, F.; et al. Mechanism and Control of Asymmetric Floor Heave in the Gob-Side Coal Roadway under Mining Pressure in Extra-Thick Coal Seams. *Energies* **2023**, *16*, 4948. https://doi.org/10.3390/en16134948

Academic Editor: Manoj Khandelwal

Received: 18 May 2023
Revised: 14 June 2023
Accepted: 16 June 2023
Published: 26 June 2023

1. Introduction

Mining roadways are seriously deformed by strong mining activities in extra-thick coal seams [1,2]. Setting coal pillars in sections is a conventional method to maintain the stability of roadways. Serious floor heave occurs when the roadway floor is weak rock and the coal pillar's width is unsuitable, which considerably hinders pedestrians as well as adequate ventilation and transportation [3,4]. The formation mechanism and control technology of roadway floor heave are a problem in the industry at present. In addition, due to tension in mining relationships, roadways are affected by multiple intense mining pressures, the stress distribution and failure law of the floor have changed significantly compared with conventional roadways, and the difficulty of floor control has further increased [5–9]. Many scholars have conducted a large amount of research on the floor deformation mechanism and control of roadways. Indraratna et al. [10] conducted research on the crack expansion process of floor strata and proposed that the main stress difference is the major factor leading to the cracking of the floor strata. Małkowski et al. [11,12] proposed two methods for simulating roadway floor heave: one is that the roadway floor is made of dry rock mass, and the other is that the roadway floor is made of water-rich rock mass. The second modeling method is more in line with the actual situation of roadways and has been applied more widely. Sakhno et al. [13] showed that the rock in the stress-increasing area of a roadway is in a state of failure and is a discrete medium that is prone to floor heave. Moreover, the grouting reinforcement of the floor can effectively control floor heave. Wang et al. [14] calculated the relative displacement of the floor by establishing a mechanical model. They proposed the main causes of floor heave as follows: tensile damage to the floor strata, the plastic flow after the damage of the floor and the sinking of solid coal gangs. Based on the basic features of floor heave, Jiang et al. [15] summarized the types of floor deformation into four categories: shear, extrusion, flexural and expansion. He et al. [16] posited that a solution to floor heave might be taking the support structure of the surrounding rock as a whole to coordinate the deformation. Bai et al. [17] showed that the weaker the strength of the rock strata of the floor, the greater the probability of floor heave, and there is a threshold for the influence of horizontal stress. When it is greater than this threshold, the deformation depth of the floor will increase. Chen et al. [18] proposed an overall control strategy of an anchor mesh cable, full-section grouting and a reverse bottom arch based on the great asymmetric deformation of the floor of a roadway. Jia et al. [19] analyzed the stress environment around a roadway and the failure patterns under the action and showed that the uneven distribution of stress is the internal cause of the asymmetric floor heave of the roadway. Yang et al. [20] used the methods of a numerical simulation and mechanical analysis to obtain the failure mechanisms of a layered floor as follows: (1) the lower rock mechanics performance of the floor; and (2) an increase in the floor load and lateral pressure. Wang et al. [21] applied Rankine's theory of earth pressure to study the reason for floor heave. He believed that the vertical stress is greater than the horizontal stress in the formation factors of floor heave, and the difference in vertical stress around a roadway is the root cause of the asymmetric distribution of floor heave.

The above research is significant in clarifying the floor heave mechanism to ensure the stabilization of roadways. However, for roadways affected by repeated mining, the stress environment is relatively complex, and the laws of floor deformation are also particular. Therefore, further research on the floor heave of such roadways is required. This paper uses many methods to study the formation mechanism and control of asymmetric floor heave in a gob-side coal roadway under multiple mining operations in extra-thick coal seams. After measuring the distribution characteristics of the roadway stress during roadway excavation and the horizontal vertical stress of the goaf of the upper working face by establishing a mechanical model, it is concluded that the internal factor of the asymmetric floor heave is due to the nonuniform distribution of vertical stress. The evolutionary trend of vertical stress during roadway excavation is studied by a numerical simulation. Additionally, the lateral vertical stress of the goaf and the coupled vertical stresses when mining of the upper working face are determined, which verify the correctness of the theoretical calculation and explain the plastic zone of the roadway and the asymmetric deformation characteristics of the floor. An asymmetric floor heave joint control scheme of "floor leveling + anchor cable support + concrete hardening" is proposed based on the joint control principle of a roof, two sides and a floor. The floor deformation monitoring results demonstrate that the asymmetric floor heave is effectively controlled during mining.

2. Overview of Geological Engineering

The object of this article is the 5209 return air roadway in the 8209 fully mechanized top coal caving working face. The 5209 return air roadway lies in the North Second Panel with the 8210 working face on the north side, and the coal pillar section is 30 m. The inclination length of the 8209 working face is about 220 m, and the strike length is about 1346 m. The main mined coal seam is the #3–5 coal seam. The average burial depth is 415 m. The coal seam is nearly horizontal with an average thickness of 15 m. The coal seam structure is complex with a serious internal fissure. The coal seam contains 0.7~3.5 m gangue, such as mudstone, carbonaceous mudstone, etc. The immediate roof of the 8209 working face is mainly composed of mudstone with a developed fissure and low strength. The main roof includes medium–coarse-grained sandstone with high strength and an average thickness of 14 m. The immediate floor is carbonaceous mudstone with low strength and an average thickness of about 5.3 m. The main floor consists of medium–fine-grained sandstone. Figure 1 is the layout of the working face and the geologic histogram.

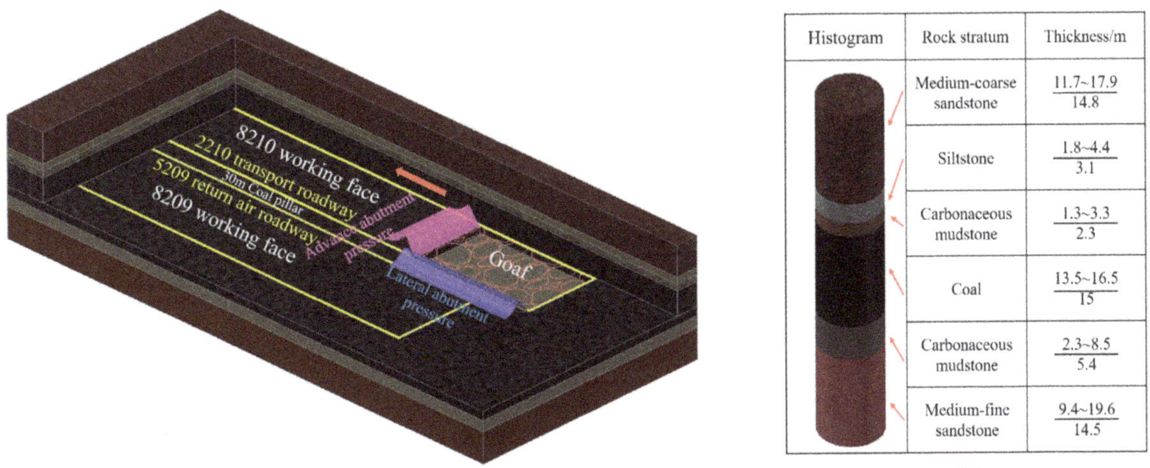

(a) (b)

Figure 1. Layout of working face and geologic histogram. (**a**) Layout of the 8209 working face; (**b**) geologic histogram.

To mitigate the tense mining succession relationship, the excavation of the 5209 return air roadway lagged behind the excavation of the 2210 transportation roadway in the mine, and the 5209 return air roadway had already been completed when the 8210 working face was excavated. The size of the 5209 roadway is 5.0 m × 3.9 m. The 5209 roadway experienced severe impacts during the mining of the 8210 working face because of the special mining and excavation succession relationship. As shown in Figure 2, the floor of roadway underwent severe deformation and damage due to the advanced and lateral support pressure formed during the mining. The roadway floor shows asymmetry, indicating that the maximum position of the floor heave is located on the coal pillar side.

(a) (b)

Figure 2. Failure characteristics of the roadway floor. (**a**) Floor heave; (**b**) Asymmetric floor deformation.

3. Monitoring the Floor Deformation

During the mining process of the 8210 working face, the asymmetric floor heave phenomenon is significant in the 5209 return air roadway. In order to specifically know the deformation laws of the floor, monitoring stations of the floor displacement were set up in the 5209 roadway. Figure 3 shows the arrangement of the monitoring stations and measurement points. Station 1 and station 2 are placed 150 m and 50 m away from the working face, respectively, as shown in Figure 3a. The locations of the measurement points at each station are shown in Figure 3b,c. Measurement points A (D), B (E) and C (F) are arranged on the solid coal side, the middle of roadway and the coal pillar side at each station, respectively. The centerline of the roadway height is used as the reference line. The height of the floor from the reference line is measured with a steel ruler. The monitoring ended when the measuring station lied at 150 m of the 8210 working face. Station 1 was monitored for 75 days, with a working face advance distance of 300 m. Station 2 was monitored for 50 days, with a working face advance distance of 200 m.

The deformation monitoring curve of the 5209 return air roadway floor is shown in Figure 3b–d. The maximum deformation of the roadway floor at Station 1 and Station 2 lies on the coal pillar side (i.e., the positions of measurement points C and F). The roadway floor shows an asymmetric deformation law in that the displacement of the coal pillar side is more than that of the solid coal side. The max. deformation of the floor near the coal pillar side at Station 1 reaches 1064 mm, the max. deformation of the middle floor reaches 713 mm, and the max. deformation of the solid coal side floor reaches 396 mm, as shown in Figure 3d. The three measurement points at Station 1 all experience two sharp increases in the deformation rate. The first time is where the roadway is about 20 m in front of the upper working face, which is subjected to the advanced support pressure of the upper working face. The deformation rate at the three measurement points significantly increases, and the deformation rate of the coal pillar side floor increases from 4.8 mm/m to 7.7 mm/m. The second time is when the upper working face is 120 m in front of Station 1, in which the roadway is subjected to lateral support pressure from the upper working face, resulting in a more severe deformation of the floor. The deformation rate of the coal pillar side grows from 4.9 mm/m to 13.8 mm/m. When the working face is 130 m from Station 1, the deformation rates of the three measurement points tend to ease off. The max. deformation of floor near the coal pillar reaches 969 mm, the max. deformation in the roadway middle reaches 633 mm and the max. deformation of the floor near the solid coal reaches 295 mm, as shown in Figure 3b. There are two sharp increases in the deformation rate at three measurement points. The first time is when the working face is 20 m behind the measuring station, in which the deformation rate on the coal pillar side increases from 3.2 mm/m to 10.3 mm/m. The second time is when the working face is 120 m in front of the measuring station, in which the deformation rate on the coal pillar side increases from 5.5 mm/m to 15.3 mm/m. It is clear that the asymmetric deformation phenomenon of the floor is significant because of the effect of the lateral and advanced support pressures of the upper-section working face. The lateral support pressure has the most influence.

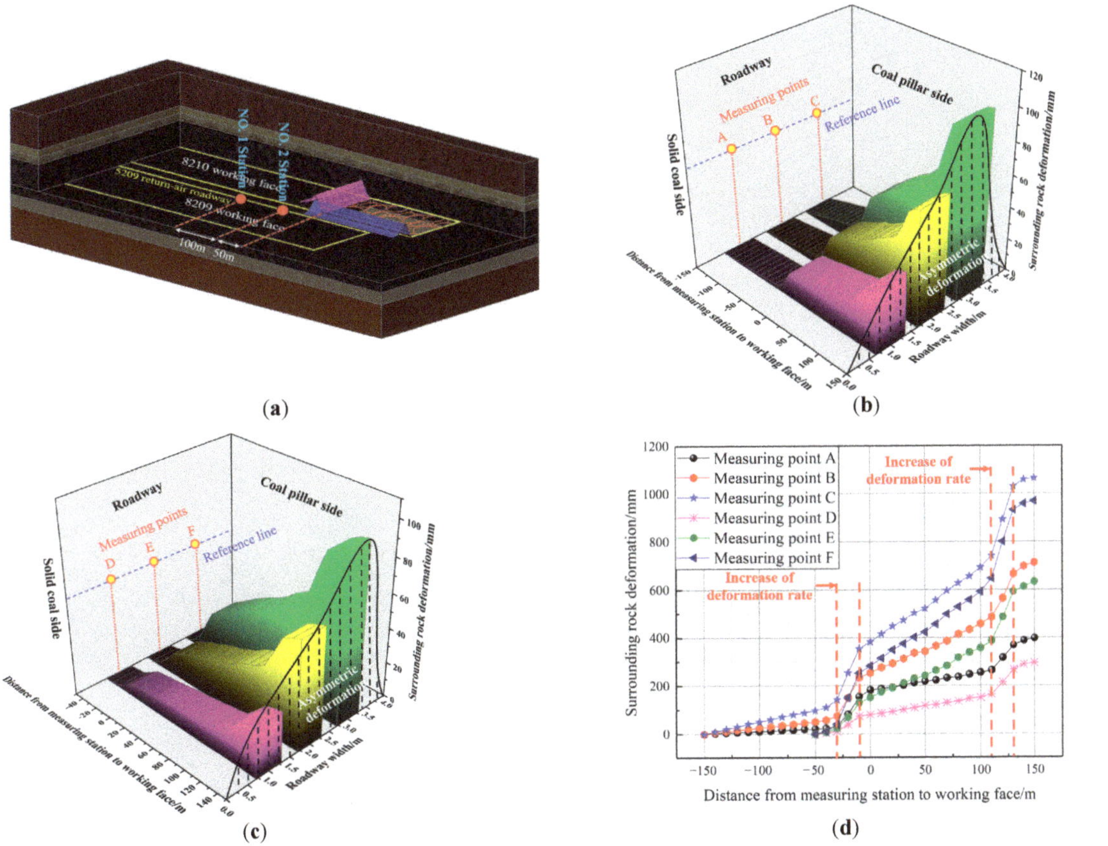

Figure 3. Layout of measuring stations and deformation curve of roadway floor. (**a**) Layout of measuring stations; (**b**) Layout of measurement points and deformation curve of the floor at Station 1; (**c**) Layout of measurement points and deformation curve of the floor at Station 2; and (**d**) Summary diagram of floor deformation curves.

4. Stress Calculation of Surrounding Rock

The deformation features of roadways are closely related to the stress environment. The 5209 return air roadway is affected by multiple mining movements caused by the roadway and the excavation of the 8210 working face, and the stress environment is complex and variable. There is a stress concentration around the roadway after the excavation. Taking a circular roadway within a biaxial isobaric stress field as an example, the stress distribution is calculated using the elastic mechanics method based on the plane strain problem. Figure 4 shows the area where the stress around the roadway is less than the strength of the rock, which is called the elastic zone (C). If the stress is more than the strength of rock, the roadway will undergo plastic deformation, forming a plastic deformation zone (B) with a radius of R. The area where the stress in the plastic zone is less than the original stress is called the fracture zone (A). The area outside the elastic zone is the original stress zone (D) [22]..

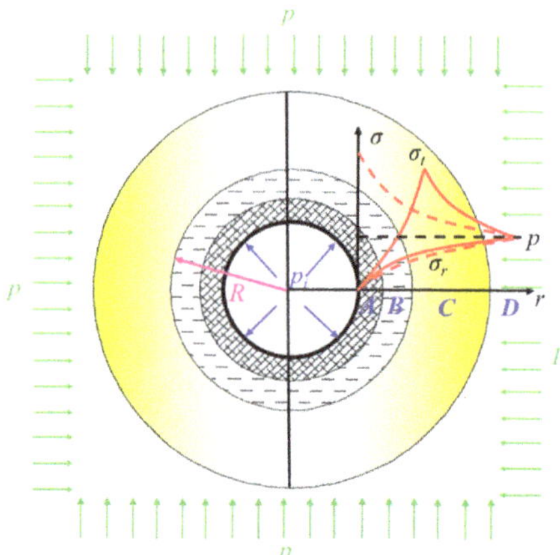

Figure 4. Surrounding rock zoning for circular roadway excavation.

The tangential stress of the plastic zone can be obtained by the following formula [23]:

$$\sigma_t = (P_i + c\cot\varphi)\left(\frac{1+\sin\varphi}{1-\sin\varphi}\right)\left(\frac{r}{r_0}\right)^{\frac{2\sin\varphi}{1-\sin\varphi}} - c\cot\varphi \tag{1}$$

In the formula, c represents the rock cohesive strength, with a value of 1.5 MPa; φ represents the internal friction angle of coal with a value of 18°; r is the radius of the studied point within the limit equilibrium zone in m; and r_0 is the radius of the roadway, with a value of 2.4 m. According to the famous modified Fenner equation, the radius of the plastic zone is obtained as follows:

$$R = r_0\left[\frac{(p + c\cot\varphi)(1-\sin\varphi)}{P_i + c\cot\varphi}\right]^{\frac{1-\sin\varphi}{2\sin\varphi}} \tag{2}$$

In the formula, p is the original stress in MPa. P_i is the support resistance of the roadway, with a value of 0.2 MPa. The tangential stress in the elastic zone can be obtained as follows:

$$\sigma_t = p + (c\cos\varphi + p\sin\varphi)\left[\frac{(p + c\cot\varphi)(1-\sin\varphi)}{P_i + c\cot\varphi}\right]^{\frac{1-\sin\varphi}{\sin\varphi}}\left(\frac{r_0}{r}\right)^2 \tag{3}$$

H represents the burial depth of the roadway, which is 415 m. The unit weight of the overlying rock γ is 2500 kN/m^3. The following formulas can be obtained through our calculations:

$$\begin{cases} \sigma_t \approx 9.15 \times \left(\frac{r}{2.4}\right)^{0.89} - 4.62\ (2.4 \le r \le 5.61\ \text{m}) \\ \sigma_t \approx 10.38 + \frac{152.81}{r^2}\ (r > 5.61\ \text{m}) \\ R \approx 5.61\ \text{m} \end{cases} \tag{4}$$

Based on the monitoring results of the roadway floor heave in the previous section, it can be concluded that the lateral support pressure due to the mining of the 8210 working face has a significant impact on the roadway floor heave. Therefore, the following section focuses on analyzing the distribution of the lateral support pressure caused by the 8210 working face. The pressure is determined by the lateral overlying rock structure of the goaf. The main roof rock blocks are difficult to form a stable masonry beam with because of the large mining height. Instead, they form a cantilever beam structure, which can only form a stable hinge structure in higher rock layers. Whether the main roof can form a masonry beam can be determined based on the following formula [24]:

$$\Delta = M + (1 - K_p)\sum h_i > h - \sqrt{\frac{2ql^2}{\sigma_c}} \tag{5}$$

M represents the thickness of the coal seam with a value of 15 m; the coefficient of the collapse and expansion K_p of the collapsed rock block is taken as 1.3; $\sum h_i$ represents the thickness of the immediate roof with a value of 5.4 m; the main roof thickness h is 14.8 m; the periodic weighting step l is taken as 25 m; the load on the overlying rock is q; and the main roof compressive strength σ_c is 78.5 MPa. It is calculated that Δ is 13.38, which is greater than 12.85, so a cantilever beam is formed on the main roof, as shown in Figure 5.

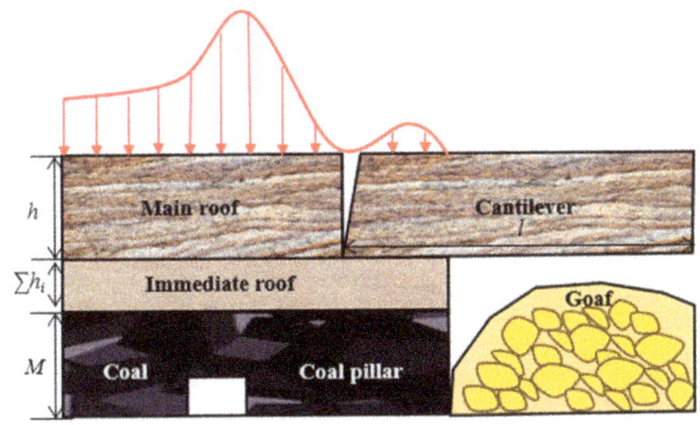

Figure 5. Overlying rock structure of roadway.

The reason for the lateral support pressure is consistent with that of the leading support pressure, and the form of the lateral support pressure is also similar to that of the leading support pressure, which can be expressed by a simplified exponential function. A model to calculate the lateral support pressure is established using the limit equilibrium theory, as shown in Figure 6 [25,26]. In the figure, P_0 represents the support resistance in MPa. x represents the distance from the inspection point to the coal wall in m. $0 - x_0$ is the plastic zone, $x_0 - x_1$ is the elastic zone, and K is the lateral support pressure concentration coefficient. From the plastic zone, a unit is taken, and the force balance in the x-direction is shown as follows [27]:

$$M(\sigma_x + d\sigma_x) - M\sigma_x - (f_1\sigma_y + c_1)dx - (f_2\sigma_y + c_2)dx = 0 \tag{6}$$

In the formula, σ_x is the horizontal stress of the coal body. f_1 and f_2 are the friction coefficients of the bedding plane; and c_1 and c_2 are the cohesion between the coal seam and the immediate roof and floor interfaces, respectively. We assume that the yield of coal in the plastic zone satisfies the Mohr–Coulomb criterion, i.e.:

$$\begin{cases} \sigma_y = S_c + \sigma_x \varepsilon \\ S_c = \frac{2c\cos\varphi}{1-\sin\varphi}, \ \varepsilon = \frac{1+\sin\varphi}{1-\sin\varphi} \end{cases} \tag{7}$$

In the formula, φ represents the internal friction angle in the coal. Substituting Formula (7) into Formula (6) yields the following:

$$\frac{d\sigma_y}{dx} - \frac{(f_1 + f_2)\varepsilon}{M}\sigma_y = \frac{(c_1 + c_2)\varepsilon}{M} \tag{8}$$

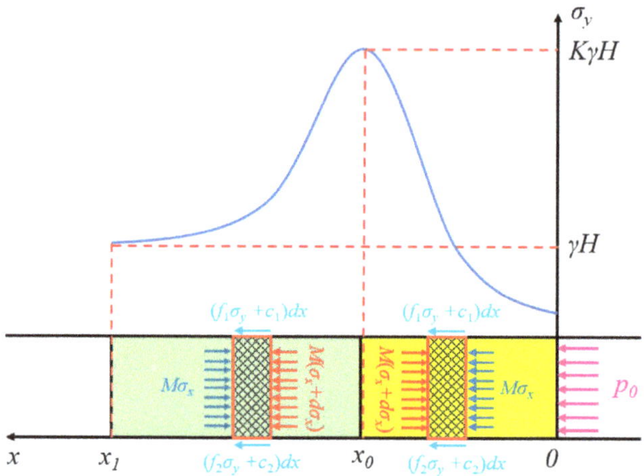

Figure 6. The mechanical model of the lateral support pressure.

Considering the boundary conditions, when x equals 0 and σ_x equals p_0, the stress in the y-direction is as follows:

$$\sigma_y\mid_{x=0} = \varepsilon\sigma_x\mid_{x=0} + S_c = \varepsilon P_0 + S_c \tag{9}$$

Solving Formula (8) can obtain:

$$\sigma_y = (S_c + P_0\varepsilon + \frac{c_1+c_2}{f_1+f_2})e^{\frac{(f_1+f_2)\varepsilon x}{M}} - \frac{c_1+c_2}{f_1+f_2} \tag{10}$$

We substitute $\sigma_y = K\gamma H$ into Formula (9) to obtain the position of the peak point of the support pressure as follows:

$$x_0 = \frac{M}{(f_1+f_2)}\ln\left(\frac{K\gamma H + \frac{c_1+c_2}{f_1+f_2}}{S_c + P_0\varepsilon + \frac{c_1+c_2}{f_1+f_2}}\right) = 23.93 \tag{11}$$

Taking a unit body from the elastic zone, the following can be inferred from the force balance in the x-direction:

$$M(\sigma_x + d\sigma_x) - M\sigma_x + (f_1\sigma_y + c_1)dx + (f_2\sigma_y + c_2)dx = 0 \tag{12}$$

In the elastic zone, it satisfies the following conditions: $\sigma_x = \lambda\sigma_y$, $d\sigma_x = \lambda d\sigma_y$, λ, which represents the coefficient of the lateral pressure. Considering the boundary conditions, our calculation results are as follows:

$$\sigma_y = (K\gamma H + \frac{c_1+c_2}{f_1+f_2})e^{\frac{(f_1+f_2)}{M\lambda}(x_0-x)} - \frac{c_1+c_2}{f_1+f_2} \tag{13}$$

We substitute $\sigma_y = K\gamma H$ into Formula (12) to obtain the farthest position under the support pressure as follows:

$$x_1 = x_0 + \frac{M\lambda}{f_1+f_2}\ln\left(\frac{K\gamma H + \frac{c_1+c_2}{f_1+f_2}}{\gamma H + \frac{c_1+c_2}{f_1+f_2}}\right) = 60.68 \tag{14}$$

Assuming that the friction coefficient of each layer is equal, the cohesion between the coal seam and the immediate roof and floor interfaces is equal, i.e., $f_1 = f_2 = f = 0.15$, $c_1 = c_2 = 1.5$ MPa. The stress concentration coefficient K is 2.5. The resistance p0 of the coal wall support is 0.2 MPa. λ has a value of 1.3. By incorporating the above formulas, the following can be concluded:

$$\begin{cases} \sigma_y \approx 14.51e^{0.038x} - 10 \ (0 \leq x \leq 23.93 \text{ m}) \\ \sigma_y \approx 35.94e^{0.015\times(23.93-x)} - 10 \ (23.93 < x \leq 60.68 \text{ m}) \end{cases} \tag{15}$$

5. Mechanism and Deformation Law of Roadway Floor Heave

The stress characteristics, the size of plastic zone and the deformation of the roadway floor under the roadway excavation and pressure disturbance due to the mining of the upper working face are analyzed by a numerical simulation analysis. The numerical calculation model is shown in Figure 7. A Flac3D numerical calculation model (x × y × z = 540 × 250 × 116 m) is established with a displacement of 0 at the bottom and side boundaries of the model based on the actual geological conditions. The Mohr–Coulomb criterion is adopted for a constitutive relationship. The applied stress on the upper boundary is 8.7 MPa with a lateral pressure coefficient of 1.2. The cross-section of the working face is 220 m × 15 m, and that of the 5209 roadway is 5 m × 4 m. Table 1 shows the mechanical parameters of each rock layer.

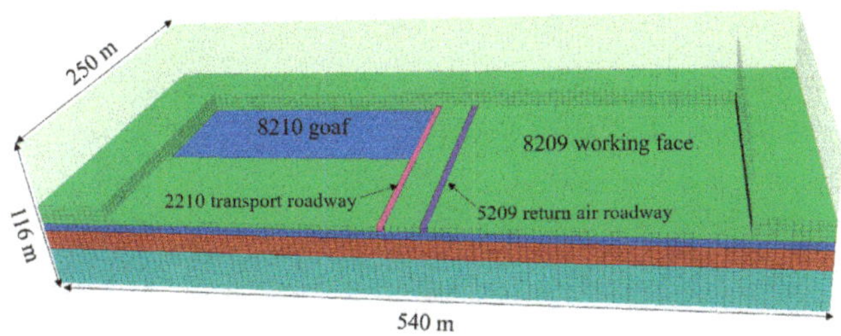

Figure 7. Numerical calculation model.

Table 1. Mechanical parameters of each rock layer and coal seam.

Lithology	*K* (GPa)	*G* (GPa)	ρ (kg/m^3)	*C* (MPa)	φ (deg.)	σ_t (MPa)
Fine sandstone	8.91	6.23	2500	2.8	32	1.67
Medium-coarse sandstone	12.58	7.75	2656	3.3	38	2.14
Siltstone	3.15	1.52	2587	1.58	33	1.52
Coal	2.56	0.51	1336	0.45	20	0.49
Carbonaceous mudstone	2.87	0.83	2615	0.77	25	1.16
Medium-fine sandstone	13.27	7.85	2630	2.9	38	2.89
Mudstone	12.15	7.23	2660	3.1	37	2.65

5.1. *Stress Distribution of Roadway Surrounding Rock*

The vertical stress distribution laws of the surrounding rock due to the roadway excavation are shown in Figure 8. The vertical stress of the rock mass on both sides centered on the central axis of the roadway presents a symmetrical distribution pattern. The vertical stress curve is obtained by setting a monitoring line in the roadway. The vertical stress on the right side of the roadway is analyzed as an example: the vertical stress is the lowest at the edge of the roadway. The vertical stress curve grows first and then decreases as it gets farther away from the central axis of the roadway. The vertical stress curve exhibits a single peak distribution pattern. At the position 6 m from the central axis of the roadway, the max. stress is 15.5 MPa, and then the vertical stress is gradually reduced until it approaches the original stress, which is 10.38 MPa. According to the vertical stress calculation formula during the roadway excavation, a theoretical curve is obtained. The vertical stress reaches its peak at a distance of 5.6 m away from the central axis of the roadway, and the max. stress is 15.4 MPa. Then the vertical stress is gradually reduced until it approaches the original stress. The theoretical curve is basically identical to the numerical simulation curve, which validates the correctness of the theoretical calculation.

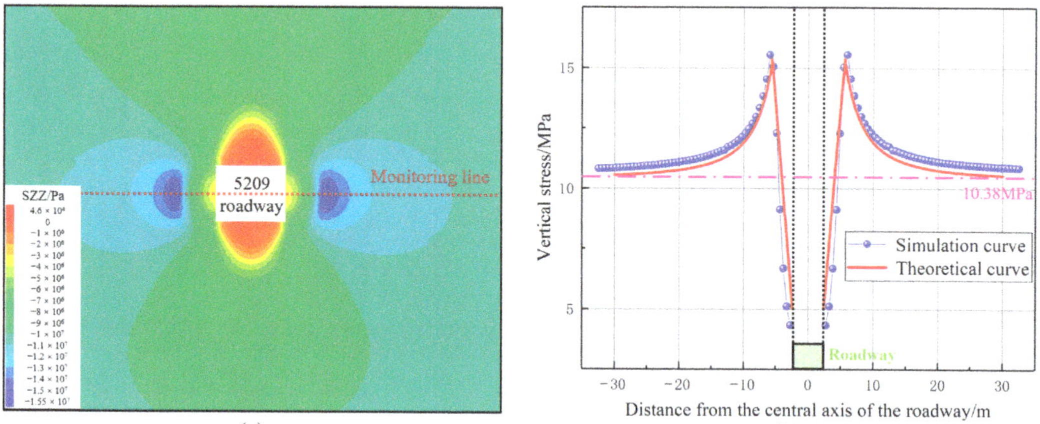

(a) **(b)**

Figure 8. Vertical stress distribution of roadway excavation. (**a**) A map of vertical stress; (**b**) Vertical stress distribution along the monitoring line.

The distribution laws for the lateral vertical stress caused by the 8210 working face are simulated for a situation in which the 5209 roadway is not excavated. Figure 9 shows that the lateral vertical stress first grows and then is reduced with the increasing distance from the goaf side coal wall. A stress core is formed at a distance of 33 m from the coal wall, and the max. stress reaches 27 MPa. The distribution curve of the horizontal vertical stress caused by the upper working face is obtained by arranging a monitoring line in the coal seam. An analysis shows that the vertical stress at the edge of the coal wall in the goaf is the lowest, and the vertical stress curve grows first and then is reduced further away from the goaf. The max. stress is 25.4 MPa at a position 5.5 m from the left of the central axis of the 5209 roadway, and then the stress is gradually reduced until it approaches the original stress with a value of 10.38 MPa. According to the above calculation formula of the lateral abutment pressure, the theoretical curve of the lateral vertical stress in the goaf is obtained. The stress reaches its peak at a position of 6 m on the left side of the central axis of the roadway for which the max. stress is 25.9 MPa, and then the stress is gradually reduced until it approaches the original stress. The theoretical curve is basically identical to the numerical simulation curve, which validates the correctness of our theoretical calculation.

The coupled vertical stress distribution laws under the influence of mining at 120 m behind the working face during the advance of the 8210 working face after the roadway excavation are simulated. Figure 10 shows the stress distribution: the left side of the roadway is a coal pillar, and the right side is solid coal. Stress cores are formed on the roadway both sides. A stress core on the coal pillar side is located at the lower left side of the roadway, while the stress core on the solid coal side is located at the upper right side of the roadway. The max. stress of stress core on the left side is 26 MPa, and the max. stress on the right side is 24 MPa. The coupled stress distribution curve is obtained by setting a monitoring line. The stress at the edges of both sides is the lowest, and the vertical stress curve grows first and then decreases as the distance from the central axis of roadway increases. The max. coupling stress on the left side of the roadway is located 12.5 m from the central axis of the roadway, and the max. stress is 26.9 MPa with a stress concentration coefficient of 2.6. The max. coupling stress on the right side of the roadway is located 11.4 m from the central axis of the roadway, and the max. stress is 19.5 MPa with a stress concentration coefficient of 1.9. Then the vertical stress is gradually reduced until it approaches the original rock stress. The stress curve on both sides of the roadway presents an asymmetric distribution pattern in which the stress of the roadway near the coal pillar is more than the stress near the solid coal. Additionally, the max. stress near the coal pillar is 1.4 times that of near the solid coal.

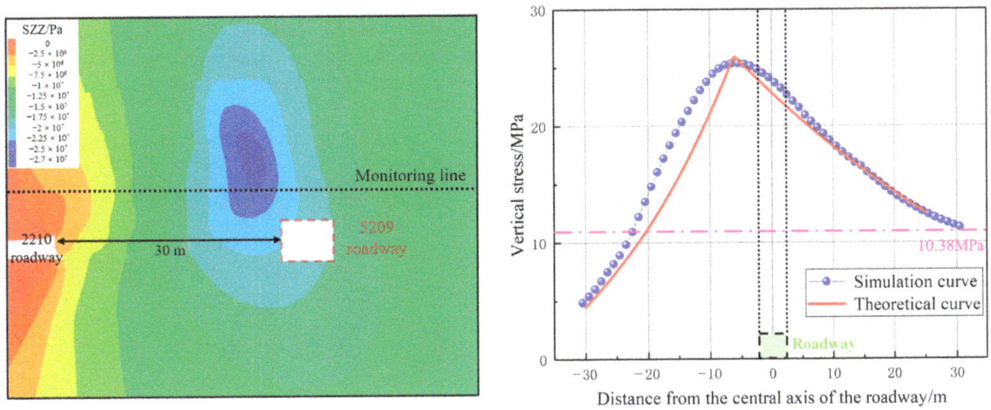

(**a**) (**b**)

Figure 9. Vertical stress distribution of 8210 working face excavation. (**a**) A map of vertical stress; (**b**) Vertical stress distribution along the monitoring line.

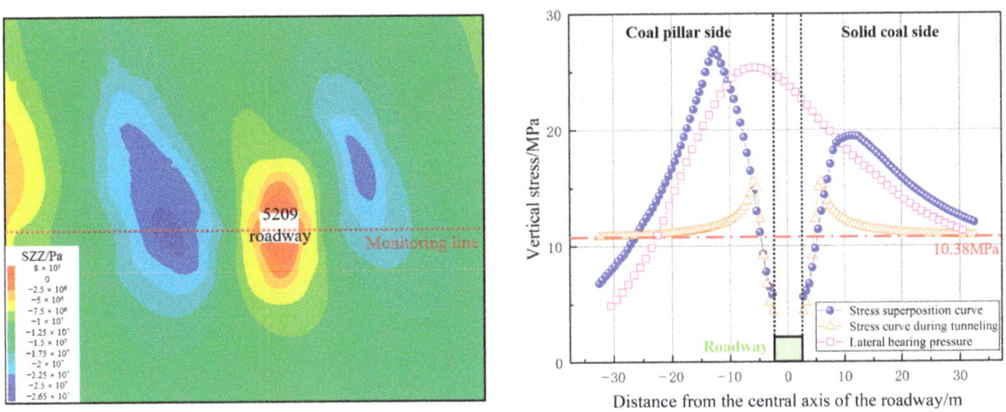

(**a**) (**b**)

Figure 10. Vertical stress distribution of 8210 working face mining after roadway excavation (**a**) A map of vertical stress; (**b**) Vertical stress distribution along the monitoring line.

5.2. Mechanism of Asymmetric Floor Heave

This article analyzes the reasons for the formation of roadway floor heave based on the slip line theory. After the stress of the floor exceeds its ultimate strength under the influence of mining, the floor will undergo elastic and plastic deformation and failure, forming a plastic zone. The failure zoning of the floor is shown in Figure 11. The plastic zone of the floor consists of three parts: the voluntary stress zone ABM, GHF, the enforced stress zone MEH, MCD, and the transitional stress zone MBC, EFH [28]. The rock in the voluntary stress area is subjected to stress compression in the vertical direction, causing an expansion deformation in the horizontal direction and increasing stress in the horizontal direction, leading to tensile failure in the horizontal direction. The deformation state of the rock is tensile strain, and the deformed rock compresses the rock in the transition stress zone. The rock is in a compressive strain state, resulting in compressive shear compression failure. Continuously compressing the rock in the passive stress area, the rock in the passive stress zone undergoes shear slip deformation. With the continuous disturbance of mining stress, the rock in this area will inevitably move towards the free face of the roadway floor, and the accumulated distortion energy in this area will be released, causing the floor to bulge. Under the mining of the upper working face, the max. value and stress range of the support pressure near the coal pillar of the 5209 roadway are significantly more than those near the solid coal. Therefore, the failure area of the floor caused by support pressure near the coal pillar is inevitably more than that near the solid coal, and the roadway floor will undergo asymmetric failure.

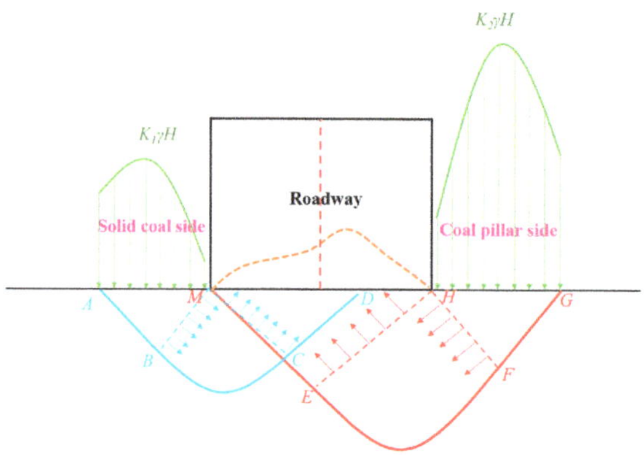

Figure 11. Schematic diagram of roadway floor failure zoning.

5.3. Distribution Laws of the Plastic Zone in the Roadway

The distribution laws of the plastic zone in the roadway and the excavation of the upper working face are simulated. Figure 12 shows that the plastic zone has a symmetrical distribution along the central axis of the roadway during the roadway excavation, and the depth of the floor plastic zone is 3 m. During the excavation of the upper working face, the plastic zone of the roadway floor 100 m in front of the working face begins to expand to the left, and the depth of the floor plastic zone is 3.5 m. During the mining of the upper working face, the scope of the plastic zone of the roadway 20 m in front of the working face is greatly expanded and interacts with the plastic zone near the coal pillar. The plastic zone of the floor presents an asymmetric distribution pattern in which the scope near the coal pillar is bigger than that near the solid coal, and the depth of the floor plastic zone is 5 m. During the excavation of the upper working face, the scope of the plastic zone 120 m behind the working face is further expanded, and serious deformation occurs in the roadway. Additionally, the plastic zone's depth increases to 5.5 m.

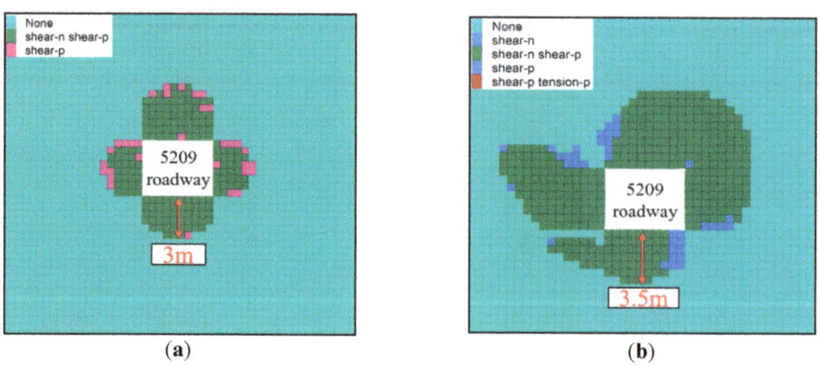

Figure 12. *Cont.*

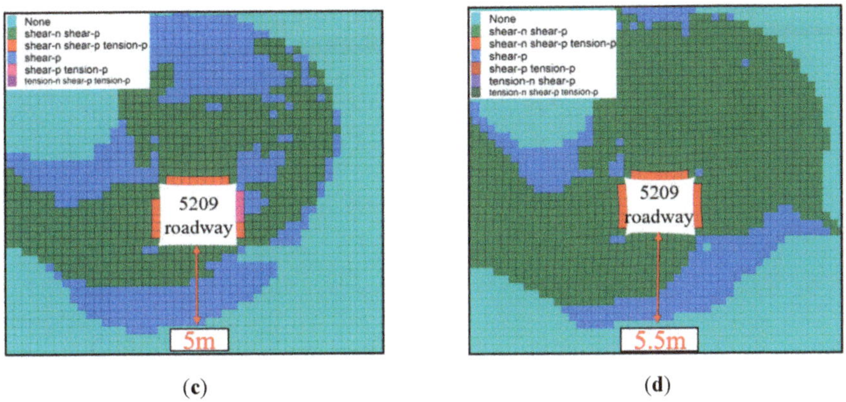

Figure 12. Plastic zones in different periods. (**a**) Roadway excavation; (**b**) 100 m in front of the upper working face; (**c**) 20 m in front of the upper working face; and (**d**) 120 m behind the upper working face.

5.4. Displacement Distribution of the Floor

The distribution of the vertical displacement of the floor during the roadway excavation is simulated. The vertical displacement distribution and displacement curves of the floor at different depths (0–4 m) are shown in Figure 13. The vertical displacement of the floor centered on the central axis of the roadway presents a symmetrical distribution after the roadway excavation. The displacement of floor decreases with the increase in the depth; this is the case for the floor on the roadway on both sides. The floor failure of the roadway is the floor heave. The central axis displacement is the largest, and the peak displacement is 49.5 mm. The deformation on both sides shows subsidence, and the max. displacement is 31.6 mm.

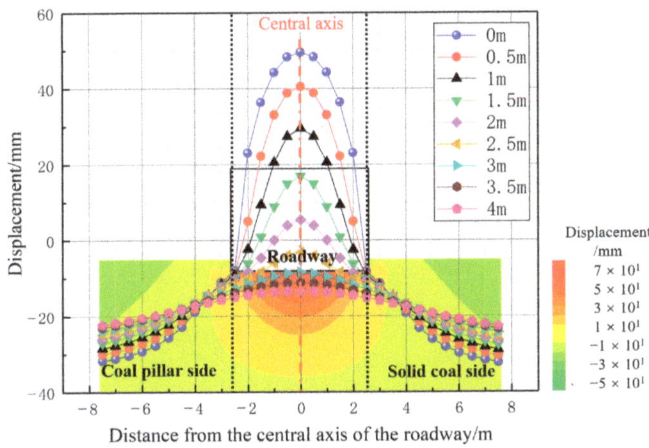

Figure 13. Curve and cloud of floor displacement during roadway excavation.

The floor displacement distribution at 120 m behind the working face during the mining of the upper working face is simulated. A vertical displacement distribution nephogram and the displacement curves of the floor at different depths (0–4 m) are shown in Figure 14. The displacement of the floor centered on the central axis of the roadway shows an asymmetric form in which the coal pillar side is relatively large under the mining of the upper working face. The displacement of the roadway floor is reduced with the increase in the depth. Floor deformation is manifested as floor heave, and the max.

displacement is reached 1 m away from the central axis on the left side of the roadway. The peak displacement is 873 mm. Floor subsidence occurs on both sides of the roadway. The maximum displacement of the floor near the coal pillar is 627 mm, and that near the solid coal is 131 mm.

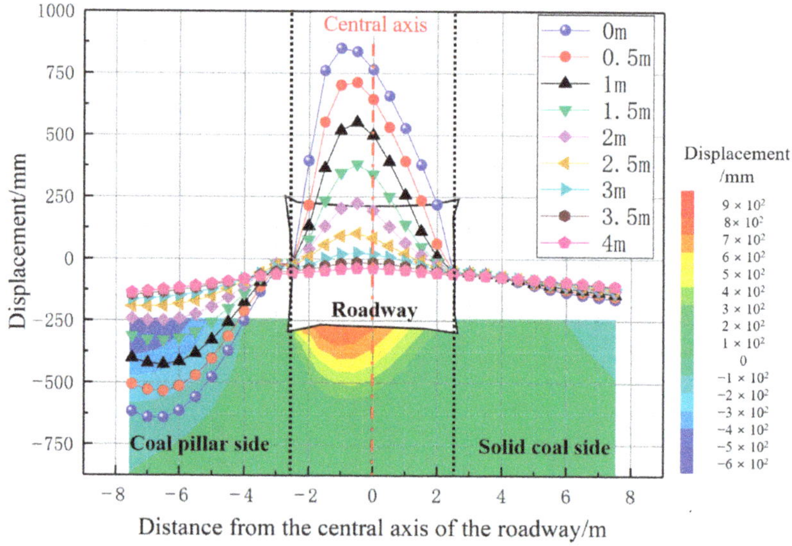

Figure 14. Curve and cloud of floor displacement during 8210 working face mining after roadway excavation.

6. Control Measures of Roadway Asymmetric Floor Heave

The roof and two sides of the coal body of the roadway have low strength, a poor bearing capacity and insufficient support strength, which lead to the instability of the roof and rock of the two sides and the transfer of stress to the floor under strong mining stress, leading further to floor failure [29,30]. Therefore, our priority in solving floor heave is strengthening the roof and two sides of the roadway, blocking or reducing the stress transfer of the roof and two sides, controlling the stability of the roof and two sides, and then avoiding floor stress concentration. Control methods are implemented to the floor based on the above.

6.1. Strengthening the Support to the Roof and Coal Wall

The instability of the roof will aggravate the failure of the two sides, and stress will be transferred to the two sides, which will indirectly lead to an increase in the supporting stress of the floor and aggravate the appearance of the floor heave. Therefore, controlling the stability of the roof is important for preventing floor heave. In the original support scheme, the strength of an anchor cable support is not enough, which leads to serious roof subsidence near the coal pillar and then causes the formation of a high stress concentration and floor heave at the coal pillar side. Therefore, an asymmetric truss anchor cable is chosen to control the roof's asymmetric deformation. As shown in Figure 15, the asymmetric truss anchor cable structure is composed of a high-strength anchor cable, reinforced ladder beam and U-bar, with a supporting pallet, a thick gasket and other auxiliary components. The anchor cable on the coal pillar side is linked with a U-bar. It is necessary to ensure that the anchoring section of the anchor cable is located in a stable rock mass deep in the roadway. The structure increases the strength and stiffness required by the supporting system, ensuring the coordinated deformation between the supporting body and surrounding rock [31–33].

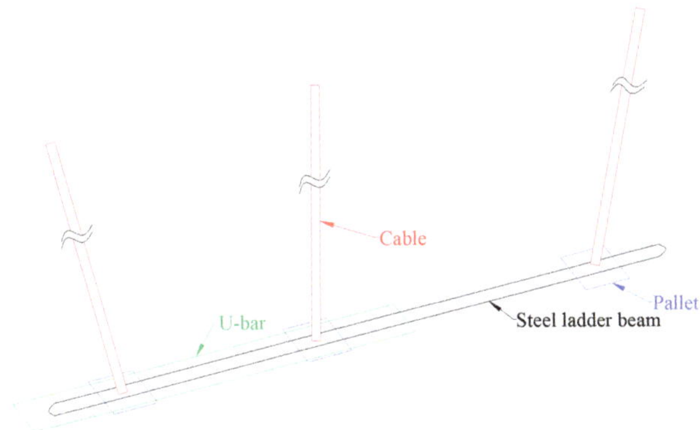

Figure 15. Asymmetric truss anchor cable structure.

It is important to increase the supporting strength of the two sides, improve their ability to resist deformation, lower the high stress transmitted on both sides and lower the high stress transmitted by the two sides of the floor. Our original support scheme of bolts on the two sides of the roadway would not be able to control the deformation. As shown in Figure 16, it is necessary to match the anchor cable with a high preload force to meet the support stiffness and strength requirements of the supporting system [34–36].

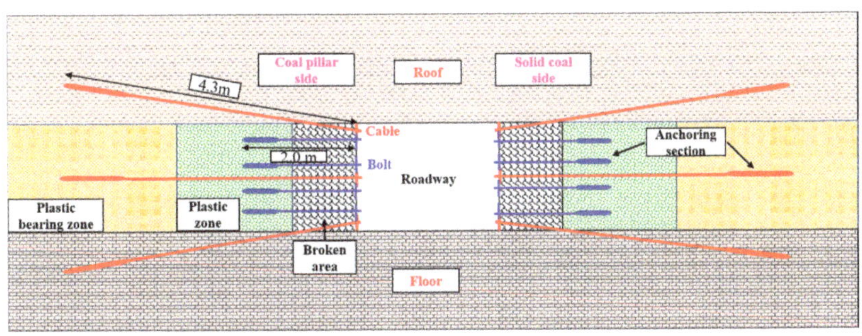

Figure 16. The support effect of anchor cables.

6.2. Floor Control Measures

Floor leveling is a common treatment method for floor heave, but it does not solve floor heave at its root. For roadways affected by repeated mining, frequent floor leveling is inefficient and uneconomic. A floor anchor cable can connect shallow, low-strength rock strata with deep, high-strength stable rock strata and prevent the sliding deformation of the weak surfaces and the floor heave phenomenon by improving the friction between the floor strata [37]. For the existing roadway floor heave, it is necessary to take floor-leveling measures first to ensure the original roadway section. After floor leveling, an anchor cable support is implemented on the floor and a high-strength concrete layer is laid on the surface to increase the floor's deformation resistance. The thickness of the concrete layer is 200 mm, with an inner lining of a Φ6 metal mesh. The concrete strength grade is C30. The combination of the above methods is able to efficiently control floor heave.

Our initial scheme to support the roadway is as follows. The roof cables are Φ21.8 mm × 6300 mm prestressed-steel strands, with three pieces in each row. The row spacing is 1800 mm × 1800 mm, the angle between the most lateral anchor cable in the horizontal direction is 60° and the roof anchor cable tray size is 250 mm × 250 mm × 16 mm. The roof bolts are Φ22 mm × 2400 mm screw steels, with six pieces in each row. The row spacing is 900 mm × 900 mm, the angle between the most lateral bolt in the horizontal direction is 75° and the roof anchor bolt's tray size is 130 mm × 130 mm × 10 mm. The two sides have Φ20 mm × 2000 mm screw steel bolt, with four pieces in each row. The upward and downward angles of the upper- and lower-side bolts are 15°. The row spacing is 1000 mm × 900 mm. The bolt tray size is 130 mm × 130 mm × 10 mm. The roof uses a Φ6 steel wire mesh with a size of 100 × 100 mm. The two sides have a #8 diamond mesh, with a size of 50 mm × 50 mm. The roof bolt support is matched with a 4800 × 280 × 3.75 mm W steel belt, the roof cable support is matched with a 3900 × 330 × 6 mm JW steel belt and the side bolt is matched with a 450 × 280 × 3.75 mm W steel belt. The scheme is shown in Figure 17.

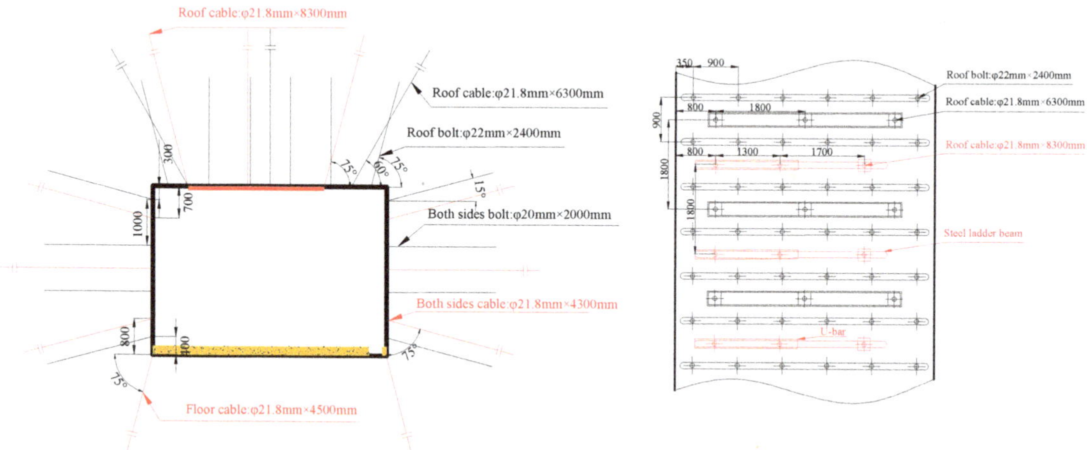

(a) (b)

Figure 17. Supporting scheme of roadway. (**a**) Front view; (**b**) Top view.

The roadway reinforcement control scheme is shown in the figure (labeled in red). The asymmetric truss anchor cables for the roof are Φ21.8 mm × 6300 mm steel strands, with four pieces in each row. The angle between the side anchor cable in the horizontal direction is 15°, and the middle one is arranged perpendicular to the roof. The row spacing is 1300 mm + 1700 mm × 1800 mm, and the range between the anchor cables near the coal pillar and coal wall is 800 mm. The bolt tray size is 250 mm × 250 mm × 16 mm. The anchor cables are linked with 3700 mm × 70 mm steel ladder beams. The roadway's middle and anchor cables at the coal pillar side are linked by a U-bar, with a length of 2200 mm. The size of the steel gasket is 250 mm × 120 mm × 16 mm. The two sides have Φ21.8 mm × 4300 mm screw steel bolts, with three pieces in each row. The upward and downward angles of the upper- and lower-side cables are 75°. The row spacing is 1100 mm × 1800 mm. The cable tray size is 250 mm × 250 mm × 16 mm. The side anchor cable support is matched with a 2600 × 330 × 6 mm JW steel belt. The floor has Φ21.8 mm × 4500 mm screw steel bolts, with two pieces in each row. The angle between the two anchor cables in the horizontal direction is 75°. The row spacing is 5200 mm × 1800 mm. The cable tray size is 150 mm × 150 mm × 12 mm.

7. Engineering Practice

The 5209 roadway was repaired after the mining of the 8210 working face to ensure normal use for the mining of the 8209 working face. The above combined control scheme of "floor leveling + anchor cable support + concrete hardening" was applied to the floor heave treatment in the roadway repair stage. To verify the effectiveness of the asymmetric floor heave control, a surface displacement monitoring station was established in the central position of the roadway alignment (at a mining position of 500 m) to monitor the floor deformation. The layout of the measurement points and the monitoring results are shown in Figure 18. Three measurement points were arranged at this station. The following is an analysis of our monitoring data of the roadway floor displacement: the first stage is the slow deformation stage, in which the deformation rate of the floor slowly increases between 60 m and 130 m in front of the working face; the second stage is the stage of the rapid deformation of the floor, in which the deformation rate rapidly increases between 20 m and 60 m in front of the working face; and the third stage is the stable deformation stage of the floor, in which the deformation rate of the roadway floor decreases and tends to stabilize between 0 m and 20 m in front of working face. The monitoring work ends when the working face pushes past the station position. The maximum floor heave near the coal pillar is 167 mm, the max. floor heave at the central roadway is 113 mm and the max. floor heave near the solid coal is 81 mm. The results show that the floor heave is obviously controlled during the mining of this working face, which proves that the asymmetric control strategy of floor heave can achieve the long-term stabilization of roadways under strong mining.

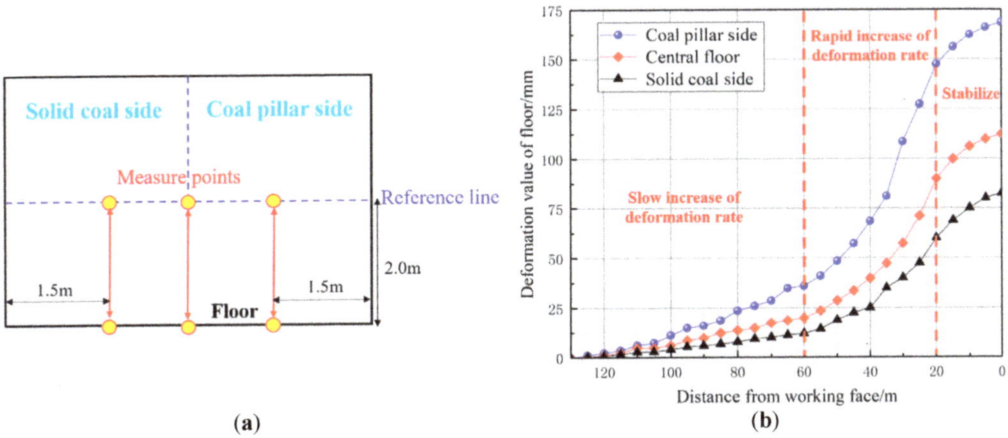

Figure 18. Layout of measure points and monitoring curve. (**a**) Layout of measure points; (**b**) Monitoring curve.

8. Conclusions

(1) The formulas to calculate the horizontal vertical stress of the roadway caused by the excavation and mining of the upper working face are derived separately. The theoretical curve is basically identical to the numerical simulation curve. The vertical stress coupling curve on both sides during the mining of the upper working face is obtained through a numerical simulation.

(2) The cause of the asymmetric floor heave is the uneven distribution of vertical stress on both sides of the floor. The stress concentration coefficient near the coal pillar of the roadway is 2.6, and the stress concentration coefficient near the solid coal is 1.9. The failure area of the floor near the coal pillar is more than that near the solid coal.

(3) Our numerical simulation shows a symmetrical distribution of the displacement curve of the floor during the roadway excavation, with a max. displacement of 49.5 mm. The displacement curve of the floor during the mining of the upper working face is asymmetrical with a max. displacement of 873 mm at a distance of 1 m from the central axis near the coal pillar. The maximum depth of floor failure is 5.5 m.

(4) According to the cooperative control principle of a roof, two sides and a floor, the asymmetric floor heave joint control scheme of "floor leveling + anchor cable support + concrete hardening" is proposed. The floor deformation monitoring results indicate that the max. floor heave at the measurement point near the coal pillar is 167 mm, and the floor heave is effectively controlled.

Author Contributions: Conceptualization, D.W. and F.H.; methodology, Y.Z. (Yajiang Zhang); software, J.S., X.W. and K.T.; formal analysis, J.Z. and F.W.; investigation, P.J.; data curation, Y.W. and B.L.; writing—original draft preparation, D.W.; writing—review and editing, Y.Z. (Yun Zheng) and F.H.; funding acquisition, Y.Z. (Yun Zheng) and F.H. All authors have read and agreed to the published version of the manuscript.

Funding: This work was supported by the National Natural Science Foundation of China (No. 52004286), the National Natural Science Foundation of China (No. 51974317), the Fundamental Research Funds for the Central Universities (2022YJSNY09) and the Fundamental Research Funds for the Central Universities (3142019010).

Data Availability Statement: The data presented in this study are available on request from the corresponding author. The data are not publicly available due to the privacy protection of research achievements.

Conflicts of Interest: The authors declare that they have no known competing financial interests or personal relationships that could have appeared to influence the work reported in this paper.

References

1. Liang, M.F.; Hu, C.J.; Yu, R.; Wang, L.X.; Zhao, B.F.; Xu, Z.Y. Optimization of the process parameters of fully mechanized top-coal caving in thick-seam coal using BP neural networks. *Sustainability* **2022**, *14*, 1340. [CrossRef]
2. He, F.L.; Liu, B.Q.; Wang, D.Q.; Chen, D.D.; Wu, Y.H.; Song, L.M.; Ma, X.; Ye, Q.C.; Jiang, Z.S.; Guo, F.F.; et al. Study on stability and control of surrounding rock in the stopping space with fully mechanized top coal caving under goaf. *Energies* **2022**, *15*, 8498. [CrossRef]
3. Wang, D.Q.; He, F.L.; Wu, Y.H.; Xu, X.H.; Zhang, J.L.; Lv, K.; Li, L.; Zhai, W.L.; Song, J.Y. Study on surrounding rock failure mechanism and rational coal pillar width of the gob-side coal roadway under influence of intense dynamic pressure. *Energy Sci. Eng.* **2023**, 1–18. [CrossRef]
4. Li, Z.X.; Zhang, Y.D.; Ma, Q.; Zheng, Y.; Song, G.Y.; Yan, W.Z.; Zhang, Y.; Hu, L. The floor heave mechanism and control technology of gob-side entry retaining of soft rock floor. *Sustainability* **2023**, *15*, 6074. [CrossRef]
5. Yang, J.; Gao, Y.B.; Liu, S.Q.; Peng, Y.H.; Zhang, H.W. Study on failure mechanism and control techniques of the preparation roadway induced by dynamic mining disturbance. *J. Min. Sci. Technol.* **2018**, *3*, 451–460.
6. Xie, S.R.; Li, H.; Chen, D.D.; Feng, S.H.; Ma, X.; Jiang, Z.S.; Cui, J.Q. New technology of pressure relief control in soft coal roadways with deep, violent mining and large deformation: A key study. *Energies* **2022**, *15*, 9208. [CrossRef]
7. Mo, S.; Tutuk, K.; Saydam, S. Management of floor heave at bulga underground operations—A case study. *Int. J. Min. Sci. Technol.* **2019**, *29*, 73–78. [CrossRef]
8. Mo, S.; Ramandi, H.L.; Oh, J.; Masoumi, H.; Canbulat, I.; Hebblewhite, B.; Saydam, S. A new coal mine floor rating system and its application to assess the potential of floor heave. *Int. J. Rock Mech. Min. Sci.* **2020**, *128*, 104241. [CrossRef]
9. Chen, Y.; Bai, J.B.; Yan, S.; Xu, Y.; Wang, X.Y.; Ma, S. Control mechanism and technique of floor heave with reinforcing solid coal side and floor corner in gob-side coal entry retaining. *Int. J. Min. Sci. Technol.* **2012**, *22*, 841–845. [CrossRef]
10. Indraratna, B.; Nemcik, J.A.; Gale, W.J. Review and interpretation of primary floor failure mechanism at a longwall coal mining face based on numerical analysis. *Geotechnique* **2000**, *50*, 547–557. [CrossRef]
11. Małkowski, P.; Ostrowski, Ł.; Stasica, J. Modeling of Floor Heave in Underground Roadways in Dry and Waterlogged Conditions. *Energies* **2022**, *15*, 4340. [CrossRef]
12. Małkowski, P.; Ostrowski, Ł.; Bednarek, Ł. The effect of selected factors on floor upheaval in roadways—In situ testing. *Energies* **2020**, *13*, 5686. [CrossRef]
13. Sakhno, I.; Liashok, I.; Sakhno, S.; Isaienkov, O. Method for controlling the floor heave in mine roadways of underground coal mines. *Min. Miner. Depos.* **2022**, *16*, 1–10. [CrossRef]
14. Wang, W.J.; Hou, C.J.; Bai, J.B.; Li, X.H. Mechanical deformation analysis and principle of floor heave of roadway driving gob in fully mechanized sub—Level caving face. *Rock Soil Mech.* **2001**, *2001*, 319–322.
15. Jiang, Y.D.; Zhao, Y.X.; Liu, W.G.; Li, Q. Research on floor heave of roadway in deep mining. *Chin. J. Rock Mech. Eng.* **2004**, *2004*, 2396–2401.
16. He, M.C.; Zhang, G.F.; Wang, G.L.; Xu, Y.L.; Wu, C.Z.; Tang, Q.D. Research on mechanism and application to floor heave control of deep gateway. *Chin. J. Rock Mech. Eng.* **2009**, *28*, 2593–2598.
17. Bai, J.B.; Li, W.F.; Wang, X.Y.; Xu, Y.; Huo, L.J. Mechanism of floor heave and control technology of roadway induced by mining. *J. Min. Saf. Eng.* **2011**, *28*, 1–5.
18. Chen, S.Y.; Song, C.S.; Guo, Z.B.; Wang, J.; Wang, Y. Asymmetric deformation mechanical mechanism and control countermeasure for deep roadway affected by mining. *J. China Coal Soc.* **2016**, *41*, 246–254.
19. Jia, H.S.; Wang, L.Y.; Liu, S.W.; Peng, B.; Zhang, W.G.; Fan, K. Mechanism and control strategy of asymmetrical floor heave of coal pillar roadway in fully mechanized caving face. *J. China Coal Soc.* **2019**, *44*, 1030–1040.
20. Yang, R.S.; Zhu, Y.; Li, Y.L.; Li, W.Y. Floor heave mechanism and control measures of layered floor in weakly cemented soft rock roadway. *J. Min. Saf. Eng.* **2020**, *37*, 443–450.

21. Wang, Z.Q.; Wang, P.; Lv, W.Y.; Shi, L.; Su, Z.H.; Wu, C.; Yu, F. Mechanism and control of asymmetric floor heave in gob-side entry. *J. Min. Saf. Eng.* **2021**, *38*, 215–226.
22. Jiang, B.S.; Zhang, Q.; He, Y.N.; Han, L.J. Elastoplastic analysis of cracked surrounding rocks in deep circular openings. *Chin. J. Rock Mech. Eng.* **2007**, *184*, 982–986.
23. Pan, Y.; Wang, Z.Q. Elasto-plastic analysis on surrounding rock of circular chamber based on strain nonlinear softening. *Chin. J. Rock Mech. Eng.* **2005**, *2005*, 915–920.
24. Xu, X.H.; He, F.L.; Li, X.B.; He, W.R. Research on mechanism and control of asymmetric deformation of gob side coal roadway with fully mechanized caving mining. *Eng. Failure Anal.* **2021**, *120*, 105097. [CrossRef]
25. Wang, Z.Q.; Wu, C.; Shi, L.; Su, Z.H.; Wang, P.; Huang, X. Analysis of surrounding rock stress and plastic zone of two-way unequal pressure circular roadway based on complex variable theory. *J. China Coal Soc.* **2019**, *44*, 419–429.
26. Vu, T.T. Solutions to prevent face spall and roof falling in fully mechanized longwall at underground mines, Vietnam. *Min. Miner. Depos.* **2022**, *16*, 127–134. [CrossRef]
27. Chang, J.C.; Xie, G.X.; Zhang, X.H. Analysis of rib spalling mechanism of fully-mechanized top-coal caving face with great mining height in extra-thick coal seam. *Rock Soil Mech.* **2015**, *36*, 803–808.
28. Xu, X.H.; He, F.L.; Li, X.B.; Lv, K.; He, C.X.; Li, L. Roadway impact mechanism and optimization design under coal pillar disturbance in large mining height working face. *J. Min. Saf. Eng.* **2021**, *38*, 547–555.
29. Lai, X.P.; Xu, H.C.; Shan, P.F.; Kang, Y.L.; Wang, Z.Y.; Wu, X. Research on mechanism and control of floor heave of mining-influenced roadway in top coal caving working face. *Energies* **2020**, *13*, 381. [CrossRef]
30. Zhou, X.M.; Wang, S.; Li, X.L.; Meng, J.J.; Li, Z.; Zhang, L.H.; Pu, D.D.; Wang, L.K. Research on theory and technology of floor heave control in semicoal rock roadway: Taking longhu coal mine in Qitaihe mining area as an Example. *Lithosphere* **2022**, *2022*, 3810988. [CrossRef]
31. He, F.L.; Xu, X.H.; Qin, B.B.; Li, L.; Lv, K.; Li, X.B. Study on deformation mechanism and control technology of surrounding rock during reuse of gob side entry retaining by roof pre-splitting. *Eng. Failure Anal.* **2022**, *137*, 106271. [CrossRef]
32. Xie, S.R.; Pan, H. Control mechanism and application of double span truss anchor cable in long span coal roadway. *J. Min. Sci. Technol.* **2018**, *3*, 148–155.
33. Zhang, G.C.; He, F.L. Asymmetric failure and control measures of large cross-section entry roof with strong mining disturbance and fully-mechanized caving mining. *Chin. J. Rock Mech. Eng.* **2016**, *35*, 806–818.
34. Chen, D.D.; Guo, F.F.; Li, Z.J.; Ma, X.; Xie, S.R.; Wu, Y.Y.; Wang, Z.Q. Study on the Influence and Control of Stress Direction Deflection and Partial-Stress Boosting of Main Roadways Surrounding Rock and under the Influence of Multi-Seam Mining. *Energies* **2022**, *15*, 8257. [CrossRef]
35. Sun, X.M.; Zhang, Y.; He, M.C.; Yang, J.K. Research of high pre-stress NPR support technology in deep shaft roadway Engineering. *J. Min. Sci. Technol.* **2023**, *8*, 50–65.
36. Xie, S.R.; Wu, Y.Y.; Guo, F.F.; Zou, H.; Chen, D.D.; Zhang, X.; Ma, X.; Liu, R.P.; Wu, C.W. Application of Pre-Splitting and Roof-Cutting Control Technology in Coal Mining: A Review of Technology. *Energies* **2022**, *15*, 6489. [CrossRef]
37. Zhao, Y.Q.; He, F.L.; Xie, F.X.; Wu, J.K.; Liu, B.Q. Study on comprehensive evaluation of surrounding rock stability grade and repair control technology of gob side entry retaining. *J. Min. Sci. Technol.* **2017**, *2*, 550–558.

MDPI
St. Alban-Anlage 66
4052 Basel
Switzerland
www.mdpi.com

Energies Editorial Office
E-mail: energies@mdpi.com
www.mdpi.com/journal/energies

www.ingramcontent.com/pod-product-compliance
Lightning Source LLC
Chambersburg PA
CBHW040822250926
51850CB00007BA/1236
9783036593166